Element Recovery and Sustainability

RSC Green Chemistry

Series Editors:
James H Clark, *Department of Chemistry, University of York, UK*
George A Kraus, *Department of Chemistry, Iowa State University, Ames, Iowa, USA*
Andrzej Stankiewicz, *Delft University of Technology, The Netherlands*
Peter Siedl, *Federal University of Rio de Janeiro, Brazil*
Yuan Kou, *Peking University, People's Republic of China*

How to obtain future titles on publication:
A standing order plan is available for this series. A standing order will bring delivery of each new volume immediately on publication.

For further information please contact:
Book Sales Department, Royal Society of Chemistry, Thomas Graham House, Science Park, Milton Road, Cambridge, CB4 0WF, UK
Telephone: +44 (0)1223 420066, Fax: +44 (0)1223 420247
Email: booksales@rsc.org
Visit our website at www.rsc.org/books

Element Recovery and Sustainability

Edited by

Andrew J. Hunt
Department of Chemistry, University of York, UK
Email: andrew.hunt@york.ac.uk

RSC Publishing

RSC Green Chemistry No. 22

ISBN: 978-1-84973-616-9
ISSN: 1757-7039

A catalogue record for this book is available from the British Library

Published by The Royal Society of Chemistry,
Thomas Graham House, Science Park, Milton Road,
Cambridge CB4 0WF, UK

Registered Charity Number 207890

For further information see our web site at www.rsc.org

Andrew J. Hunt would like to dedicate this book to his loving fiancée Nontipa, who has been a constant source of support and encouragement.

Preface

The 21st century has seen a growing recognition of the importance of resource management including the more careful use of resources, an awareness of their lifecycles and a move towards a more closed-loop approach to manufacturing. In the context of sustainable chemistry, while the 1990s saw the emergence of the green chemistry movement, with a strong focus on clean production and waste minimization in manufacturing, the new century has seen a move towards a much more holistic view. While clean and efficient manufacturing remains key and arguably the cornerstone of chemical sustainability, we now understand that this must be accompanied by a greater use of renewable resources as well as shift to environmentally compatible products. The emphasis to date in renewable resources has been on carbon – we have all become familiar with the challenges of declining traditional fossil feedstocks and the need for a smaller carbon footprint, yet only now are concerns over other elements and elemental cycles being taken seriously.

Society uses a very large number of the elements in its everyday activities. Carbon, nitrogen, oxygen, hydrogen and phosphorus, in particular, are fundamental and commonplace in life. Others including iron, chlorine, bromine, selenium and potassium are also vital to natural processes if perhaps less widely appreciated. In addition to these we have chosen to build a society around many other elements – nickel, tungsten, chromium, copper and vanadium alongside boron, gold, silver, palladium, platinum and many others are found in a multitude of articles from buildings to electronics and from sophisticated equipment to complex drug molecules. Ironically the move to a low-carbon economy – a laudable goal in principle, if not in practice, has made the problem worse by consuming large amounts of main group and rare earth elements that are not normally used to any great extent, in batteries, wind turbines, hybrid cars and other increasingly popular alternative, low-carbon technologies. We know where these elements occur and we know a lot about

RSC Green Chemistry No. 22
Element Recovery and Sustainability
Edited by Andrew J. Hunt
© The Royal Society of Chemistry 2013
Published by the Royal Society of Chemistry, www.rsc.org

their chemistries (entire book series have been written about some of them), yet we know very little about their elemental cycles which, given our almost obsessive interest in carbon cycles, seems surprising. How does our use of fluorine for example, in an increasingly high proportion of pharmaceuticals, agrochemicals and electronic chemicals affect the natural lifecycle of fluorine (which includes large quantities of static minerals and frequent emissions of hydrogen fluoride from volcanoes)? We are very aware of the dangers of halocarbons and compounds such as sulfur hexafluoride in the atmosphere but what about less volatile organofluorine molecules in the terra-sphere? The importance of achieving a better understanding of the elements in terms of their use, recovery and interaction with the environment is discussed in Chapter 1 and at various other stages in the book, in particular in Chapter 7 where the special case of platinum group metals – with applications from the decorous (jewellery) to the essential (catalysts for emission control and for the manufacture of pharmaceuticals) are discussed in terms of anthropospheric losses. The book also focuses on both tradition and novel methods of metal recovery (Chapters 2 and 3).

Fluorine, like carbon, has a number of important volatile compounds that we and nature produce, but most elements do not. Palladium, for example, is a very valuable element frequently used in catalysts for pollution control and has no significant volatile compounds. Nature provides it in the form of ores, we then transform it, for example, to a solid (but soluble) chloro compound, use it and then discard it when it is no longer "fit for purpose" typically within a solid waste or an aqueous waste stream. This traditional but totally unsustainable *linear* model for resource "management" (extract, process, use, dispose) disperses what was originally a concentrated ore over a very wide area in a way that makes it very difficult to recover. Actually, palladium is one of the few and precious elements that we do make an effort to recycle, for example by returning wastes to the supplier company which has the appropriate expertise to recover and reuse the element. Unfortunately, the majority of elements are not recycled at all and largely go into landfill sites where they can leach into the soil and waterways adding an environmental hazard problem to that from a loss of limited resource. Elemental recovery and waste management are discussed in Chapter 9.

What do we need to do to redress this alarming situation? In a limited ecosphere like our planet it is obvious that the linear economy model will lead to market uncertainties, decreasing availability and increasing reliability not only on declining resources but also on resources in distant regions of the planet. We worry in Europe about getting gas from the east, what about metals from South Africa? Can we assume that a country with major social issues or political instability will continue to supply materials and not use them as political levers?

If we can learn to recover elements from waste, if we can process more efficiently and if we can design articles (or processes) better so that they use less critical elements and enable recovery of those elements when the article (or process) reaches its end of life, then we are making massive steps towards elemental sustainability even in resource-deficient regions such as western Europe. The use of benign and more efficient processes and the recovery of otherwise wasted or lost valuable elements are described in Chapters 4, 5, 6, 8 and 9. These cover bio- and non bioprocesses and also look at metals and non-metals. They include a chapter on the especially important challenge and opportunity for WEEE mining, the enormous amount of waste electronics we produce each year, which is growing at a staggering rate – surely we cannot continue to treat our slightly out-of-date phone, television and other devices which such lack of respect for the elements that went into them. This balance in favour of practical technologies is deliberate: politicians can and do spend long hours debating these issues but as scientists we have a responsibility to develop practical solutions and these must include effective ways of recovering valuable elements both to reduce environmental harm and to recover and re-use the precious and limited resources our planet gives us. To put the words of Mahatma Gandhi into a modern consumer society context: "…we can satisfy all of our needs if we are *resource intelligent*".

James Clark
York, UK

The editors would like to thank, Peter Hunt, Nontipa Supanchaiyamat and Helen Parker for their help in the preparation of this book.

Contents

RSC Green Chemistry No. 22
Element Recovery and Sustainability
Edited by Andrew J. Hunt
© The Royal Society of Chemistry 2013
Published by the Royal Society of Chemistry, www.rsc.org

CHAPTER 1

Elemental Sustainability and the Importance of Scarce Element Recovery

ANDREW J. HUNT,* THOMAS J. FARMER AND
JAMES H. CLARK

Green Chemistry Centre of Excellence, Department of Chemistry,
The University of York, Heslington, York, YO10 5DD, UK
*Email: andrew.hunt@york.ac.uk

1.1 The Issue of Elemental Sustainability

Important topics including climate change and peak oil have been making headlines with increasing intensity over the past decade. The subject of green, clean sustainable energy, fuels and chemicals is an important topic of focus for the scientific community and is a fundamental component of the long term wellbeing of planet Earth.[1] The necessity to be carbon neutral is well known and as a consequence solutions are being sought to lessen our dependence on fossil resources. New legislation and a growing movement towards the development of "low carbon technologies" are driving this technological change towards a sustainable carbon future. Unfortunately there is a serious problem, as many "low carbon technologies" including wind turbines, electric cars, energy saving light bulbs, fuel cells and catalytic converters, require rare and precious metals for their production and use.[2] Traditional supplies of such elements are "running out", thus creating other challenges in the form of a resource deficit. In fact, such elements are not running out or being destroyed

RSC Green Chemistry No. 22
Element Recovery and Sustainability
Edited by Andrew J. Hunt
© The Royal Society of Chemistry 2013
Published by the Royal Society of Chemistry, www.rsc.org

but are being quickly dispersed throughout our human environment or what has been referred to as the technosphere.[3,4] This makes recapture of these unique elements both highly problematic and costly. Such challenges must be tackled through the development of multidisciplinary partnerships and a sustainable holistic approach to the extraction, processing, use and recovery should be adopted for all elements within the periodic table. The only exception to this would be radioactive materials which cannot be recovered in the initial state once decay has occurred. As such, it is essential to develop new sustainable routes and strategies for the recovery and reuse of these elements.

Elemental sustainability is a concept whereby the sustainability of each element in the periodic table is guaranteed. For an element to be sustainable, its use by this current generation should not impair or restrict future generations from also utilising that same element. Within these constraints, it is also important to consider the triple bottom line of sustainability, that is, the environmental, societal and economic effects of these elements and their use.[5,6] All elements within the Earth's crust are available in finite amounts, although some, like aluminium, iron and silicon, are available in many orders of magnitude higher abundances than others, like platinum, silver and selenium.[7] Each element in the periodic table also has varying levels of demand. This demand varies as new technological advances come on-stream and others become obsolete. Rising demand for some elements is caused by both developed and developing nations which require advanced materials for consumer goods products (*e.g.* mobile phones and flat screen televisions) and the level of demand for each element often varies from nation to nation and region to region. As the world's population continues to rise, the growing middle classes will continue to demand a higher standard of living, fuelling a need for consumer goods and cleaner energy.[8] This combination of known availability of certain elements and their current level of demand has caused some to have been flagged up with concern. Although we should endeavour to use all elements in the periodic table sustainably, those whose current rates of use risk depleting known reserves in the near future should be of greatest focus in the short to mid-term. Reserves are known tonnages of metals that can be economically and legally extracted using existing technologies.[9] These reserves represent only a small proportion of the element compared to the significant abundance in the Earth's crust, while the resources of elements are represented in the locations or concentrations of that element or ore that have reasonable prospects of being recovered in the future.[9]

As shown in Figure 1.1 numerous elements fall into the range where current known reserves will be consumed in less than 50 years if current rates of extraction are retained. Some of these are at high risk as a result of exceptionally low crustal abundances and these include the precious metals where the annual production of the majority is below 200 tonnes.[10,11] It is not only elements with low crustal abundances and small annual productions that are of concern as known reserves of both strontium (Sr) and manganese (Mn)would be consumed in less than 50 years at current annual production levels of 380 000 and 16 000 tonnes, respectively.[10] However, both the

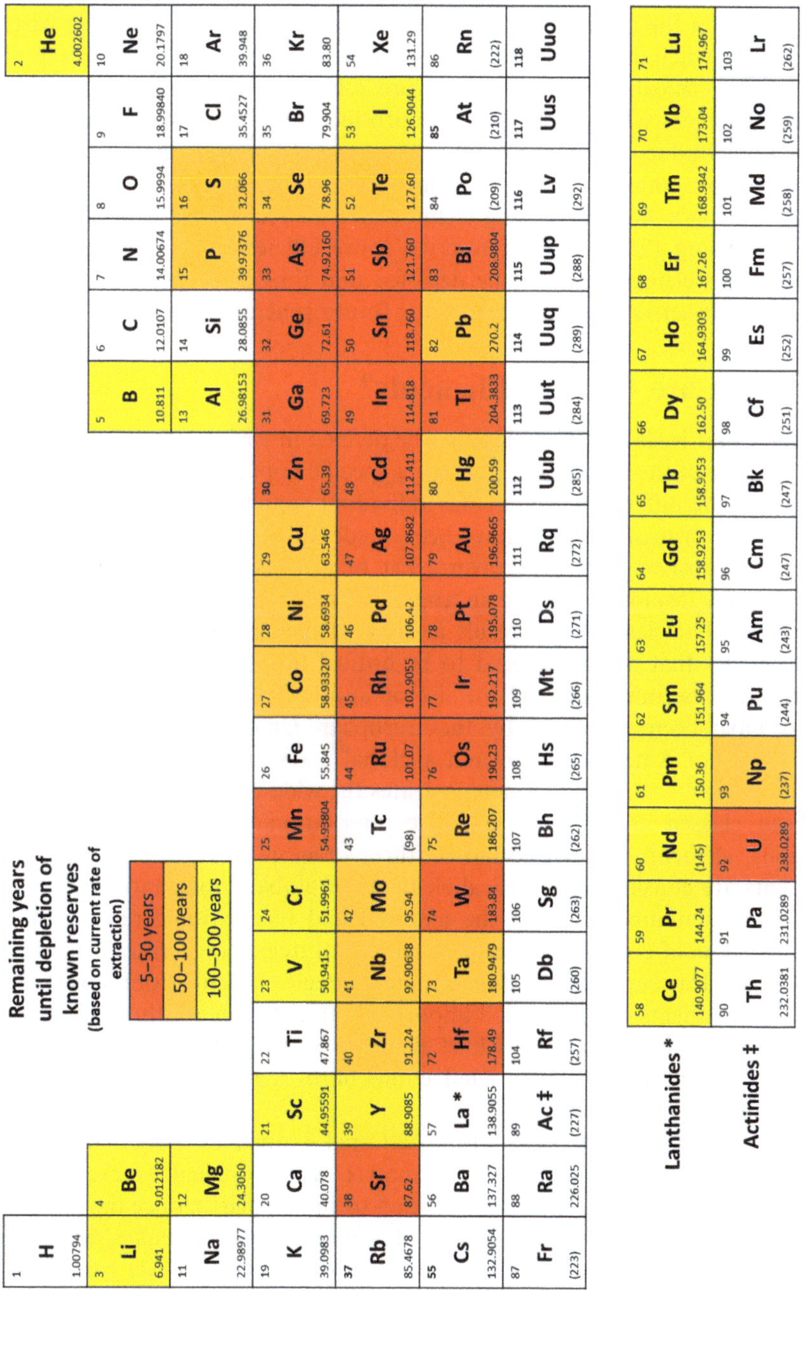

Figure 1.1 Number of years remaining of rare and precious metal reserves if consumption continues at present rate (original data adapted from Salazar,[10] Brown *et al.*[11] and Rhodes[12]).

consumption and reserves of these finite elements are continually changing in response to movements in markets, discovery of new mineral deposits, development of new applications, advances in extraction technologies and improvements in the efficiency of use, recovery and recycling.[13] As such, care must be taken when using the rate of consumption versus known reserves as a metric for the criticality of elements (Figure 1.1). Current known reserves of indium, an element which is vital for the production of display devices, solar cells and semiconductors, may run out in as little as 13 years at the current rate of consumption, thus fuelling concerns over the security of supply.[12] If investment is made into developing technologies for recovery at end-of-life, in addition to using the remaining reserves more efficiently, it is hoped that supplies of this element will not be depleted and thus by utilising them in a sustainable manner reserves will be left for future generations.

1.2   What are Critical Elements?

An element that is classed as critical can be defined in various ways depending on the purpose of the assessment (*e.g.* for a specific application such as mobile phones) and the different needs of the individual country or territory. Many assessments of critical elements have been made, all of which apply different criteria and as such generate a diverse list of critical elements (Table 1.1), although in all instances there is an appreciation that current and projected demands for that element will result in rapid depletion of known reserves.[9] Often elements that are considered to be critical in one territory, nation or company may be omitted from the list of another. Assessments of critical elements have been generated by the European Union, United Kingdom, United States of America, Japan and a global assessment was made by the United Nations.[9,13–19] Table 1.1 demonstrates the wide variety of elements that have been classed as "critical" in these international assessments. In our discussions, elements that are found on three or more of these international assessments are considered as critical elements of global importance.

A significant number of both the national and international reports that discuss elements of concern use terms such as "strategic" or "critical" for the raw materials. It is important to differentiate strategic elements from those that are critical. Elements of "strategic" importance to a nation are those vital for defence or military applications, whilst elements with significant international supply risk issues, which if restricted could harm a nation's economy, are considered to be "critical".[9]

The National Research Council of the National Academies (NRCNA, USA) developed a two-dimensional "criticality matrix" as a graphical representation of critical elements.[17] Elements can be added to such a matrix once the impact of supply restrictions and the supply risk of a mineral have been assessed. This study highlighted elements including platinum group metals (PGM), rare earth elements (REE) and indium, manganese and niobium as being most critical as they represent elements with both high supply risks and high impact if the supply of that element is impaired.[17] For a true assessment of elemental

Table 1.1 Critical elements as highlighted in international assessment.

Element	Symbol	Global UNEP (2009)[18]	Global OECD (2010)[16]	Japan NISTEP (2008)[19]	USA NRCNA (2007)[17]	USA DOE (2010)[15]	EU JRC (2011)[13]	EU E & I (2010)[9]	UK BGS (2012)[14]
Critical elements of global importance (*)									
Rare earths	REE	*		*	*	*	*	*	*
Gallium	Ga	*		*	*	*	*	*	*
Indium	In	*		*	*	*	*	*	*
Platinum group metals	PGM	*		*	*	*	*	*	*
Tantalum	Ta	*	*	*	*			*	*
Cobalt	Co	*		*		*		*	*
Niobium	Nb			*	*		*	*	*
Antimony	Sb		*	*				*	*
Beryllium	Be		*	*				*	*
Lithium	Li	*		*	*	*			
Tellurium	Te	*		*		*	*		
Germanium	Ge	*		*				*	*
Vanadium	V			*	*		*		
Tungsten	W			*			*	*	*
Molybdenum	Mo			*			*		*
Selenium	Se			*			*		*
Critical elements of multinational importance (*)									
Hafnium	Hf			*			*		
Nickel	Ni			*			*		
Bismuth	Bi			*					*
Strontium	Sr			*					*
Barium	Ba			*					*

Table 1.1 (*Continued*)

Element	Symbol	Global UNEP (2009)[18]	Global OECD (2010)[16]	Japan NISTEP (2008)[19]	USA NRCNA (2007)[17]	USA DOE (2010)[15]	EU JRC (2011)[13]	EU E & I (2010)[9]	UK BGS (2012)[14]
Magnesium	Mg								*
Manganese	Mn			*	*			*	
Titanium	Ti			*	*				
Copper	Cu				*				
Cadmium	Cd						*		
Silver	Ag						*		
Tin	Sn						*		
Mercury	Hg								*
Thorium	Th								*
Arsenic	As								*
Yttrium	Y					*			
Rubidium	Rb			*					
Caesium	Cs			*					
Zirconium	Zr			*					
Chromium	Cr			*					
Rhenium	Re			*					
Boron	B			*					
Thallium	Tl			*					

Critical elements of national importance (*)

UNEP - United Nations Environment Programme (UN), OECD - Organisation for Economic Cooperation and Development (UN), NISTEP - National Institute of Science and Technology Policy (Japan), NRCNA - National Research Council of the National Academies (USA), DOE - Department of Energy (USA), JRC - Joint Research Centre (EU), E & I - Enterprise and Industry (EU) and BGS - British Geological Survey (UK).

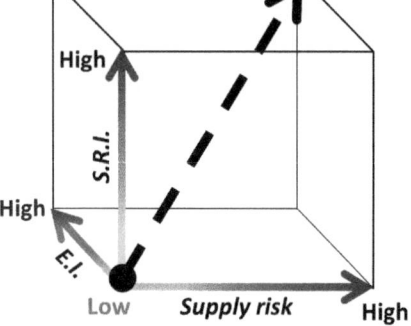

Figure 1.2 Diagram of criticality of elements, where the three axes are represented by the supply risk, the supply risk impact (S.R.I.) and the environmental impact of element extraction and use (E.I.).

sustainability, the environmental impact of an element's extraction and use should also be taken into account. Figure 1.2 demonstrates this in a graphical illustration giving a three-dimensional representation of elemental sustainability which inherently includes criticality. As criticality increases so the sustainability of the element decreases.

1.3 Why is there a Growing Security of Supply Issue?

In many cases developing economies have significantly greater potential reserves and production capacity of critical elements than nations with established economies (Table 1.2). Frequently these developing economies are adopting aggressive trade restrictions, thereby obtaining exclusive domestic use of their elemental reserves.[9] These strategies include the systematic tightening of exports through the application of taxation, implementation of strict quotas, subsidies, price-fixing or restrictive investment, distortion of international trade, elevated export duties and investment policies which are frequently at odds with international trade agreements. The combination of export restrictions, political factors and the manipulation of markets has resulted in REE supply shortages and significant price increases.[9] Consumption patterns, materials efficiency and applications all change over time, thus demand for these raw materials is also in a state of flux. The ability to respond quickly to such changes can restore balance to markets. The technical and economic obstacles of increasing production (by opening new mines) are considerable and take many years to achieve. To alleviate supply restriction issues, a number of new REE mines are due to be opened in South Africa, Kazakhstan, Malaysia, Burma and Canada.[21] In cases where the production of minerals or elements is limited to a small number of countries, these producers can gain political or commercial advantages by influencing supplies and markets. In some cases these aggressive trading strategies have been used to provide a competitive advantage for domestic industries over their international competitors.[9]

Elements such as antimony, gallium, germanium, indium, magnesium, REE and tungsten now have a high supply risk (Table 1.2). In 2010, China controlled 95% of the world's supply of rare earth metals.[22] When one nation has a monopoly in terms of production of an element it results in issues relating to the longer term security of supply (Table 1.2).[9] Countries whose manufacturing or technology base depends on imported critical elements are beginning to look

Table 1.2 Critical elements of global importance and percentage of current production via hitch-hiker recovery (original data adapted from references 14 and 29–33).

Element	Symbol	Relative supply risk[14]	Major producing nation	[a]Total global mined production/tonne	Global production as hitch-hiker/%	Attractor metal
Rare earths	REE	9.5	China	114 000	47	Fe
Tungsten	W	9.5	China	61 000	—	—
Antimony	Sb	9.0	China	126 000 (*40 000*)[b]	No data[b]	Au, Cu, Pb[b]
Molybdenum	Mo	8.6	China	221 000	100	Cu
Germanium	Ge	8.1	China	84 (*36*)	100	Zn
Beryllium	Be	8.1	USA	190[c]	—[c]	—[c]
Gallium	Ga	7.6	China	106 (*55*)	100	Al
Indium	In	7.6	China	574	100	Zn
Platinum	Pt	7.6 (all PGMs)	South Africa	192 (*57*)	100[d]	Cu, Ni
Palladium	Pd	7.6 (all PGMs)	South Africa	202 (*158*)	100[d]	Cu, Ni
Rhodium	Rh	7.6 (all PGMs)	South Africa	29 (*7*)	100	Pt/Pd
Ruthenium	Ru	7.6 (all PGMs)	South Africa	35	100	Pt/Pd
Cobalt	Co	7.6	DRC	88 000	85	Ni (50%) Cu (35%)
Niobium	Nb	7.6	Brazil	63	3	Sn
Selenium	Se	7.1	Japan	3250	100	Cu
Tantalum	Ta	7.1	Brazil	772 (*244*)	13	Sn
Lithium	Li	6.7	Australia	25 300[e]	—[e]	—[e]
Vanadium	V	6.7	South Africa	56 000[f]	74[f]	Fe (59%) Al, U[f]
Tellurium[g]	Te	N/A	USA	475	100	Cu

[a]Numbers in parentheses are additional production from recycling.
[b]Sb from Roskill Report 2011.[33]
[c]Be data from USGS 2011 report.[31]
[d]Pt and Pd are primarily dispersed in Cu and Ni ores, but in the instance of South Africa these ores are mined for the Platinum Group Metal (PMG).
[e]Li data from USGS 2011 report.[30]
[f]V data from Pacific Ore website. (Pacific Ore Mining Corp, 2013).[32]
[g]Te is not in the BGS Risk List (2012)[14] as it is solely produced as a by-product from other metal processing and thus supply issues are difficult to assess.

for alternative sources, whilst other countries and companies that are dependent on rare earths elements are racing to secure control of mineral rights in Australia, South Africa and Greenland.[22] In 2009, a diplomatic dispute between China and Japan resulted in a temporary halt in REE exports to Japan.[9] Such restrictions could be devastating for the Japanese economy which currently consumes 30% of the world's annual rare metal production.[23] Similar risks are also true for PGM, where South Africa generates 89% of the world's production, 40% of the world's cobalt is from the Democratic Republic of Congo (DRC) and 90% of the niobium is produced in Brazil.[9,15,22]

By using the list of critical elements of global importance we can surmise that in a global sense all of the elements are of concern and could be significant supply risks in the coming decades.[14]

Many of the critical elements are located in parts of the world that are often viewed as politically unstable (Figure 1.3). Niobium and tantalum are extracted from tantalite ore, of which significant quantities come from the Democratic Republic of Congo.[24] In recent years, the illegal mining of these ores in the Democratic Republic of Congo has significantly contributed to instability and aided in financing conflict in the region.[25,26] A 2003 UN Security Council report highlighted that a significant amount of ore was being smuggled out of the country by militia.[27,28] Such illegal uncontrolled mining can have serious environmental and social impacts, whilst also damaging local wildlife. Eastern mountain gorilla and elephant populations in the Congo have been severely reduced through hunting by miners and militia.[25]

A key concern regarding the availability of these elements in the future is their abundance and ease of accessibility. Currently, the majority of critical elements are mined and extracted from primary ore in highly energy intensive processes that require a sufficient concentration of the element of interest. As already indicated, the geological abundance of most mineral resources is potentially high, however, the concentration of elements within the ores compared to industrial or base metals (such as iron) can be very low.[9] The exploitation of these lower grade ores and the challenges of having to mine in geographically and politically hostile locations can have a significant bearing on the cost and could have a potentially negative impact on the surrounding environment. This is compounded by the fact that the production volumes of elements of critical global importance are much smaller than industrial metals and frequently require difficult extractive metallurgy.[1,9] Mining will continue to contribute a significant proportion of critical element supply in the future. Consequently, improving mineral detection in order to locate new deposits and focussed investment on research into sustainable mining are of vital importance to the long term security of supply of these elements.

A holistic approach to elemental use must be adopted throughout the life cycle, including processing, manufacture, recycling and substitution. A significant impact on future availability of certain metals will be achieved if more efficient processing methods are developed. These will improve the yields of elements that are mined as "by-products" of or "hitchhikers" on primary or "attractor" metal deposits.[29] The Institut Européen d'Administration des

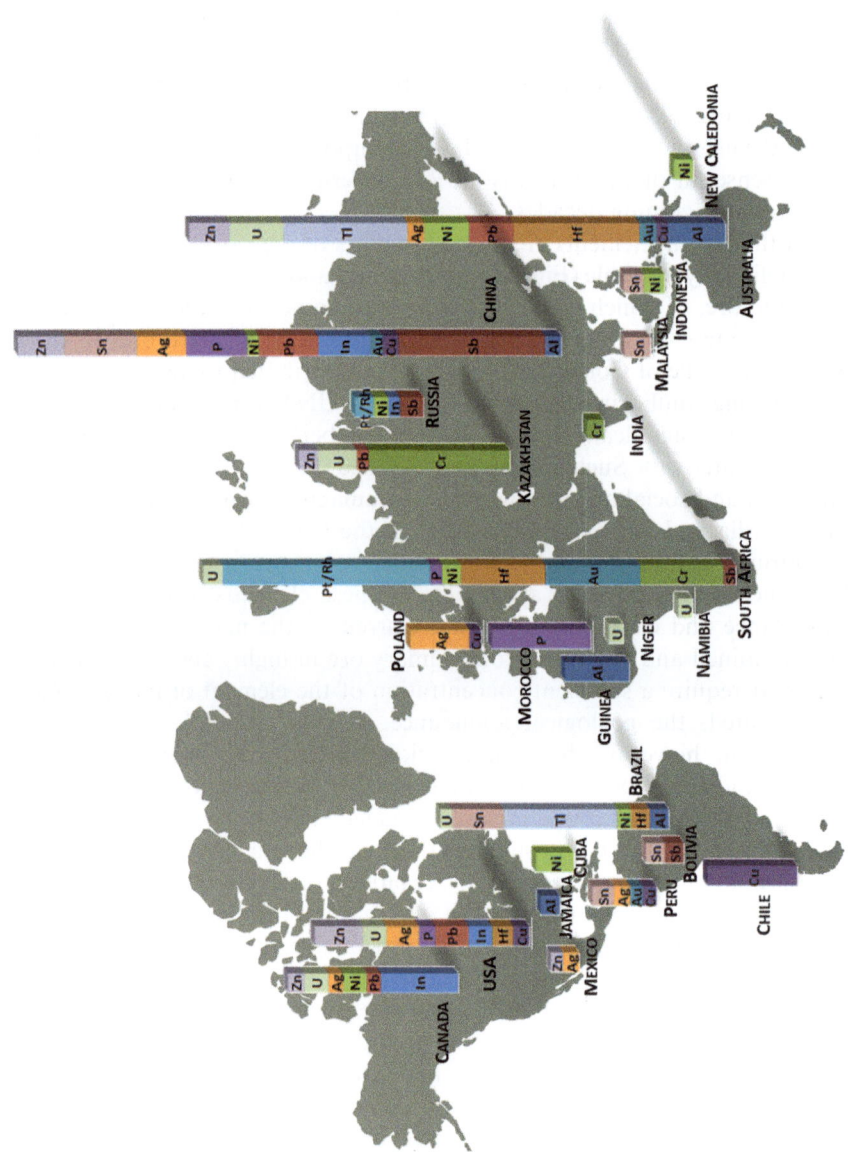

Figure 1.3 Distribution of rare and precious metal reserves around the world (original data adapted from Brown *et al.*[10]).

Affaires (INSEAD) report (2012) highlights the current scenario whereby many of these critical elements are predominately produced as a by-product of mining and processing of base "attractor" metals (*e.g.* Zn, Cu and Pt) and as such are dubbed "hitch-hiker elements" (*e.g.* In, Co and Ru).[29] This scenario could have major implications for future increases in demand for the specific critical element. Either the rate of mining of the attractor metal would have to increase or the rate of recovery of the critical metal from the base metal ore would have to improve or new resources of the critical metal, independent of a base metal, would have to be found. The first option above is unlikely to make economic sense unless demand for the base metal also increases, while the latter two options require technological advances and increased initial capital expenditure. The only other viable alternative is for further technological advances to be implemented that result in significant improvements in the rates of critical hitch-hiker metal recycling. Table 1.2 demonstrates the percentage of current production via hitch-hiker recovery and from what base metals these are predominately derived from the previously highlighted list of globally significant critical elements.[29–33] More efficient use of resources, recycling and also substitution with alloys or more abundant elements, can be very effective in alleviating pressure on existing diminishing reserves.

1.3.1   Trends in Elements

You have only to study the recent trends in the price and production of many elements to realise that the consumption of reserves is increasing at an alarming rate (Figure 1.4). Markets for critical elements are highly volatile and are frequently influenced by supply risk. This can be observed in the price of REE, which has risen dramatically following growing concerns over security of supply from China (Figure 1.4).[34] Production and consumption rates of many critical elements such as gallium are also rising at a dramatic rate from a relatively low base level (Figure 1.4).

The price of elements can also be influenced by a rapidly growing demand for use in new applications. Indium prices rose a staggering 800% in 6 years from approximately US$85/kg in 2002 to US$685/kg in early 2008 (Figure 1.4), paralleling a growth in large screen televisions sales.[35,36] As many of these elements are "hitch-hikers" it means that increases in production cannot be easily achieved and if demand for attractor elements decreases, so too will the production of the rare by-products. This has resulted in predictions that the demand for some elements will soon outstrip supply.[1,12] The key question is what will happen to the prices next and how will demand for all these metals that underpin our technologically advanced lifestyles be met sustainably in the future.

The critical element markets are vulnerable to production concentration and price volatility. There has been a significant drop in the production and sale of tantalum from African mines since the dramatic spike in the price of this metal in the year 2000.[37] This unprecedented spike in the market was a result of supply fears caused by nervous dealers scrambling to lock themselves into

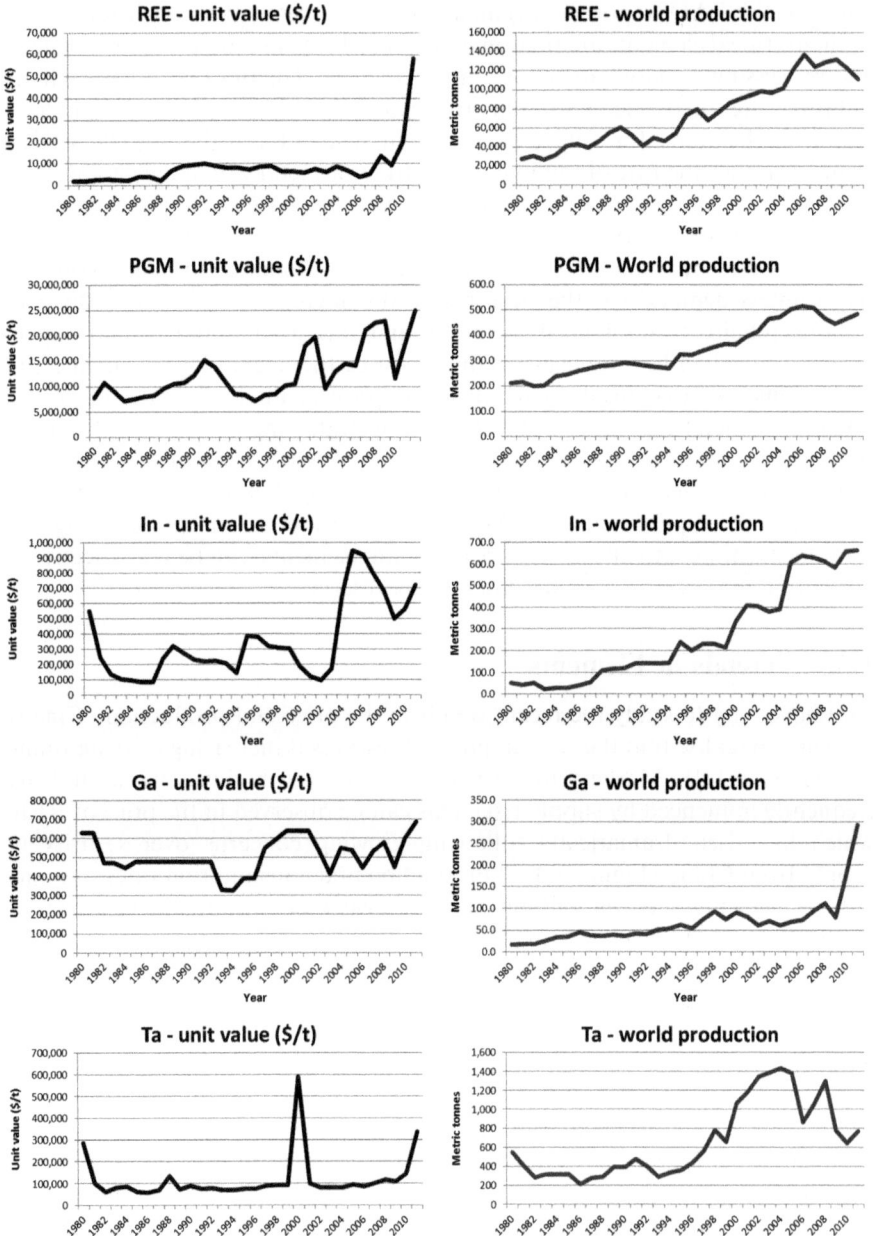

Figure 1.4 Market price (US$/tonne) and world production in metric tonnes for REE, PGM, In, Ga and Ta (original data adapted from Kelly and Mattos[38]).

long-term contracts at dramatically high market prices.[37] In September 2001 the US House of Representatives passed a resolution banning the purchase of tantalum from the war ravaged Democratic Republic of Congo, further fuelling the frenzy to secure future delivery from legitimate sources.[25] Since 2010 similar panic buying and increases on the markets have been observed in response to growing concerns over security of supply.

There is a significant negative societal impact which is associated with the increased global demand and supply of critical elements. This issue will need to be addressed on a global scale. The development of effective recycling methods, strategies for greater materials efficiency in manufacturing and the pursuit of new substitute raw materials from technological innovations are imperative to meet these challenges. In order to mitigate the risk of supply and become more self-reliant in terms of elements, developed nations should look at their waste as a valuable resource ready for exploitation.

1.4 Current Uses of Critical Elements

A great concern of the government bodies of developed nations investigating critical elements is about the impact of the scarcity on emerging energy technology.[9,20] Rare earth metals are extensively used in the modern solution to energy and pollution woes such as in the batteries of electric cars or energy efficient light bulbs (Table 1.3). Their use also extends to applications in photovoltaic cells, generators (*i.e.* for wind, tidal and wave turbines) and catalytic converters for the abatement of emissions from automobile exhausts.[9,20] Other critical elements are also used extensively in consumer goods, typically electronics (LCDs, hard drives of computers, lasers *etc.*), while their application has also spread to specialist alloys for engineering (*e.g.* platinum aluminde for aircraft turbines), resistant glass, ceramics, direct medical applications (*e.g.* cancer treatment with *cis*-platin derivatives) and extensively as catalysts in the chemical and related industries.[9,20] In 2008 alone, 1300 million new mobile phones produced worldwide, consumed approximately 12 000 tons of copper, 4900 tons of cobalt, 325 tons of silver, 31 tons of gold and 12 tons of palladium.[39] 300 million new personal computers and laptops were produced which consumed 150 000 tons of copper, 9100 tons of cobalt (Li-ion batteries for laptops), 300 tons of silver, 66 tons of gold and 24 tons of palladium.[39] Seemingly the types and volumes of applications where critical metals are used are ever increasing and thus demands for these elements are also rising.

1.4.1 Critical Elements in Catalysis

Many of the critical elements mentioned above are used as catalysts (*e.g.* PGMs, Co, Ce, Ge, Sb and In) and reagents (Li) used in the chemical and related industries. Therefore their scarcity would also have an impact on a plethora of chemical transformations used in the generation of pharmaceuticals, plastics, fuels, home and personal care products and packaging.

Table 1.3 Major uses of critical elements of global importance.

Element	Symbol	Major uses
Antimony	Sb	Flame retardant, semiconductors, alloys, pharmaceuticals, catalyst, flame and PET catalysts
Beryllium	Be	Electronics
Cobalt	Co	Superalloys, catalysts and batteries
Gallium	Ga	Semiconductors, solar cells, MRI contrast agent, electronics (integrated circuits) and solar cells
Germanium	Ge	Semiconductors, solar cells, catalyst, infrared optics, PET catalysts, solar cells
Indium	In	Flat-panel displays, alloys, photocells and touch screens
Lithium	Li	Batteries, ceramics and glass
Molybdenum	Mo	High performance stainless steel
Niobium	Nb	HSLA steel (high strength low alloy steels)
Palladium	Pd	Alloying agent, industrial catalyst, fuel cells and catalytic convertors for automobiles
Platinum	Pt	Alloying agent, industrial catalyst, fuel cells and catalytic convertors for automobiles
Rare Earths	REE	Magnets, batteries, ceramics and catalysts
Rhodium	Rh	Alloying agent, industrial catalyst, fuel cells and catalytic convertors for automobiles
Ruthenium	Ru	Hard dish drives, catalysts and electrochemistry
Selenium	Se	Glass, photovolteics and infrared optics
Tantalum	Ta	Capacitors for electronics
Tellurium	Te	Steel additive, solar cells and thermoelectronics
Tungsten	W	High strength cutting tools
Vanadium	V	HSLA steel (high strength low alloy steels)

The above only shows globally significant Critical Elements highlighted in Table 1.1.

A variety of these transformations have been regarded as green (*e.g.* the atom economy of metathesis) and even applied in the valorisation of waste, a key move towards a more sustainable chemical industry.[40] Statistical data from the *Green Chemistry Journal* (RSC) and *ChemSusChem* (Wiley) show that since their inception in 2003 and 2008, respectively, there have been numerous articles demonstrating the use of some of these critical metals, typically as catalysts (Figure 1.5).

However, one of the principles of green chemistry is that reagents are renewable and it is a disappointing oversight that many researchers do not carry this through to consideration of the renewability of their catalysts. For elements that have high criticality there are major concerns about both their high consumption and long-term supply. As such, high criticality elements can also be viewed as elements with major sustainability concerns (our non-renewable use of them now would prevent future generations having access to these metals). It could therefore be further argued that any transformation, even of bio-derived chemicals, is no longer sustainable if critical elements have been used for catalysis. A requirement for any process claiming to form a sustainable product should be that all aspects and auxiliaries of that process are themselves sustainable and this would therefore require the development of

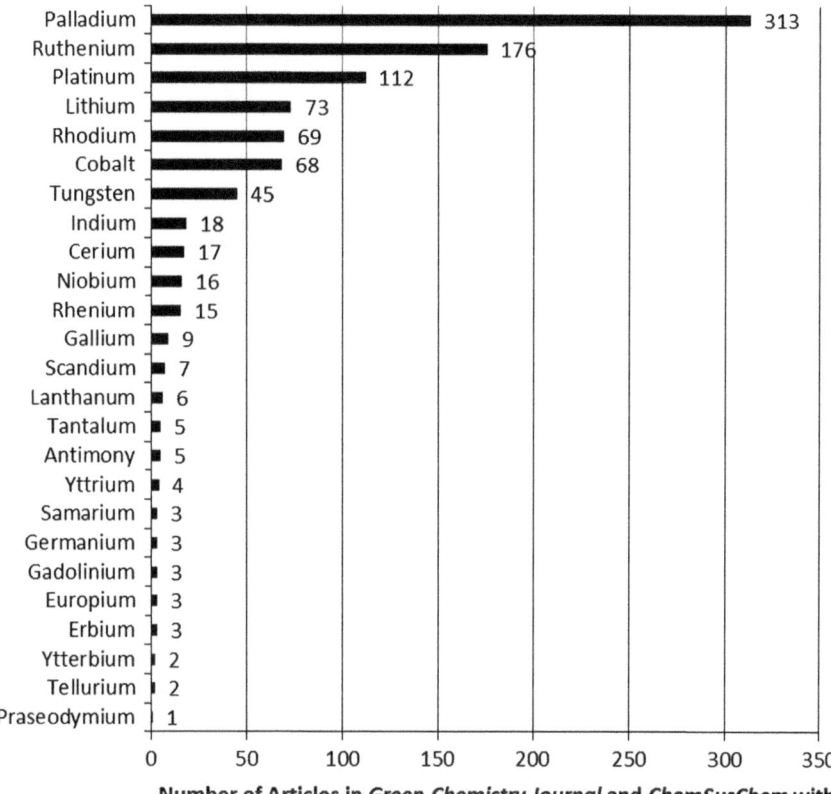

Number of Articles in *Green Chemistry Journal* and *ChemSusChem* with Reference to Critical Element in Topic

Figure 1.5   Statistical data from the *Green Chemistry Journal* (RSC) and *Chem-SusChem* (Wiley) indicating metal use (each element's name was searched in Web of Knowledge under Topic, February 2013).

truly sustainable reagents, solvents and, of course, sustainable catalysts. For critical elements to be used as truly sustainable catalysts their recovery, ideally total, after use followed by subsequent reuse is required, thus progressing to an ideal circular economy.

An alternative to the total recovery of critical metals from catalytic processes is to seek replacements for active catalysts, either via the use of non-metallic catalysts (*e.g.* enzymes) or by using metals with less criticality. Many critical elements used in catalysis today, such as palladium, ruthenium and cerium, have replaced lesser or non-critical elements from earlier versions of catalytic processes. One such example of this is the metathesis reaction (Figure 1.6), including olefin cross-metathesis (CM), ring-closing metathesis (RCM) and ring-opening metathesis polymerisation (ROMP).[41]

Metathesis has long been touted as a key "green" reaction owing to its versatility, atom efficiency (if all products are useful) and applicability to emerging and established bio-derived commodity chemicals (*i.e.* unsaturated

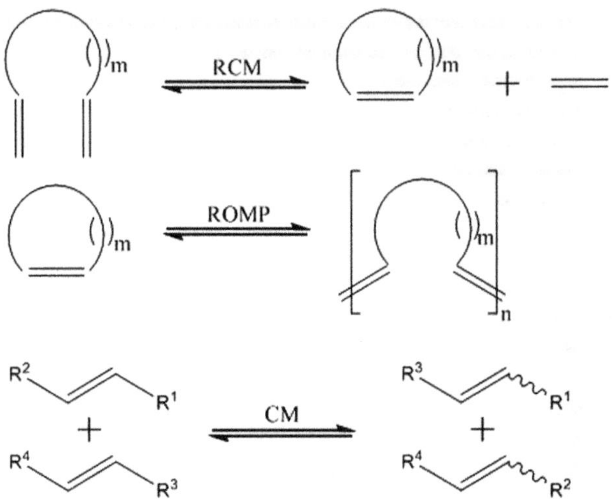

Figure 1.6 General schemes for ring-closing metathesis (RCM), ring-opening metathesis polymerisation (ROMP) and cross metathesis (CM).

fatty acids, cinnamic acid and fumaric acid).[42–44] Metathesis is also well known for the 2005 Nobel Prize won by Chauvin, Grubbs and Shrock, although key breakthroughs in this reaction protocol date back several decades earlier.[45] Modern understanding and application of olefin cross-metathesis is dominated by ruthenium catalysts such as those developed by Grubbs and Hoyveda. From the discussions above it is evident that ruthenium is an element of global critical importance, with major issues in supply security, recyclability and high usage in current and emerging technologies. Clearly for olefin cross-metathesis to play a role in sustainable catalysis, alternatives to the currently preferred Ru catalysts must be found. Early research into olefin cross-metathesis made use of tungsten and molybdenum carbene catalysts with these often being found to have higher activity then Ru equivalents but with poor water, air and functional group tolerance.[46,47] Although W and Mo are highlighted as critical metals in some reports, they are not as unanimously perceived as critical in the way that Ru is (Table 1.1). At first glance it may then seem that W and Mo could be exploited as candidates for more sustainable metathesis catalysts, especially when considering the extensive reserves of both elements compared to Ru. However, there are major issues with both W and Mo, which are:

- Recycling rates for W are only comparable to Ru (10–25%), Mo is a little better (25–50%), although this is predominately via recycling and reuse of steel alloys containing Mo.[10]
- Both W and Mo are considered by the BGS to have greater supply risk than Ru (Table 1.2).
- Mo is currently primarily produced as a hitch-hiker in Cu production (Table 1.2) and therefore its price and availability are linked directly to the demand for another element.

- Although W is primarily produced from mines that focus on it as an attractor metal (*i.e.* not a hitch-hiker), consumption of currently known reserves will be exhausted in less than 50 years if current demand is maintained (Figure 1.1).
- Reserves of W and Mo are 3.2 and 11 million tonnes respectively,[10] this being considerably higher than the 5000 tons for Ru,[48] though rates of use of W and Mo are significantly higher.

Various metals other than Ru, W and Mo have also been utilised as metathesis catalysts (Figure 1.7); these include Re, Os, Ir, Ti, Cr, Co, Nb, Rh and Ta.[45] Unfortunately, all these elements have been cited nationally or even internationally as being critical (Table 1.1). When taking each of these metals individually, as for the Mo and W catalysts above, major issues around supply risk, recycling or global reserves can be found. Some, such as Cr, also suffer from concerns regarding toxicity and are thus unviable as a result of international legislation such as REACH.[49] At first glance Ti seems to offer a glimmer of hope. Although Ti was highlighted by two national reports as being critical,[18,19] it has high recycling rates (91%),[50] has global mineral reserves totalling more than 2 billion tonnes and is not regarded as having a supply risk (Table 1.1).[51] To date, the utilisation of Ti catalysts for metathesis seems to be limited to ROMPs and occasional examples of cross-metathesis, such as those using Tebbe's reagent (Figure 1.7), but reagents typically need to be devoid of carbonyls to prevent Wittig type methenylations (the original purpose for Tebbe's reagent) and quantitative amounts of the titanium "catalyst" are required.[52-54]

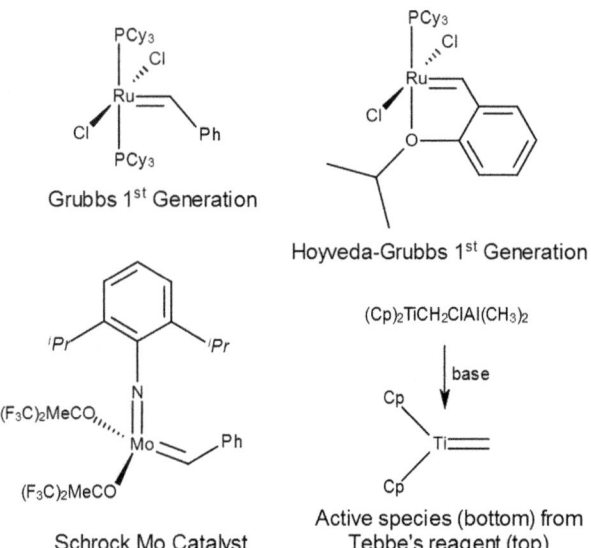

Figure 1.7   Metathesis catalysts based on Ru, Mo and Ti centres.

Perhaps the answer to sustainable metathesis catalysts lies in the development of ligands for Ti centred catalysts that both improve the air and moisture stabilities of the catalyst and also prevent carbonyl methenylations, while still promoting the desired cross-metathesis. Above all, the example of metathesis highlights how the approach of replacing an efficient but critical metal, such as Ru, with others that are less critical is not always feasible. It is certainly not simple and therefore methods of more effective metal recovery and reuse are a must. Concerns about the application of critical metals in catalysis should not only focus on sourcing, but also on using them (or other metals) in a more sustainable manner.

1.4.2   Growing Need for Greener Elemental Recovery

Sustainable and green methods for the recovery of elements are imperative. Hyperaccumulation of metals by plants has become a significant research focus in recent years owing the significant potential for metal extraction by phytoremediation. Hyperaccumulation refers to the accumulation of elements at 100 times greater concentrations (normally toxic to plants) than typically observed for traditional plants growing in the same location (hyperaccumulation is discussed in detail in Chapter 5).[55] These plants can be harvested, combusted (thus producing energy) and then smelted to enable production of pure metal.[56]

Phytoremediation opens up the opportunity not only to restore degraded land and generate energy by burning the biomass but also allows elements to be recovered from low grade ores, contaminated soils, mine tailings and wastewaters that would typically be uneconomic to exploit. Plants need to be able to tolerate elements, be capable of rapid growth and have high biomass yields. At present, no plant is known that has all of the above attributes.[57] Genetic modification could hold the key to opening up this technology in the future, but this must be done in way that would not introduce negative ecological impacts. Recovery of waste metals via hyperaccumulation (bioremediation) and subsequent re-use as catalysts or catalyst precursors is of growing interest. This is of great potential value in a sustainability and critical metal circular economy sense.[58,59] For example it has been demonstrated that Friedel–Craft alkylation and acylation Lewis acid catalysts can be prepared from extracts of metal hyperaccumulating plants.[60]

Harnessing the ability of bacteria to solubilise elements selectively could become a major new technology owing to industrial interest focused on copper, nickel, cobalt, zinc, gold and silver.[61] This bioleaching process has been successfully used to treat over 160 000 tonnes of copper ore daily.[62] Bioleaching has also been studied extensively for the recovery of Pd from waste streams, with some bacterium (*Desulfovibrio desulfuricans* and *Shewanella oneidensis*) found to generate Pd(0) nano-particles with a narrow size distribution, located on the outer parts of the cell and adequate cell adherence for use in catalysis.[63] However, these bacteria suffer from Cu(II) inhibition which is a major issue for

Pd recovery from printed circuit boards as Cu(II) often makes up $>25\,wt\%$ of the solid in this waste leachate.

A potential method for metal recovery and catalyst preparation from wastewaters can be the use of biosorbents. Polysaccharides are ideal for such processes, as they interact with metal ions by electrostatic interactions, ion exchange or to form complexes with metal ions.[64] It is likely that bioremediation processing will result in mixed-metal recovery as the location where bioremediation occurs contains waste with an array of mixed metallic ions. This will either require separation of the various metals, or alternatively may result in the growth of catalysis via mixed metals.

There is increasing interest in the field of bimetallic (or alloy) catalysts, as this often accesses improved efficiency compared to monometallic catalysts.[65] The synergistic effect of metal combinations could offer vast potential for the future of catalyst design. A greater understanding of bimetallic synergistic effects and mixing patterns will ultimately lead to greater possibilities for predicting suitable applications for metals derived from mixed-metal wastes and alloys of base or low criticality metals may give more active catalyst than those currently derived from pure high critical elements.

Ultimately we have to seek 100% recovery of critical metals from catalytic processes and with this also understand better the synergistic effects of mixed metals from waste, whilst also maximising the metals catalytic activity. In the mid-term, finding non-critical metals, such as Fe, for a wider range of catalytic processes is a possibility, but the topics addressed in this book focus on the ideal scenario whereby even scarce critical metals can be used as sustainably catalysts long into the future if total recovery from anthropogenic cycles is guaranteed.

1.5 New Sources of Critical Elements

As we disperse materials throughout the environment there are several potential new sources of elemental recovery including municipal and industrial solid waste, electrical and electronic products (WEEE-waste), landfill sites, low grade ores, mine tailings and aqueous wastewaters.[1] These offer reductions in hazardous waste and supply an alternative to virgin resources.

The amount of waste electrical and electronic equipment (WEEE) produced globally is increasing rapidly. Global WEEE generation annually is estimated to be 50 million tonnes per year.[66] This waste is a significant potential resource for the supply of scarce and valuable metals. However, WEEE also frequently contains toxic and hazardous materials and therefore requires careful or special treatment. Globally, only a small portion of well separated e-waste is treated; two good examples of industrial practice are at Umicore in Belgium and New Boliden in Sweden.[39,67] Illegal trading of e-waste from the western world to developing countries causes serious environmental problems and a high risk to human health. The effective recycling and refining of WEEE has become a global issue. Chapter 8 highlights economic, technological, social and environmental concerns relating to WEEE recovery.

The annual estimated worldwide generation of municipal solid waste (MSW) is 1636 million tonnes and rising.[68,69] As such, MSW offers a great potential for the recovery of elements.[69] There are further discussions on the potential for recovery from MSW in Chapter 9.

Ultramafic soils from weathered mineral landscapes are relatively high in nickel, chromium, manganese, cobalt, titanium, iron and other metals, as are industrially contaminated soils.[56] Not only is there a growth in saving energy by recycling elements but also new sources of recovery must be sought, including that of roadside dust as a source of PGM (up to 1.5 ppm of platinum) from catalytic converter emissions.[70,71]

Mining processes can utilise aqueous/organic-based metal extraction systems which generate large volumes of wastewater containing low concentrations of dissolved metals.[72] At source recovery of valuable and environmentally hazardous elements can prevent damage to local ecosystems. The nuclear industry, electroplating and metal processing operations also produce wastewaters containing Cr, Ni, Cd, Zn, Cu, U and precious metals.[73] Exploitation of these new sources of elements at low concentrations could lead to further recovery and enhance security of supply but as previously discussed it is of significant importance that green and sustainable recovery methods are employed.

1.6   Could a Circular Economy Hold the Answer?

Reuse, recovery and recycling are an answer to avoiding a future resource deficit and improving the sustainability of all elements. The recovery and recycling of metals from wastes and end-of-life products is not new. The recovery and recycling of metals from waste streams can be an efficient, economical and environmentally beneficial route to valuable materials. In fact in recent years, significant quantities of steel,[74] aluminium,[75,76] zinc,[77] copper[78] and lead,[79] supplied to the market have been produced from secondary resources. Significant energy saving can be made through the recycling of metals compared to processing ores to generate these metals.[80,81] Steel recycling is carried out alongside primary steelmaking using basic oxygen furnace (BOF) or electric arc furnace (EAF) steelmaking processes. These processes save 74% energy, 90% of virgin materials, reduce 86% of air pollution, 40% of water use, 76% of water pollution, 97% of mining waste and a considerable amount of consumer wastes generated, when compared with production from virgin materials.[82] However, these metals are still not recovered to their maximal extent and such strategies should be adopted for all elements (metals). Development of efficient physical and metallurgical recovery technologies is vital to ensure that the ideal of 100% recovery is achieved.

On the other hand, there is little or no recycling of many of the elements highlighted in the numerous national and international reports that are viewed as critical, even those that are required for the emergence of renewable energy sources (Figure 1.8). Figure 1.8, prepared from data given in the UNEP 2011 report, shows a mixed range of recycling rates of the critical elements.[50] As stated in the report, numerous difficulties in data collection were encountered

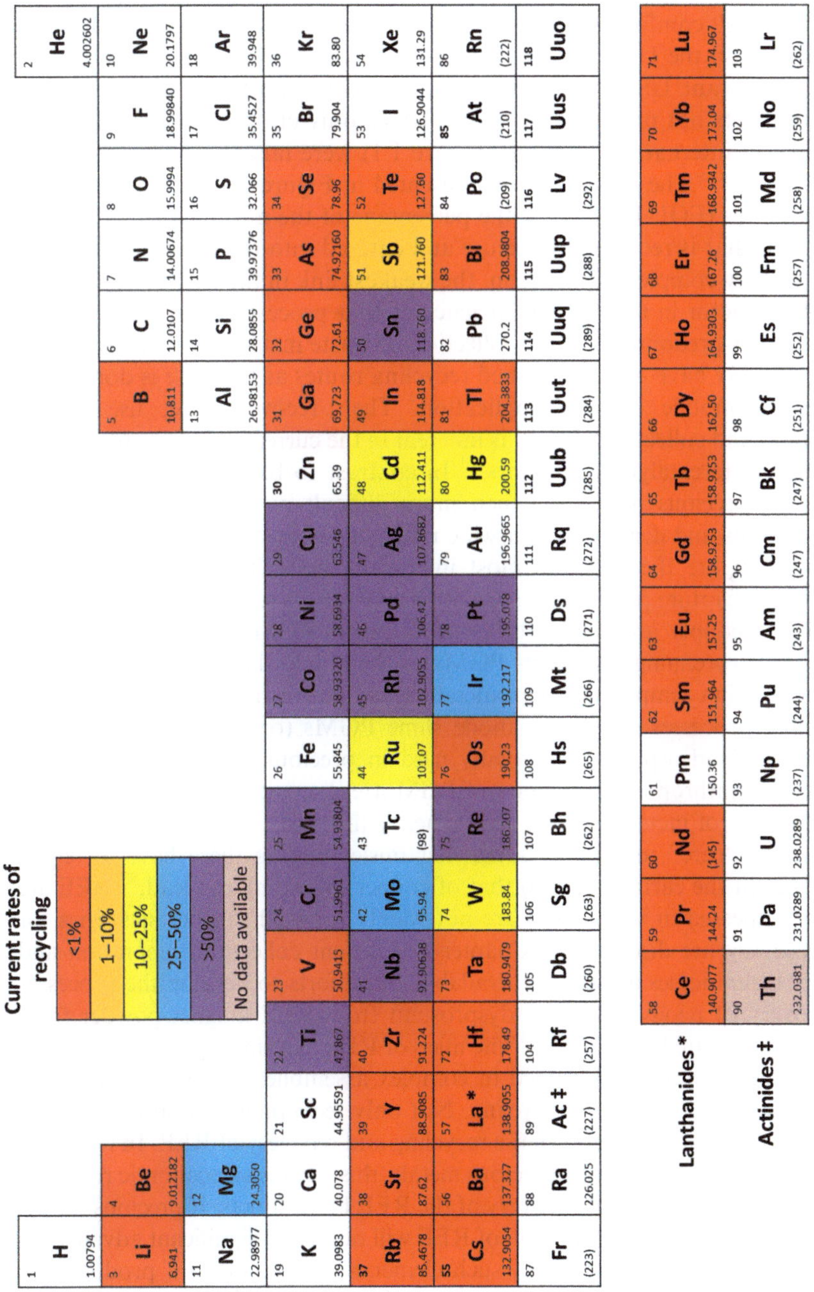

Figure 1.8　Recycling rates for those elements highlighted as "critical" in Table 1.1 (original data adapted from Salazar[10] and Graedel *et al.*[50]).

and as such the figures given are predominately from the years 2000–2005 and thus are likely to ignore the ripple effect of recent supply and demand concerns for many of these elements. Perceived error margins in the data also mean that UNEP only felt comfortable quoting ranges rather than exact figures. Elements not included in the UNEP report, such as Rb and Cs were added using data from the relevant USGS reports.[10,50] The UNEP report also included additional elements not presented in Figure 1.8 as only critical elements from the national and international reports (Figure 1.1) were taken as relevant for this discussion.[9,13–18] The range of rates presented in Figure 1.8 are the end-of-life recycling rates (EOL-RR), being the percentage of the metal, either in pure or alloy form, that is recovered after use and entered into the recycling chain.[50]

As with any complex flow system, the weakest link in the chain is always the controlling factor in the overall efficiency of that process and in the EOL-RR process this is predominately the collection of waste metal directly after use. Pt and Pd already have well established recycling routes as their use is dominated by the automobile catalyst application. Their recovery after use is well understood and collection of these is inherent in the current processes for dealing with end-of-life catalytic convertors.[50] In contrast to base metals, recovery of scarce and precious metals is much more difficult owing to their very low concentrations and dispersion in a wide range of waste streams. Collection and concentration are the first and most important steps. The current industrial practice is to capture precious metals with heavy metals such as copper and lead. Smelting of copper and lead with e-wastes brings PGMs into the base metal stream. Electrorefining transports PGMs to the anode slime. Further leaching, roasting (a high temperature pyrometallurgical conversion process) and electrowinning processes enrich the anode slime PGMs (or precious metals). Ion exchange and adsorption are broadly used in precious metals recovery from solutions.[82] Anthropospheric losses of PGM are discussed in Chapter 7, which maps the flow of metals through the element's lifecycle.

The term "rare" in REE is somewhat misleading as these have an equal abundance in the earth's crust to that of copper (50 ppm) and lead.[83] REE are scarcely dispersed in wastes and are not widely recycled, further strategies for REE recovery are discussed in Chapter 6. Efficient collection, separation and recovery technologies are not yet available, but efforts to develop the processes for recovery of neodymium and dysprosium from REE magnet powder from hard disc drives are being made.[84] The mass of REE in a single product is usually very low and is mostly embedded in complex assemblies combined with other elements, which complicates recycling. Not only technology but also economic factors play an important role in the recycling and recovery of REE. In recovery, if the REE are combined with precious metals, this creates an economic recycling incentive, which can lead to recovery of the REE by-products. If precious metals are absent, the economic attraction of REE will not exist. Significant advances in the recovery and sustainability of such elements may be made if products are designed for disassembly or component reuse.

Some metals, such as Mn and Mo have decent levels of recycling as a result of their recovery and recycling via alloys, while other current applications, including catalysis, may pose significant technological difficulties to be overcome to improve EOL-RRs further.[50] Although ELO-RRs are below 1% for 35 of our highlighted critical elements, this shows that there is great potential in the future for reducing dependence on virgin resources if these rates can increased. In many cases increased rates of recycling would also be highly desirable to counteract current fears about supply security (Table 1.1 and 1.2).

At the end of their life products containing these elements, such as mobile phones, televisions and computers, are ending up in landfills or being incinerated and the elements are being lost. Japan has already highlighted that waste resources will become increasingly important for its economy as concerns over securing supplies increase.[23] Analyses of the ash from waste incinerators in Japan has revealed the composition to include zinc, lead, copper, silver, indium, palladium, nickel, chromium, tantalum, vanadium, zirconium, potassium, calcium, antimony and sodium. The sources of many of these scarce elements in incineration residues are expected to be WEEE or flame retardant materials in the case of antimony.[85,86] In developed countries, it is estimated that WEEE can contribute 8% by volume of MSW.[60] New approachs are being tried to recover all elements and to re-use them in close-looped systems by designing the direct recycling of elements through intelligently designed disassembly of materials at their end of life. These measures should limit the demand for new supplies of elements and increase the lifetime of our reserves infinitesimally.

In the example shown in Figure 1.9, increased rates of recycling would move towards a circular economy and would reduce reliance on hitch-hiker element production. It should also be noted that following the discussion above

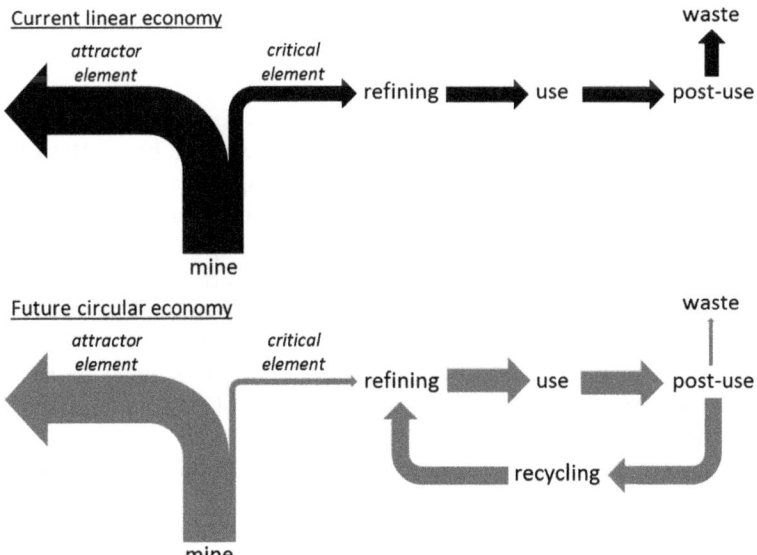

Figure 1.9   Diagram highlighting that increased recycling could move towards a circular economy.

regarding the difficulty of increasing critical element production when that element is predominately produced as a hitch-hiker, recycling in a circular economy is the only viable way to deal with further increases in demand for that element. The concept of the circular economy is discussed in greater detail in Chapter 9.

It is evident from all of the above discussions that for these critical elements to be used in a sustainable manner, recycling rates must increase and this includes the recovery of these metals from wastes such as those in landfill sites and mine tailings, a true resource for the future. Included in this book is an in-depth assessment of current greener approaches to critical metal recovery along with future prospects and ideas about where research should be directed in the coming decades. The concept of elemental sustainability or being sustainable for all elements, not just those regarded as being critical, is likely to become increasingly important in the future. Now is the time for producers and users alike to progress to circular economies and embrace elemental recovery and sustainability.

References

1. J. R. Dodson, A. J. Hunt, H. L. Parker, Y. Yang and J. H. Clark, *Chem. Eng. Process*, 2012, **51**, 69.
2. M. Z. Jacobson and M. A. Delucchi, *Sci. Am.*, 2009, **November**, 58.
3. British Geological Survey, http://www.bgs.ac.uk/research/highlights/2010/peakMetal.html, accessed 24th March 2013.
4. House of Commons Science and Technology Committee, *Strategically Important Metals, Volume I: Report, together with formal minutes, oral and written evidence*, 4 May 2011.
5. R. A. Sheldon, *Chem. Ind.*, 1997, **1**, 12.
6. J. H. Clark, *Green Chem.*, 1999, **1**, 1.
7. D. R. Lide, *Handbook of Chemistry and Physics*, CRC Press, Boca Raton, Florida, 85th edn, 2005. Section 14, Geophysics, Astronomy, and Acoustics; Abundance of Elements in the Earth's Crust and in the Sea.
8. S. E. Kesler, *Proceedings for a Workshop on Deposit Modeling, Mineral Resource Assessment, and Their Role in Sustainable Development*, J. A. Briskey and K. J. Schulz, USGS, USA, edn 1, 2007, 55–62.
9. European Commission Enterprise and Industry, *Critical Raw Materials for the EU, Report of the Ad-hoc Working Group on Defining Critical Raw Materials*, 2010, http://ec.europa.eu/enterprise/policies/raw-materials/files/docs/report-b_en.pdf, accessed 24th March 2013.
10. K. Salazar, *Mineral Commodity Summaries 2013: US Geological Survey (USGS)*, US Geological Survey, 2013.
11. T. J. Brown, R. A. Shaw, T. Bide, E Petavratzi, E. R. Raycraft and A. S. Walters, *World Mineral Production 2007–11*, British Geological Survey, 2013.
12. C. Rhodes, *Chem. Ind.*, 2008, **16**, 21.

13. R. L. Moss, E. Tzimas, H. Kara, P. Willis and J. Kooroshy, *Critical Metals in Strategic Energy Technologies*, European Commission, Joint Research Centre (JRC) Institute for Energy and Transport, 2011.
14. British Geological Survey (BGS), *Risk List 2012*, http://www.bgs.ac.uk/mineralsuk/statistics/riskList.html, accessed 24th March 2013.
15. US Department of Energy (DOE), *Critical Element Strategy*, 2010.
16. OECD, *Critical Metals and Mobile Devices*, OECD Environment Directorate, 2010.
17. National Research Council of the National Academies, *Minerals, Critical Minerals, and the US Economy*, 2007.
18. United Nations Environment Programme (UNEP) & United Nations University, *Critical Metals for Future Sustainable Technologies and their Recycling Potential*, 2009.
19. H. Kawamoto, *Japan's Policies to be adopted on Rare Metal Resources*, 2008, http://www.nistep.go.jp/achiev/ftx/eng/stfc/stt027e/qr27pdf/STTqr2704.pdf, accessed 24th March 2013.
20. APS Panel on Public Affairs and the Materials Research Society, *Energy Critical Elements: Securing Materials for Emerging Technologies*, 2011.
21. D. Merriman, *A Review of Global Supply of Rare Earths*, RSC Environmental Chemistry Group, Distinguished Guest Lecture and Symposium, 20th March 2013, London, UK.
22. D. Kramer, *Phys. Today*, 2010, **63**, 22.
23. K. Halada, K. Ijima, M. Shimada and N. Katagiri, *J. Jpn. Inst. Metals*, 2009, **73**, 151.
24. J. F. Papp, *Mineral Commodity Yearbook 2011: US Geological Survey, Niobium (Columbium) and Tantalum*, US Geological Survey, 2012, 52.1–52.14.
25. R. Hunziker, *Eng. Min. J.*, 2002, **November**, 20.
26. M. Taka, *New Perspectives on Human Security*, M. McIntosh and A. Hunter, Green Leaf Publishing, Sheffield (UK), 2010, 159–173.
27. M. Kassem, *United Nations Security Council, S/2003/1027*, 2003, http://www.un.org/Docs/journal/asp/ws.asp?m = S/2003/1027, accessed 24th March 2013.
28. United Nations Security Council, *Press Release SC/7057, Security Council Condemns Illegal Exploitation of Democratic Republic of Congo's Natural Resources*, 3rd May 2001, http://www.un.org/News/Press/docs/2001/sc7057.doc.htm, accessed 24th March 2013.
29. L. Talens Peiró, G. Villalba Méndez and R. U. Ayres, *Rare and Critical Metals as By-products and the Implications for Future Supply*, 2011, INSEAD http://www.insead.edu/facultyresearch/research/doc.cfm?did = 48916, accessed 24th March 2013.
30. B. W. Jaskula, *Mineral Commodity Summaries 2013: US Geological Survey, Lithium*, US Geological Survey, 2011, 94–95.
31. B. W. Jaskula, *Mineral Commodity Summaries 2013: US Geological Survey, Beryllium*, US Geological Survey, 2011, 28–29.

32. Pacific Ore Mining Corp., http://www.pacificoremining.com/vanadium/ where-is-it-produced, accessed 22nd March 2013.
33. J. Chegwidden, *ANCOA Ltd, Study of the Antimony Market*, Roskill Consulting Group Ltd, 2011. http://www.ancoa.com.au/RoskillCRT.pdf, accessed 24th March 2013.
34. C. Zhanheng, *J. Rare Earths*, 2011, **29**(1), 1.
35. J. D. Jorgenson, *Mineral Commodity Summaries 2002, US Geological Survey, Indium*, US Geological Survey, 2002.
36. A. C. Tolcin, *Mineral Commodity Summaries 2008, US Geological Survey, Indium*, US Geological Survey, 2008.
37. C. Kelly, *Tantalum Industry in Dire Need of New Resources*, Reuters, Washington, 22 Oct 2009, http://www.reuters.com/article/2009/10/21/ markets-tantalum-idUSN2154079620091021, accessed 24th March 2013.
38. T. D. Kelly and G. R. Matos, *Historical Statistics for Mineral and Material Commodities in the United States: US Geological Survey Data Series 140, 2011*, http://pubs.usgs.gov/ds/2005/140/, US Geological Survey, accessed 24th March 2013.
39. C. Meskers, *Umicore Precious Metals Refining*, Guest lecture presentation for e-waste recycling at Delft University of Technology, 8 September 2009, Delft, Netherlands.
40. P. Anbarasan, Z. C. Baer, S. Sreekumar, E. Gross, J. B. Binder, H. W. Blanch, D. S. Clark and F. D. Toste, *Nature*, 2012, **491**, 235.
41. A. Fürstner, *Angew. Chem., Int. Ed.*, 2000, **39**, 3012.
42. E. L. Scott, J. Spekreijse, J. L. Nôtre, J. Haveren and J. P. M. Sanders, *Green Chem.*, 2012, **14**, 2747.
43. M. J. Burk, P. Pharkya, S. J. Van Dien, A. P. Burgard and C. H. Schilling, *Methods for the Synthesis of Olefins and Derivatives*, US Patent, 8026386, 2011.
44. J. P. M. Sanders, E. L. Scott, J. Spekreijse, J. L. Nôtre and J. Haveren, *Bio-derived Olefin Synthesis*, EP Patent, 2269974, 2011.
45. O. M. Singh, *J. Sci. Ind. Res.*, 2006, **65**, 957.
46. R. R. Shrock and A. H. Hoveyda, *Angew. Chem., Int. Ed.*, 2003, **42**, 4592.
47. R. R. Shrock, S. C. Marinescu, P. Müller and A. H. Hoveyda, *J. Am. Chem. Soc.*, 2009, **131**, 10840.
48. J. Emsley, *Nature's Building Blocks, An A–Z Guide to the Elements*, Oxford University Press, New Edition, 2011.
49. REACH, *Registration, Evaluation and Assessment of Chemicals*, http:// ec.europa.eu/enterprise/sectors/chemicals/reach/index_en.htm, accessed 24th March 2013.
50. T. E. Graedel, J. Allwood, J.-P. Birat, B. K. Reck, S. F. Sibley, G. Sonnemann, M. Buchert and C. Hageluken, *Recycling Rates of Metals – A Status Report, A Report of the Working Group on the Global Metal Flows of the International Resource Panel*, UNEP, 2011.
51. G. M. Bedinger, *Mineral Commodity Summaries 2013: US Geological Survey, Titanium*, US Geological Survey, 2013, 174–175.
52. J. D. Rainier and K. Iyer, *J. Am. Chem. Soc.*, 2007, **129**, 12604.
53. T. Takeda, T. Fujiwara and Y. Kato, *Tetrahedron*, 2000, **56**, 4859.

54. T. Takeda, *Chem. Rec.*, 2007, **7**, 24.

55. R. Brooks, J. Lee, R. Reeves and T. Jaffre, *J. Geochem. Explor.*, 2001, **7**, 49.

56. V. Sheoran, A. S. Sheoran and P. Poonia, *Miner. Eng.*, 2009, **22**, 1007.

57. K. Shah and J. M. Nongkynrih, *Biol. Plant.*, 2007, **51**, 618.

58. B. D. Pandey and J. Lee, *Waste Manage.*, 2012, **32**, 3.

59. F. Beolchini, V. Fonti, A. Dell'Anno, L. Rocchetti and F. Vegliò, *Waste Manage.*, 2012, **32**, 949.

60. G. Losfeld, V. Escande, P. V. de La Blache, L. L'Huillier and C. Grison, *Catal. Today*, 2012, **189**, 111.

61. R. Widmer, H. Oswald-Krapf, D. Sinha-Khetriwal, M. Schnellman and H. Boni, *Environ. Impact Assess.*, 2005, **25**, 436.

62. G. J. Olson, J. A. Brierley and C. L. Brierley, *Appl. Microbiol. Biotechnol.*, 2003, **63**, 249.

63. N. Boon, S. D. Corte, T. Hennebel, B. D. Gusseme and W. Verstraete, *Microb. Biotech.*, 2012, **5**, 5.

64. T. A. Davis and B. Volesky, *Water Res.*, 2000, **34**, 4270.

65. G. J. Hutchings, M. Sankar, N. Dimitratos, P. J. Miedziak, P. P. Wells and C. J. Kiely, *Chem. Soc. Rev.*, 2012, **41**, 8099.

66. United Nations University, *Review of Directive 2002/96 on Waste Electrical and Electronic Equipment (WEEE): Final Report*, 2008.

67. Boliden, http://www.boliden.com/Documents/Press/Publications/Broschures/Atervinning_Ronnskar_eng.pdf, accessed 24th March 2013.

68. OECD, *OECD Environmental Outlook to 2030: Consequences of Policy Inaction*, 2008.

69. E. Baker, E. Bournay, A. Harayama and P. Rekacewicz, *Vital Waste Statistics*, UNEP/DEWA/GRID-Europe, 2004.

70. D. Cohen, *New Sci.*, 2007, **2605**, 34.

71. C. Rhodes, *Chem. Ind.*, 2010, **1**, 26.

72. C. Mack, B. Wilhelmi, J. R. Duncan and J. E. Burgess, *Biotechnol. Adv.*, 2007, **25**, 264.

73. B. Volesky, *Hydrometallury*, 2001, **59**, 203.

74. World Steel Association: Steel and Raw Materials: Fact Sheet 2011, http://www.worldsteel.org/dms/internetDocumentList/fact-sheets/Fact-sheet_Raw-materials2011/document/Fact%20sheet_Raw%20materials2011.pdf, Assessed 24th March 2013.

75. K. Tsesmelis, *Recycling – an important part of the aluminium story, Aluminium Recycling*, OEA Congress, Duesseldorf, 25–26 February (2013), http://www.world-aluminium.org/media/filer_public/2013/02/27/aluminium_recycling_-_an_important_part_of_the_aluminium_story.pdf, accessed 24th March 2013.

76. *The Aluminium Story*, http://www.thealuminiumstory.com/home.html, accessed 24th March 2013.

77. International Zinc Association: *Zinc Recycling*, http://www.zinc.org/basics/zinc_recycling, accessed 24th March 2013.

78. C. R. Risopatron, (ICGS), *Global Copper Scrap Market – Report for China: Trends and Outlook,* in China International Copper Conference & ICSG

China Copper Market Seminar, 6–7 November 2009, Wuhan, China, 2009, http://www.icsg.org/index.php/meetings-and-presentations-2/finish/118-2009-china-international-copper-conference-2009-icsg-china-copper-market-seminar/684-icsg-global-copper-scrap-market-report-for-china, accessed 24th March 2013.

79. D. E. Guberman, *Mineral commodity yearbook 2011: US Geological Survey, Lead*, US Geological Survey, 2013, 42.1–42.18.
80. J. Cui and L. Zhang, *J. Hazard. Mater.*, 2008, **158**, 228.
81. J. Johnson, B. K. Reck, T. Wang and T. E. Graedel, *Energy Policy*, 2008, **36**, 181.
82. J. Cui and E. Forssberg, *J. Hazard. Mater.*, 2003, **B99**, 243.
83. US Geological Survey, *Rare Earth Elements—Critical Resources for High Technology*, http://pubs.usgs.gov/fs/2002/fs087-02/, accessed 24th March 2013.
84. M. Tanaka, T. Oki and T. Akai, *Recycling of Rare Earth Elements, Rare Metals*, AIST (National Institute of Advanced Industrial Science and Technology) Publication, 2008, 8–9.
85. C. H. Jung and M. Osako, *Resour. Conserv. Recycl.*, 2009, **53**, 301.
86. C. H. Jung, T. Matsuto, N. Tanaka and T. Okada, *Waste Manage.*, 2004, **24**, 381.

CHAPTER 2

Integration of Traditional Methods for Elemental Recovery in a Zero-waste Recycling Flow Sheet

XUAN WANG,[a,b] TOM VAN GERVEN*[a] AND
BART BLANPAIN[b]

[a] Department of Chemical Engineering, KU Leuven, de Croylaan 46, 3001
Leuven, Belgium; [b] Department of Metallurgy and Materials Engineering,
KU Leuven, Kasteelpark Arenberg 44, 3001 Leuven, Belgium
*Email: tom.vangerven@cit.kuleuven.be

2.1 Introduction

Metals exist in the earth's crust in different forms, particularly as oxides, sulfides and carbonates. A few more noble metals, such as gold and platinum, can be found in metallic form, but in very limited amounts.[1] Since antiquity various techniques and processes have been developed to extract metals from natural minerals and this development will be continuously stimulated by the growing demands of modern society. On the other hand, the decreasing grade of natural ores and increasing environmental burden of industrial residue and scrap metal make recycling a strong driving force of economic growth. The traditional methods of extracting steel, aluminium, copper, zinc and lead, which are some of the basic materials for human activities, are briefly described in this chapter. In a case study of copper slag, these traditional methods of

RSC Green Chemistry No. 22
Element Recovery and Sustainability
Edited by Andrew J. Hunt
© The Royal Society of Chemistry 2013
Published by the Royal Society of Chemistry, www.rsc.org

extractive metallurgy are integrated with utilisation of the residual material after extraction, in order to design a complete recycling flow sheet.

2.2   Metal Production Processes

2.2.1   Basic Stages in Metal Production

There is a wide variety of techniques and processes available for metal production from original resources to final products. According to their purpose, all these techniques and processes can be divided into five stages, as summarised by Hayes.[2]

- *Separation* refers to the removal of unwanted elements or compounds from the feed materials. This can be achieved through mineral processing, which includes size reduction of particles and selection of desired phases. The selection of desired phases is carried out based on their characteristics, such as size (sieving and screening), density (cyclones), magnetic properties (magnetic separation), electrical properties (electrostatic separation) and surface chemistry properties (flotation). The physical or chemical properties of the individual phases are not changed during mineral processing. Separation can also be achieved by changing the chemical or structural properties of the feed materials, such as roasting and leaching followed by precipitation.
- *Compound formation* is the production of a material chemically and/or structurally different from the feed material. According to the aim of the process, the compounds obtained can be feedstocks for following metal extracting steps, as well as end-products themselves.
- *Metal extraction* involves the production of an impure metal product. Pyrometallurgical processes, such as in a blast furnace and flash smelter, can be applied in metal extraction. It is more preferable to treat other feedstocks by hydrometallurgical methods, such as solvent extraction and chemical precipitation.
- *Metal purification* refers to the removal of undesired elements from the impure metal product obtained by metal extraction. It can be achieved through pyrometallurgical processes, such as carbon removal using a basic oxygen furnace, or electrochemical processes, such as electrorefining.
- *Product preparation and processing* is the final stage of the metal production ensuring the product is suitable for sale and marketing. It may constitute shaping, blending, packaging, and so on.

The hierarchy of these stages is demonstrated in Figure 2.1. For a specific metal production flow sheet, it is not necessary to go through these five stages. On the other hand, there are also possible interconnections and similarities between different stages. For example, during the sintering of lead sulfide concentrate, sulfur transforms into SO_2 (compound formation) and is then separated from lead oxides (separation).

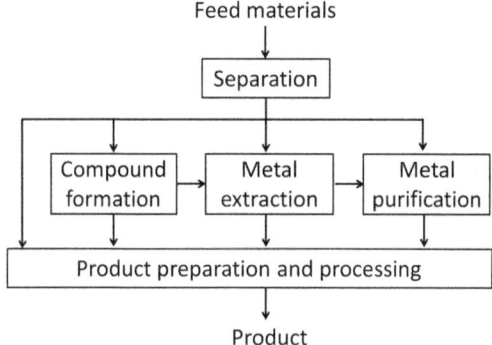

Feed materials

Separation

Compound formation → Metal extraction → Metal purification

Product preparation and processing

Product

Figure 2.1 Hierarchy of processes in metal production.[2]

2.2.2 Extractive Metallurgy

Extractive metallurgy is the industrial activity that focuses on the production of metals from primary (ores and concentrates) or secondary (home or old scrap) raw materials. The production schedule or flow sheet of a metallurgical plant consists of a sequence of individual processes or unit operations. These operations have been described in detail in a number of text books, dating back to the work of Georgius Agricola's *De Re Metallica* in 1556. Recent extensive descriptions of the field may be found, for instance in Hayes.[2] Here we will give a concise overview. Unit processes in the field of extractive or chemical metallurgy are generally classified in pyrometallurgical, hydrometallurgical and electrometallurgical operations.

Pyrometallurgical processes take place at an elevated temperature, wherein the metals and metal compounds are treated in concentrated streams or batches. The typical temperature range for these processes is between 500 °C and 1700 °C, although much higher temperatures can be achieved, for instance in the arc of an electric arc furnace or much lower temperatures, such as in the Mond process for refining nickel, which runs at temperatures as low as 250 °C. We list hereafter some typical pyrometallurgical processes.

- *Calcination* is the thermal decomposition of the process material to remove volatile species. It is used frequently for hydrates and carbonate, such as the calcination of $Al(OH)_3$ to Al_2O_3 in the last step of the Bayer process.
- *Roasting* is the reaction of the solid process material with a gas phase. One distinguishes between oxidative, reductive, chlorination and sulfation roasting processes. An example is the oxidative roasting of zinc concentrate as a preparatory step in the predominant roasting-leaching-electrowinning process for zinc production.
- In *smelting* the process material is rendered liquid to allow separation based on elemental and compound solubility in and density differences between the metallic phase, the oxidic phase (slag) and/or the sulfidic

phase (matte). Depending on the processing conditions, the smelting operation can be neutral, reducing or oxidising. An example of a reducing smelting operation is the blast furnace production of iron (hot metal) from iron ore concentrate (sinter or pellets), cokes and lime.

- In *converting* processes an oxidising gas is injected in the liquid process material (metal or matte) to remove impurities as oxide to the slag phase or off-gas. Examples are the converter process used in integrated steel production where technically pure oxygen is injected into hot metal to produce steel or blowing matte to blister copper in a Peirce–Smith convertor.
- *Metallothermic* processes use a more reactive metal to extract the process metal from its compound, *e.g.* the Kroll process for the extraction of Ti from $TiCl_4$ with Mg.

Hydrometallurgical processes involve using aqueous and/or organic solutions to extract, concentrate, separate or refine the metal. These processes typically require large amounts of dilute solutions in which the temperature remains relatively low. Typical hydrometallurgical processes are:

- *Leaching* is the dissolution of metal or metal compounds in aqueous or organic solutions. Leaching operations can be very large (*e.g.* heap leaching) or more compact (percolation or agitation leaching). An example is the leaching of copper from oxidic ores in an aqueous sulfuric acid solution.
- *Precipitation* is used to separate species from a solution as, for example hydroxides or sulfides. An example is the precipitation of goethite or jarosite to remove Fe from the aqueous solution in the zinc production flow sheet. In *cementation*, a specific type of precipitation reaction, the precipitant is a (less noble) metal.
- *Solvent extraction* uses immiscible liquids to concentrate and separate metal species in a solution. It typically consists of an extraction and stripping stage, as in the case of the solvent extraction of copper from the pregnant sulfuric acid solution from a heap leach operation prior to the electrowinning process.

Electrometallurgical processes are based on electrochemical reactions in an electrolytic cell. Both aqueous and molten salt solutions may be used as electrolytes. Organic solvents are currently considered to be too expensive. Processes in which electrical energy is used only in order to supply process heat, such as an electric arc furnace, are generally regarded as pyrometallurgical processes. Electrowinning and electrorefining are the main electrochemical processing operations:

- In *electrowinning* the metal ions in the solution are reduced at the cathode. Examples are the production of aluminium using molten salts (cryolite) in the Hall–Héroult process or the production of zinc cathodes from a zinc sulfate aqueous solution.

- In *electrorefining* the metal is anodically dissolved in the electrolyte and cathodically reduced to refine the metal. Less noble elements remain in the electrolyte, more noble metals are collected as anode slimes. An example is the electrorefining of copper anodes to refine the copper and collect precious metals in the slimes as a final step in copper production.

Flow sheets are composed of a sequence of these unit processes. In practice, a distinction is made between the predominantly pyrometallurgical and hydrometallurgical nature of the treatment schedule and the flow sheet is thus indicated as a pyrometallurgical or hydrometallurgical schedule, respectively. Electrometallurgical unit operations are generally counted as hydrometallurgical for this purpose. For many metals, both hydro- and pyrometallurgical schedules have been developed and used industrially. An example of a pyrometallurgical flow sheet is the production of steel from iron ore through the blast furnace/converter route. An example of a hydrometallurgical flow sheet is the production of zinc from zinc concentrates through roasting–leaching–electrowinning.

It is useful to examine the overall characteristics of pyrometallurgical and hydrometallurgical processing schemes. These are some general considerations for pyrometallurgical processing:

- Concentrated process streams are used as a prerequisite because of the high temperatures. Therefore, they are generally preceded by beneficiation.
- High reaction rates are achieved, which are mainly determined by mass transport. Because of the high temperatures, the reaction kinetics in itself is not usually a barrier.
- High specific productivity is reached. The size and investment costs for dedusting and off-gas treatments are usually larger than those for the actual reactor or furnace.
- High temperatures also mean that the processes tend not to be very selective given the increasing solubility of the phases with increasing temperatures.
- The high temperatures do not always lead to a high additional energy requirement, in some cases the exothermicity of the reaction reduces the energy consumption or even results in net energy production.

Here are some key features of hydrometallurgical flow sheets:

- Dilute solutions and low reaction temperatures are used, with the exception of molten salt processing, where elevated temperatures are essential to render the salts liquid.
- In consequence, reaction rates are low. The reactors used are therefore large, because of the low specific production capacity.
- Ample attention should also be given to solid–liquid separation (thickeners, filters, *etc.*). This means that the footprint of a

hydrometallurgical plant is generally larger than that of an equivalent pyrometallurgical plant.

- Owing to the high heat capacity of water, a significant amount of energy is required to reach the process temperature. The low temperatures allow low quality heat, for example waste heat from a pyrometallurgical process, to be used.
- In comparison with pyrometallurgical processes, hydrometallurgical processes are usually much more selective.
- In electrometallurgical operations the reagent, namely electrons, does not cause additional contamination.
- The selectivity of the reagent also means that very poor raw materials can be treated.

The selection of a hydrometallurgical or pyrometallurgical treatment therefore is the result of numerous considerations, such as the available raw materials, the available energy vectors, the desired products and possible synergies with existing installations and flow sheets.

2.2.3 Important Flow Sheets

The annual world production of several metals, which have high economic value, from natural ores in 2011 has been listed in Table 2.1.[3]

The major commercial extractive processes of the earlier mentioned metals are listed in Table 2.2 together with the corresponding principal primary sources.[4–8]

The details of the processes can be found in the references mentioned above. The extractive flow sheets are briefly introduced in this chapter together with the major by-products generated during the processes, which have the potential to be treated as secondary resources.

2.2.3.1 Iron and Steel

After crushing and possibly beneficiation and agglomeration into pellets, iron ore is fed into a blast furnace together with coke and limestone. Air is blown from the bottom of the blast furnace. CO, which acts as a reductant for iron

Table 2.1 Annual world production of different metals in 2011.

Metal	World production (kt)
Steel	1 500 000
Aluminum	44 100
Copper	16 100
Zinc	12 400
Lead	4500

Table 2.2   Metal extractive processes and feeding materials.

Metal	Natural resource	Process
Steel	Iron ore (20–70 wt% Fe)	Blast furnace and basic oxygen furnace
Aluminium	Bauxite ore (20–30% Al)	Bayer process and Hall–Héroult process
Copper	Sulfide ore (0.5–2% Cu)	Smelting and/or converting and electrorefining
	Oxide/sulfide ore (~0.5% Cu)	Leaching and solvent extraction/electrowinning
Zinc	Sulfide ore (5–10% Pb, 10–30% Zn)	Electrolytic process and imperial smelting process
Lead	Sulfide ore (5–10% Pb, 10–30% Zn)	Lead blast furnace, Imperial smelting process, Kivcet process and QSL process

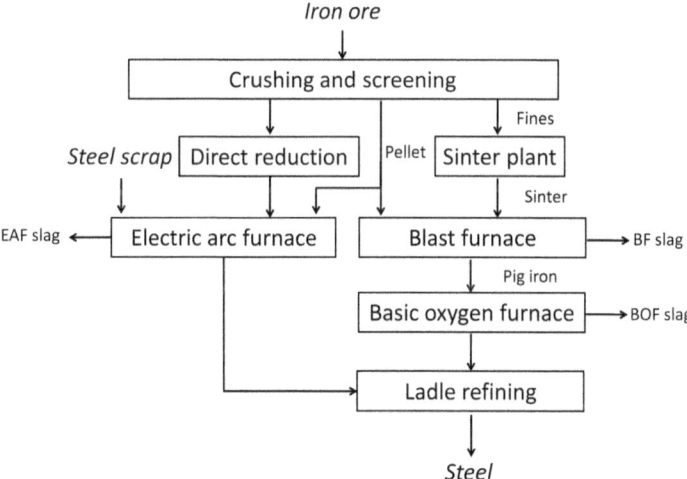

Figure 2.2   Main process flow sheet for steel making.

ore, is generated by the combustion of coke. The reduced pig iron (~94% Fe) is then refined in a basic oxygen furnace for decarburisation to finalise the steel production. Modern steel plants will also have a secondary refining operation to remove impurities and adjust the composition. The main process flow sheet is shown in Figure 2.2.

The major by-product streams during iron and steel making are blast furnace (BF) slag, basic oxygen furnace (BOF) slag and electric arc furnace (EAF) slag. The current annual world BF slag and steel slag (including EAF and BOF slag) production is estimated to be 260–310 million tonnes and 130–210 million tonnes, respectively.[3] The major element components and phases of these slags are listed in Table 2.3. Valuable metals such as Fe, Co, Mo, Cr, Mn and V can be found in steel slag in the form of oxides or a metal phase.[9]

Table 2.3 Chemical composition and major phases of iron and steel slag.[4,10–13]

| | Chemical composition (wt%) | | | | | |
Slag	Ca	Si	Al	Mg	Fe_{tot}	Major phase
BF	24–31	16–17	4–8	2–8	~1	Gehlenite ($2CaO \cdot Al_2O_3 \cdot SiO_2$), akermanite ($2CaO \cdot MgO \cdot 2SiO_2$)
BOF	30–39	6–8	<2	2–5	14–20	Larnite (Ca_2SiO_4), merwinite ($Ca_3Mg(SiO_4)_2$), lime (CaO)
EAF	18–28	5–8	2–4	2–9	18–29	Larnite (Ca_2SiO_4)

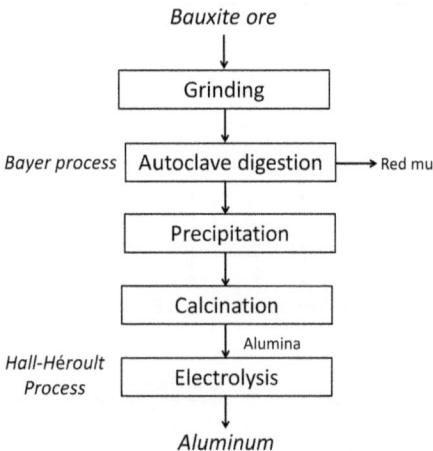

Figure 2.3 Flow sheet for aluminium production.

2.2.3.2 *Aluminium*

The commercial production of aluminium from ores is divided into two stages. The first stage is to obtain alumina from bauxite ores via the Bayer process. The second stage is the electrolysis of alumina by the Hall–Héroult process.[4] The major steps of the process are shown schematically in Figure 2.3.

The bauxite residue, also called red mud, generated from Bayer process is the major by-product stream from aluminium production. The major element components and mineral phases in the residue are listed in Table 2.4. The annual world production of bauxite residue is estimated to be 120 million tonnes and the global inventory of historical residue is estimated to be 2.7 billion tonnes.[14]

2.2.3.3 *Copper*

Most of the sulfide copper ores (containing 80% of the world's copper-from-ore), found mainly in chalcopyrite ($CuFeS_2$) form, are fed into

Table 2.4 Composition and major phases of bauxite residue.[14,15]

	Chemical composition (wt%)						
	Al	Fe	Si	Ti	Ca	Na	Major phase
Red mud	5–13	7–35	1–12	2–11	0–18	0–7	Hematite (Fe_2O_3), goethite (FeOOH), magnetite (Fe_3O_4), alumina (Al_2O_3)

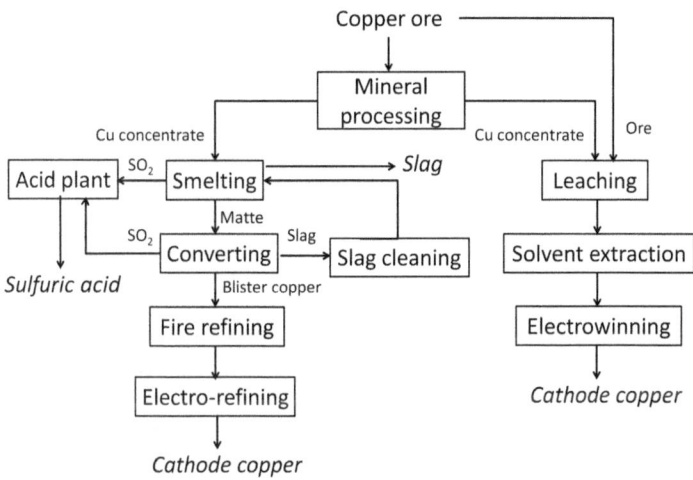

Figure 2.4 Main flow sheet for copper production from ore.[6]

pyrometallurgical processes, while ores containing oxidised minerals and chalcocite (Cu_2S) are treated by hydrometallurgical methods. These two streams are shown in Figure 2.4.

The major by-product stream from the copper extractive process is copper slag. More details of copper slag including various recycling methods will be introduced in the case study further on in this chapter.

2.2.3.4 Zinc and Lead

Zinc and lead are often mined in association with each other in the sulfide forms. They can also be produced together by the imperial smelting process. The separately produced zinc and lead concentrates can also be treated by the electrolytic zinc process and lead blast furnace, respectively. The processes are illustrated in Figure 2.5.

As can be observed from Figure 2.5, lead and zinc can be produced by various methods. Slag generated from an imperial smelting furnace (ISF) is one of the by-products of lead and zinc production. 13.2% of the world's annual primary zinc is produced from ISF, which brings the annual production of ISF

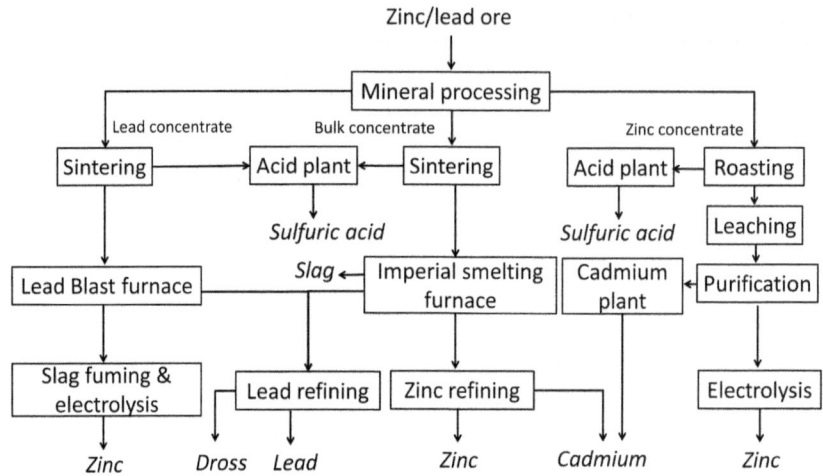

Figure 2.5 Main processes of zinc and lead production.[6]

Table 2.5 Composition and major phases of ISF slag.[16,17]

	Chemical composition (wt%)							
	Al	*Ca*	*Si*	*Fe*	*Pb*	*Zn*	*S*	*Major phase*
ISF slag	3–4	9–11	5–9	21–28	1–7	2–13	3–10	Iron(II) oxide (FeO), zinc sulfide (ZnS)

slag up to approximately 975 000 tonnes.[16] The major element components and phases of ISF slag are listed in Table 2.5.

2.3 Metal Recovery From Metallurgical Waste

2.3.1 Introduction

Despite continuous improvement in the efficiency of extractive productions over generations, the environmental burden and shortage of materials has become ever more severe owing to the rapid growth of human society. Industry has now started to "take another look" at the materials which have previously been considered as waste.

Numerous studies have been performed on recycling the various industrial residues listed above. The application of grounded blast furnace slag in the cement industry has beensuccessful.[11] Traditional processing techniques have been proven to have potential in treating residual materials. In the following section, a case study of copper slag recycling will be highlighted to demonstrate the potential of metal recovery using traditional extractive techniques and to integrate these with the utilisation of the residual waste from the processes.

2.3.2 Case Study: Metal Recovery from Copper Slag

2.3.2.1 Current Copper Slag Production and Treatment

Copper is one of the most important and widely used materials in modern society. Total copper production in recent years was between 16 million tonnes (2012)[3] and 20 million tonnes (2010).[8] Although it can be recycled from used objects and scrap, copper is mostly produced from primary natural ores. Fayalite ($2FeO \cdot SiO_2$)-based slags are by-products generated during the pyro-metallurgical production of copper. Calcium ferrite slag ($CaO \cdot Fe_2O_3$) and olivine slag ($2(Fe,Ca)O \cdot SiO_2$) are also produced in some continuous copper making processes, such as the Mitsubishi continuous converting and "direct-to-blister" smelting, but in limited amounts.[18,19] Each year, a huge amount of fayalite slag is produced worldwide. It has been estimated that about 2.2 tonnes of slag is generated for every tonne of copper. This resulted in an annual worldwide slag production of approximately 24.6 million tonnes as reported by Gorai *et al.* at the beginning of this century.[20] By 2009, this amount increased to above 30 million tonnes. China is the largest primary copper producer in the world (Table 2.6).[21] Every year, more than 8 million tonnes of copper slag is generated in China. Chile accounts for over one-third of the world copper mining production and has 10% of the world's copper smelter output.[22] Around 4 million tonnes of copper slag is disposed annually in Chile and around 40 to 45 million tonnes is estimated to be historically cumulated.[23] Japan has a similar copper smelter production as Chile (11% of world production) and a similar annual copper slag output (4 million tonnes) can be estimated. The total amount of copper slag produced from 1900 to 2004 is estimated to be 842 million tonnes.[24]

Most metallurgical slags are just piled up without being used. These heaps of slag are a burden on the environment and the metallurgical plants themselves. Fayalite slag normally contains a notable amount of heavy metals or even toxic elements (such as As and Pb).[26,27] Dumping and land filling do not comply with sustainable practice. Alternative approaches are required to deal with these slags.

Interest in new and cost-effective waste management techniques has increased significantly in recent years owing to strict environmental regulations,

Table 2.6 Copper smelter production and estimated copper slag production of the world's major copper producing countries in 2009.[22,23,25]

Country	Copper smelter production (million tonnes)	Copper slag generation (million tonnes)
China	3.4	8
Japan	1.6	4
Chile	1.5	4
Russia	0.8	1.8
India	0.7	1.6
USA	0.6	1.3
Germany	0.5	1.1

high waste treatment costs and the limited availability of disposal sites. Considering the chemical, mineralogical and physical properties of fayalite slags, current management options are recycling as aggregates, recovery of metals, production of value added products such as abrasive tools and disposal in slag dumps or stockpiles. Management practices vary throughout the world. In densely populated industrialised regions like Belgium (EU), land filling is not acceptable any more. Most fayalite slag still contains a certain amount of valuable metals such as copper, nickel and cobalt. The iron content (up to 45%) is especially considerable when taking the total quantity of slag into account. Air-cooled and granulated fayalite slags have a number of favourable mechanical properties for use as aggregate, including high hardness, good abrasion resistance and good stability. In general, land filling fayalite slags is a loss of metal resources and leads to environmental problems.[20]

Several papers review the metal recovery and utilisation of metallurgical slags.[20,27,28] This case study focuses on recent progress in research and achievements with respect to the valorisation of fayalite slag. The chemical and mineralogical compositions of fayalite slag are first introduced. Then, techniques to recover metals and the utilisation of slag are described. In order to achieve full valorisation, the combination of metal recovery and the utilisation of the residue is essential.

2.3.2.2   Chemical and Mineralogical Composition

In nature, copper most commonly exists as copper–iron sulfide and copper sulfide minerals such as chalcopyrite ($CuFeS_2$), bornite (Cu_5FeS_4) and chalcocite (Cu_2S). Approximately 90% of the world's copper-from-ore production originates from sulfide ores. Copper–iron sulfide ores are not suitable for a direct hydrometallurgical process since they do not dissolve easily. The majority of copper extraction from minerals is carried out by a pyrometallurgical treatment. The concentration of the sulfide minerals in the ore body is too low (0.5–2 wt% Cu) for economic direct smelting. Several steps are required, typically concentration, smelting, converting, fire refining and electrorefining.[29] In modern processes, such as Noranda and Teniente smelting, some specific copper slags still contain a large amount of copper (up to 10 wt%). Therefore, a slag cleaning step can be included in the process flow sheet. Fayalite slag is generated by smelting, converting and the slag cleaning processes.

The main constituents of a copper slag are FeO and SiO_2, each present at about 20–55 wt% (Table 2.7). Normally, fayalite and magnetite are the two dominant crystalline phases accompanied by a certain amount of amorphous phase as shown in Figure 2.6.[37–40] The converter slag contains a larger amount of magnetite in comparison with smelter slag.[41] Together with iron oxides, other oxides such as silica, alumina, lime and magnesia constitute 95% or more of the total composition. In most cases, Ni and Co are present in the form of oxides but the mineralogy of copper-containing phases differs in different slags. Cu in the form of oxides, metallic copper and various copper sulfides hs been

Table 2.7 Typical chemical composition of copper slag.

Slag	Composition (wt%)									
	Cu	Ni	Co	Fe	Si	Pb	Zn	Mg	Al	Ca
Cu smelter slag 1[a]	1.43	—	0.72	20.70	15.37	—	8.90	2.53	2.56	6.26
Cu smelter slag 2	1.35	0.04	4.09	28.43	15.38	1.16	1.70	2.15	3.22	5.13
Cu smelter slag 3	1.01	0.02	0.10	39.65	18.50	—	0.72	1.69	1.27	2.80
Cu converter slag 1[b]	4.36	—	0.45	52.18	—	—	0.64	—	—	—
Cu converter slag 2	6.35	—	—	44.53	14.27	0.21	2.66	—	—	—
Cu converter slag 3	3.25	—	0.05	44.17	22.4	—	1.56	—	—	—
Cu cleaning slag[c]	0.77	—	—	43.38	13.21	0.11	—	0.45	1.56	1.43

[a]Cu smelter slag: 1. Lubumbashi, DR Congo,[30] 2. Guandong, China,[31] 3. Etibank Ergani, Turkey[32]
[b]Cu converter slag: 1. Ergani Copper Plant, Turkey,[33] 2. Karadeniz Bakir Isletmeleri, Turkey,[34] 3. Black Sea Copper Works, Turkey.[35]
[c]Cu cleaning slag: ENAMI, Chile.[36]
—, data not available.

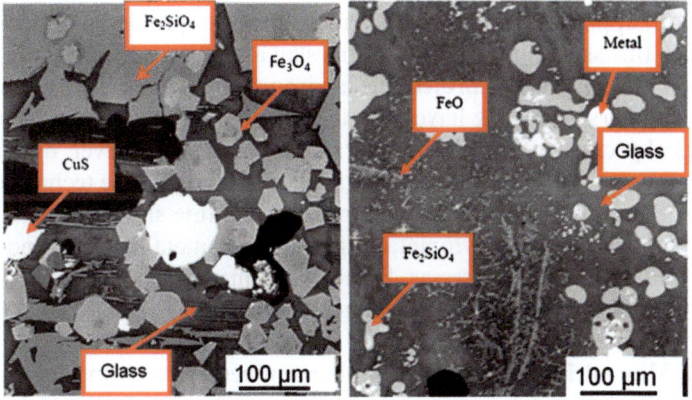

Figure 2.6 Microstructure of slowly cooled (left) and rapidly cooled (right) copper smelter slag.

identified.[20,27,39,42] The copper content of a smelter slag is normally around 1 wt% while converter slag contains in general much more Cu (2–25 wt%), which is much higher than that of copper ores, owing to over oxidation and entrapment of metal droplets. By 2011, for a typical large operating copper smelter, a profit of over six million US dollars can be estimated through a reduction of only 0.1 wt% copper in slag over an operation year.[42] Therefore, Cu recovery from the slag before disposal is necessary.[43] The commonly used slag cleaning procedures are slag reduction/settling (precipitation of metal or matte droplets) and slag minerals processing (comminution and flotation). The cleaning process also generates a fayalite slag, with a lower Cu content. In some plants, converter slag is also directly recycled to the smelter.[29] Some Cu slag also contains Co and/or Ni at levels of interest for recovery. Finally, molybdenum can also be found in copper slag associated with iron silicate and spinel phases at amounts up to 0.4 wt%.[44]

The mineralogical composition of slag generated from different origins is quite diverse owing to many factors such as ore types, processing techniques and cooling rates (Figure 2.6). Slow cooling of slag may result in significant crystallinity, while fast cooling may produce amorphous slag which exhibits a more homogeneous distribution of metals. Microscopic observations indicate that most of the slowly cooled copper slag is well crystallised. Slow cooling in the initial periods is imperative for coalescence and crystallisation.[45] Copper recovery from slag is therefore largely dependent on the cooling history of the slag. Higher copper recovery, separation efficiency and grindability of the slag can be obtained using a slow cooling rate.[46]

2.3.2.3 Metal Recovery

Copper slag usually contains a certain quantity of valuable metal and therefore can be treated as a secondary resource for metal extraction rather than as waste. Furthermore, the release of these metals upon disposal or use in construction applications can cause environmental problems. Recovery of metals from this slag and utilisation of the slag are important, not only to save metal resources, but also to protect the environment.[27] The existing techniques for metal recovery from slag have a lot in common with the current extractive metallurgical processes, described earlier in this chapter. However, owing to the different properties of slag from ore, it is necessary to adjust the techniques. The various methods of recovering metals from the fayalite slags can be classified into mineral processing, pyrometallurgical, hydrometallurgical and mixed pyro-hydrometallurgical processes. Some of the methods are discussed below, together with relevant pretreatment mineral processes.

2.3.2.3.1 Mineral Processing. By applying mineral processing technologies, such as crushing, grinding, magnetic separation, eddy current separation and flotation, followed by hydro- and/or pyrometallurgical processes, it is possible to recover metals such as Fe, Cr, Cu, Al, Pb, Zn, Co, Ni, Nb, Ta, Au, Ag from the slags.[27,47]

Isolation of copper and matte grains from the slag by comminution is crucial for the following flotation or leaching process. The finer the grind, the more efficient will be the liberation and recovery of Cu from copper slag.[38,48] It has been reported that the slag sample should be ground to a particle size below $100\,\mu m$.[49] On the other hand, ultrafine particles ($<5–10\,\mu m$) float more slowly than particles in a mid-size range.[50] In addition, crushing and grinding are energy intensive processes. Therefore, an optimum particle size should be aimed for.[29] Owing to the presence of glassy phases or the presence of metallic phases that plastically deform during grinding, the grindability of the granulated slag is much lower than that of sulfide ores.[49] Slow cooling of slag increases the size of the particles of copper containing minerals through diffusion and coagulation of the small particles. Higher copper recovery, separation efficiency and grindability of the slag can be obtained as a result of a slow cooling rate.[22,46]

The grinding conditions can significantly affect the subsequent flotation of copper sulfide. The iron ions present in the slurry can form an oxide layer on the sulfide surface and prevent the oxidation and adsorption of xanthate, thus hindering the flotation process. This could be avoided by using lined mills and non-ferrous or corrosion resistant grinding media (*e.g.* ceramic, stainless steel). A mildly oxidising environment is favourable to improve adsorption of xanthate oxidised species and enhances the self-induced floatability of the copper minerals.[51]

Slag flotation can be applied in slag recycling with a wide range of slags from different processes.[52] In principle, it is the same as sulfide ore flotation. Therefore only metallic copper and sulfide minerals can be concentrated effectively, while the oxides of Cu, Co and Ni will not be recovered. The obtained concentrate will be recycled as a feed material for the smelting furnace. Good selectivity of copper sulfides is achieved in an alkaline environment in the presence of xanthate molecules owing to the depression of iron sulfides at high pH.[49] Xanthogen species are reported to have advantages over the common thiol collectors (*e.g.* xanthates). Stability in acid conditions and selectivity between copper sulfides and pyrite seems to be better.[53] The typical collectors, frothers and experimental conditions applied are listed in Table 2.8.

In a flotation slag cleaning plant, the slag eventually ends up in the form of a fine crystalline (due to the needs of flotation) waste material which is disposed of in a tailings pond. A tailings pond is less chemically stable than a slag dump because of the increased surface area and crystalline form of the flotation tailings. Moreover, the fine material can be blown into the atmosphere if care is not taken in managing the tailings pond.[52] Hence, cleaning a slag by means of flotation and storing the tailings in a pond is maybe not the best solution to tackle environmental concerns.

In smelter slags coexisting with mattes containing less than 55 wt% Cu, nearly 90 wt% of the copper content of the slag consists of copper sulfide. For mattes containing over 60 wt% Cu, the copper oxide content increases strongly reaching as much as 70 wt% of the copper content of the slag.[54] In converter slag, dissolved Cu is mostly present as Cu_2O.[29] Therefore, the application of flotation in metal recovery from slag can be limited.[27]

2.3.2.3.2 Pyrometallurgy. According to the differences in the process conditions and reduction rates, it is theoretically possible to recover different metals in a multiple-stage process. Recovery of metals from fayalite slags with hydrocarbonaceous, hydrogeneous gases or coke as reductant and pyrite (FeS_2) as sulfidising agent to form a matte have been studied. The addition of pyrite promotes the removal of the base metals from the slag, decreases the liquidus temperature of the system, provides sufficient volume of the matte/alloy product to allow separation from the slag and produces a matte which is desirable for further treatment in existing plant flow sheets. Recovery rates of Co, Cu, and Ni up to 97.1%, 91.2% and 99.4%, respectively, have been reported.[55,56]

Table 2.8 Typical experimental conditions in copper slag/ore flotation.

Reference	pH	Solid	Collector	Dosage	Frother	Dosage
46	11.5	30 wt%	Mercapto benzo thiazole, R407	30 g t^{-1}	Methyl isobutyl carbinol (MIBC)	35 g t^{-1}
51	—	35 wt%	Potassium amyl xanthate Sodium dithiophosphate	100 g t^{-1} 25 g t^{-1}	Polyglycol alcohol	60 g t^{-1}
49	9	25 wt%	Sodium isopropyl xanthate	600 g t^{-1}	Methyl isobutyl carbinol (MIBC)	75 g t^{-1}
18	10.5–11.5	—	Potassium ethyl xanthate (KEX)	150–200 g t^{-1}	Polypropylene glycol	0.25% w/v
50	10.5	38 wt%	Sodium isopropyl xanthate: SF-113 SF-508	30 g t^{-1} 8 g t^{-1}	Propylene glycol DF-250	25 g t^{-1}
53	5–10.5	—	Xanthogen formates	5–10 mol L^{-1}	Methyl isobutyl carbinol (MIBC)	—
49	9	60 wt%	Sodium isopropyl xanthate	600 g t^{-1}	Methyl isobutyl carbinol (MIBC)	75 g t^{-1}

—, data not available.

The recovery possibility of valuable metals and oxides from copper smelter slag by a high temperature direct reduction process has been investigated by González et al.[57] An electric furnace for metal recovery from copper smelter slag may also be used. This technology is attractive owing to its flexibility in reducing the Fe_3O_4 content of the slag to 5 wt% or less, reducing the copper oxides and recovering small droplets of metal in the bath. However, low priced electric power is the prerequisite of the application.[52] González carried out research in which a mixture of copper slags from smelter, coke and calcium oxide as a flux was melted under a nitrogen atmosphere in alumina crucibles in an electric furnace at a temperature around 1460 °C. The major reactions are:

$$Cu_2O(l) + CO(g) = 2Cu(l) + CO_2(g) \tag{2.1}$$

$$2Fe_3O_4(s) + 3SiO_2(s) + 2CO(g) = 3Fe_2SiO_4(l) + 2CO_2(g) \tag{2.2}$$

$$Fe_2SiO_4(l) + 2CO(g) = 2Fe(l) + SiO_2(s) + 2CO_2(g) \tag{2.3}$$

$$CO_2(g) + C(s) = 2CO(g) \tag{2.4}$$

The predominance area diagram of the system was calculated as indicated in Figure 2.7. The black dot indicates the coexistence of the obtained Fe–Fe$_3$C pig iron melt saturated with carbon and silica, slag with high silica content and a gas phase with a low oxygen partial pressure (P_{co2}/P_{co}). Under these smelting conditions, copper and molybdenum oxides are reduced to their metallic form and dissolve in a Fe–C pig iron together with the precious metals. The pig iron matrix contains about 5 wt% C. After solidification, the Fe–Cu–C alloy with its precious metals content is suitable for metal production.

Busolic et al.[23] studied the pyrometallurgical recovery of iron from copper smelter slags. Up to 80% of iron and almost 100% of copper can be recovered by melting coke and copper smelter slag at 1450 °C. With sufficient coke, the copper content decreases from 2.3 wt% to 0.5 wt% during the first 15 min. Iron oxide reduction starts after 15 min. The selective reduction of copper and iron can be achieved by adjusting the coke additions resulting in a copper rich alloy and an iron rich alloy in two steps. Carbon dosage and reduction time have a significant effect on the recovery of metal.

During reduction, the correct slag composition is crucial to maintain the slag in a molten state in order to run the reduction process properly. Fayalite slags saturated with silica are highly viscous and foaming can be caused by the gases (CO_2 and/or CO) generated from the reduction of oxides. Viscous slags inhibit both the rates of reduction and the subsequent settling of the reduction products. In order to decrease the viscosity of the slag, an oxidic flux is added to adjust the slag composition. A desirable ratio of SiO_2 to CaO is between 2.0 to 3.3 and the correct content of Al_2O_3 is about 5–10 wt%, which permits the slag to remain molten at about 1400 °C.[58]

Although high metal recovery of metals from copper slags can be achieved using pyrometallurgical methods, the high energy and carbon input of the techniques could hinder development in the industrial scale.

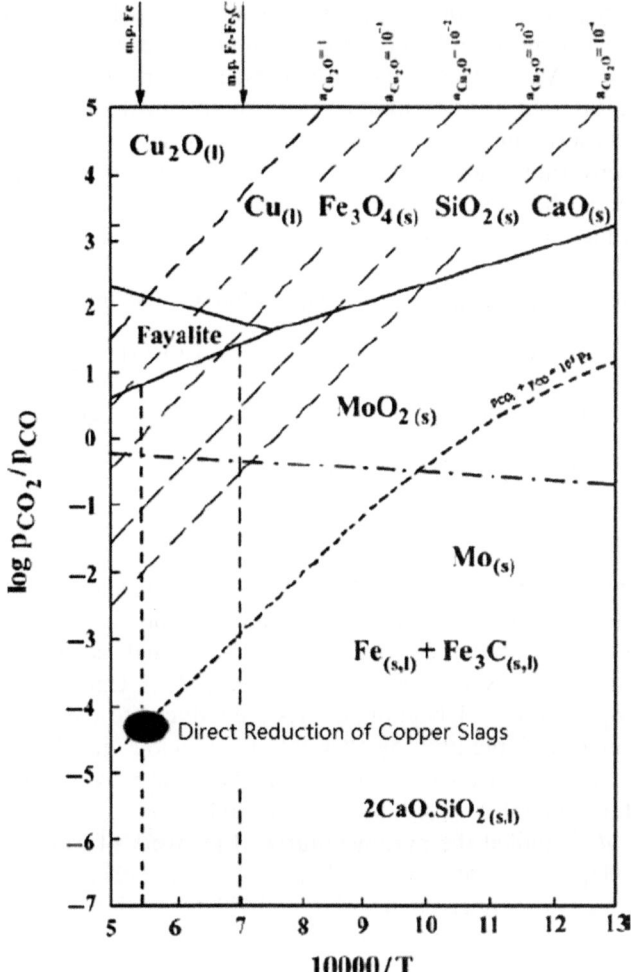

Figure 2.7   Predominance area diagram of the Cu–Mo–Fe–Si–Ca–O–C system.[57]

2.3.2.3.3   Pyro-hydrometallurgy.   Pyro-hydrometallurgical recovery of metal from fayalite slags normally consists of two steps: roasting and leaching. The principle of roasting is to convert the metals in a slag into a desired form which can be separated from the slag. One effective method is sulfation roasting, in which metals in various forms are converted into soluble sulfates by a series of chemical reactions at 150–600 °C with the addition of sulfatising agents. The obtained soluble sulfates are dissolved in water and thus separated from the slag. Up to 98% of the Cu can be recovered from converter slag using optimised parameters. Commonly used sulfatising agents in roasting are H_2S, H_2SO_4, $(NH_4)_2SO_4$, $Fe_2(SO_4)_3 \cdot xH_2O$ and pyrite.[27] Results obtained from roasting copper slag with H_2SO_4 and leaching the product with hot water showed that the recovery of Cu, Co, Zn and Fe can be higher than 80%. During sulfuric acid roasting, Cu extraction increased with increasing roasting time (up to 4 h) and roasting temperature (up to 300 °C) up to 100%. Application of thermal treatment at a relatively high temperature ($\sim$600 °C) after acid roasting leads to the decomposition of ferric sulfate ($Fe_2(SO_4)_3$) and the formation of iron oxide (Fe_2O_3). This is favourable from the perspective of recovering non-ferrous metals from solutions dowing to the absence of iron in the leachate.[59] Compared to water, application of sulfuric acid in leaching of roasted material could markedly improve the cobalt and nickel recoveries since they are trapped in fayalite matrices which can be dissolved by the acid.[60] Roasting can also promote the oxidation of iron silicate, forming hematite and silica. The entrained Mo can also be oxidised and recovered by acid leaching.[44]

2.3.2.3.4 Hydrometallurgy. The essential characteristic of hydrometallurgy is to dissolve the metals in liquids and then process them in subsequent solvent extraction and/or electrowinning operations. Compared to pyrometallurgy, hydrometallurgy is more accurate, more predictable and easier to control. In combined-cycle plants, it can be used to reduce the energy demands and environmental impact of pyrometallurgical plants.[61] On the other hand, hydrometallurgical processes require relatively long reaction times and therefore large reactors. Moreover, the requirement of corrosion resistant reactor material due to the highly corrosive reagents and the following waste water treatment should also be taken into account. Comprehensive research has been performed with respect to the hydrometallurgical recovery of metals from fayalite slags.[27]

The most commonly used leachants are sulfuric acid, hydrochloric acid, ferric chloride, ammonia and cyanide. Cyanide is no longer used owing to health concerns and environmental problems. Sulfuric acid is popular as a leaching reagent since it is available as a by-product of the smelting process. Metal oxides are leached directly by acid solutions while leaching of sulfides requires an oxidant as well as acid:

$$M_xO + 2H^+ = M^{x+} + H_2O \qquad (2.5)$$

$$2M_yS + 5O_2 + 4H^+ = 2M^{y+} + 2SO_4^{2-} + 2H_2O \qquad (2.6)$$

Direct leaching of copper slag with sulfuric acid involves huge amounts of acid (over 1 tonne acid per tonne slag) and a lengthy process time (more than one month). This leads to high operational costs and the recovery of copper is only similar or even lower than with the pyrometallurgical methods.[54] Addition of H_2O_2, leaching under pressure or leaching in a Cl_2/Cl^- system can oxidise the metal sulfides into metal oxides which dissolve more easily in acid. Hence, increased leaching efficiency can be achieved.[27]

Leaching of metals from fayalite slag with sulfuric acid (H_2SO_4) results in a low recovery rate of Cu but high recovery rates of Co, Zn and Fe. Increasing the H_2SO_4 concentration yields better extraction but the Cu extraction rate generally remains low. This can be attributed to the different mineralogical forms of the elements. Most of the Cu in the slag is in sulfide form, which is hard to dissolve in acid, while Co, Zn and Fe are present in iron-based mineral phases which can be more easily dissolved. The oxidants used in previous research to oxidise metal sulfides include hydrogen peroxide (H_2O_2), sodium chlorate ($NaClO_3$), potassium dichromate ($K_2Cr_2O_7$), chlorine (Cl_2) and gaseous oxygen (O_2). In the case of oxidising leaching, goethite (FeOOH), which is commonly used as iron ore, will form in the residue.[30]

Increasing the temperature (up to 60–80 °C) during the acid leaching of copper smelter slag can improve metal recovery. The effect of a further temperature increase is negligible. Addition of hydrogen peroxide (H_2O_2) as oxidant considerably decreases the dissolution of iron while it increases the copper recovery and improves the filtration of pulp. Selective copper and cobalt–zinc solvent extraction can be achieved with organic reagents. Overall recoveries of about 80% Cu, 90% Co and 90% Zn in separate solutions have been reported. The recoveries of the metals from their respective solutions can be carried out by electrowinning or salt precipitation.[30] Selective extraction of cobalt, copper and zinc from copper smelter slag can be performed by leaching with sulfuric acid and sodium chlorate and neutralisation with calcium hydroxide. In this way, silica gel formation and iron extraction are avoided during the process. Recovery of up to 95% of Co and Ni and 85% of Cu can be achieved.[62] Addition of dichromate compounds in H_2SO_4 leaching has a significant positive effect on the rate of copper dissolution and decreases the extraction yields of Co, Zn and Fe. This may be due to the oxidation of copper sulfides and passivation of the iron-based silicate slag particle surface by dichromate.[33] During chlorine leaching of copper smelter slag, the particle size and initial chlorine concentration are the most significant parameters. Fast (<5 min) leaching of sulfide phases can be reached with a small particle size (<20 μm) and a Cl_2 concentration of 10^{-2}–10^{-3} M at ambient temperature.[38] During high pressure oxidative acid leaching of copper and nickel smelter slag, temperature and acidity are the key factors. Relatively high temperature (~ 250 °C) and low acid addition (15%) are favourable for selective extraction of nickel, copper and zinc from iron. More than 95% of these metals can be recovered by optimised procedures.[39,63] The optimal experimental conditions and metal recoveries of the research work mentioned are summarised in Table 2.9.

Table 2.9 Optimal experimental conditions and metal recoveries of sulfuric acid leaching tests with different oxidants.

Reference		*30*	*31*	*33*	*38*	*39,63*
Oxidant		H_2O_2	$NaClO_3$	$K_2Cr_2O_7$	Cl_2	O_2
Temperature (°C)		70	95	25	20	250
Time (h)		2	3	2	0.1	2
Metal	Cu	80	89	81	80–90	99
recovery (%)	Ni	—	—	—	—	99
	Co	90	98	12	—	99
	Zn	90	97	10	—	94
	Fe	—	0.02	3	4–8	2.2

—, data not available.

The mineralogical phases present in the slag can affect the leaching rate of metals. Pavéz *et al.*[36] mention in their research that a vitreous phase is more insoluble and inert to chemical attack. A crystalline fayalite slag yields large amounts of extracted metal via oxidative sulfuric acid.[39,63]

The leaching process of the slag can be enhanced by applying different technologies. Ultrasonic energy can provide positive effects on the dissolution of base metals in copper slag. A noticeable increase in extraction of copper, zinc, cobalt and iron from the slag is obtained during ultrasound enhanced leaching, in which no special attributes except an ultrasonic source are needed.[35] Biotechnology has also been applied in research into recovery of copper from slag. Bacteria are used as oxidants in acid leaching in the biooxidation of Fe(II) producing Fe(III).[48]

The economics of the metal recovery from fayalite slags should be kept in mind. After separation from slag, subsequent purification steps could also be costly. Combination of metal recovery from slag with existing equipment and processes is suggested. The utilisation of residues generated from the metal recovery process is also worth consideration and could have a positive effect on the overall economy of the recycling process.

2.3.2.4   Utilisation

The use of the fayalite slag in cement, aggregates, abrasive tools, tiles pavement, ceramic-glass and wastewater treatment applications have been investigated and can have economic importance for the whole recovery and recycling process. Instead of spending on disposal costs and potentially causing environmental problems, the utilisation of fayalite slag can be economically beneficial. On the other hand, utilisation of such a metallurgical by-product can ease the natural resource shortage and carbon emission either directly or indirectly.

2.3.2.4.1   Cement.   Utilisation of slag in cement and concrete production is one of the greatest potential outlets for fayalite slags. The production of cement is an energy intensive process which also generates a significant amount of CO_2. Applying fayalite slag as a partial replacement for cement not only reduces the energy consumption and CO_2 output of cement making (only grinding is needed) but also beneficially recycles the by-product.

Cement making takes up 92% of the energy demand of concrete production. Therefore, the application of copper slag as supplementary cementitious materials (SCM) has a higher value than its use as aggregates. Owing to its high Fe content, copper slag has been used as an iron source in cement clinker production. Meanwhile, the melting point of copper slag is lower than the calcination temperature for cement making ($\sim 1500\,^{\circ}C$). Therefore, the application of copper slag in cement production can lower the energy consumption of the process.[28]

Secondary fayalite based slag produced from battery smelting has been studied with respect to its use as a substitute for cement and/or as aggregate in the production of concrete blocks with superior properties. In the early ageing stage, the compressive strengths of the slag-containing samples are lower than that of samples in which only OPC (ordinary Portland cement) was used. After 90 days ageing, the sample containing 40 wt% of slag had a higher compressive strength (7 MPa) than those made with 100 wt% OPC (6.2 MPa). Pb leaching from the samples with a high slag content (30% cement substitution, 100% aggregate substitution) is much lower (around 0.06 ppm) than the Thai hazardous waste disposal standard (5 ppm).[64]

The high content of silica and iron oxides of fayalite slag indicates its potential for use as high quality pozzolanic material in blended cement.[65] It has been reported that the compressive strength of mortars made with copper smelter slag (up to 10% cement replacement) is slightly lower than that of mortars made only with commercial Portland cement.[66] However, in other research, it was indicated that the use of up to 10–15% copper slag in concrete mixtures resulted in a significant (up to 50%) increase in the compressive strength (90 days of ageing).[37,67] The increase is considered to be due to the densification of the microstructure in the capillary pore region. A decrease has been observed in the water absorption rate by capillary suction and in the absorption and carbonation depth in the copper slag concrete, compared to conventional concrete, which indicates an improved durability.[68]

The optimal experimental conditions and corresponding results from different studies are summarised in Table 2.10. Although a lot of research has been performed on the use of fayalite slag as cement replacement, industrial applications are still limited.

2.3.2.4.2 Aggregate. The rapid growth of the construction industry globally results in an increasing consumption of natural resources and destruction of the environment.[70] Aggregates take up more than 70% of the volume of concrete. In order to reduce the dependency on natural aggregates, suitable alternative materials should be found.[65] Fayalite slags have similar properties to those of natural basalt (crystalline) or obsidian (amorphous).[71] Using fayalite slag as an aggregate is an application with large volume throughputs which will save the space used in piling up the slag and produces economic benefits at the same time.

Table 2.10   Optimal experimental conditions and results of different slag cement tests.

Reference	Slag type	Slag replacement (wt%)	Fineness of slag	Water/cement weight ratio	28-day compressive strength (MPa) Slag cement	Portland cement
69	CSS	5	$1261\,cm^2\,kg^{-1}$	0.5	48	46
66	CSS	10	$80\% < 40\,\mu m$	0.5	38	42
67	CSS	15	$3000\,cm^2\,kg^{-1}$	0.4	52	42
68	CSS	20	$27.2\,\mu m$	0.4	40	39

Owing to its high hardness and good mechanical properties, copper slag has the potential to be used as fine and coarse aggregates.[72] When compared with common concrete, the 28 day compressive strength and splitting tensile strength of the concrete using copper smelter slag as coarse aggregate are improved by 10–15% and 10–18%, respectively.[73] Other research, indicates that the addition of copper smelter slag as sand replacement yielded comparable or higher strength compared to concrete mixed only with sand. An improvement of 70% in the 28 day compressive strength of mortars is achieved by 50% copper slag substitution.[74] Further addition of copper smelter slag would reduce the strength of the concrete owing to the presence of excessive free water in the mix.[65]

Similar to the composition of fayalite slag, the results of leaching tests performed on fayalite slags differ from site to site or furnace to furnace. The leaching results obtained from several abandoned mine sites in the USA indicated that the Cu concentration and Zn concentration exceeded US-EPA (United States Environmental Prtoection Agency) toxicity standards (10 and $100 \, \mu g \, L^{-1}$, respectively). Thus, the copper slag may be a source of toxic elements.[75] Leaching tests have also been performed on copper slag generated in modern processes from various origins. The obtained results indicated that very little of the heavy metal content (As, Cd, Cr, Pb and Se) was removed in an aggressive laboratory test and the values were well below US regulatory levels of drinking water quality standards.[76] It has been reported that the heavy metal (Cu, Ni, Pb and Zn) leachability of a mortar containing up to 10 wt% of copper slag is below Malaysian environmental quality orders.[31] Results obtained from leaching copper smelter slags are illustrated in Table 2.11. The leaching values of relevant elements are below the US-EPA regulatory level.[76]

2.3.2.4.3 Abrasive Tool. Copper slag is widely used as an abrasive media to remove rust, old coating and other impurities in dry abrasive blasting owing to its high hardness (6–7 Mohs), high density (2.8–3.8 g cm^{-3}) and low free silica content. Working pressure, feed rate and surface contamination have significant effects on productivity and consumption of copper slag in

Table 2.11 Results of leaching tests on copper smelter slags.

Reference		77	78		76	USEPA regulatory level
Test		TCLP^a	TCLP	^{b}MEPT	TCLP	
Leachate	Ni	0.37	0.07	1.2	—	—
concentration	Cu	33.6	2.4	3.01	6.23 ± 7.44	—
(mg L^{-1})	Pb	0.51	0.08	0.6	<0.84 ± 0.765	5.0
	Zn	18.5	0.07	2.2	0.84 ± 0.543	—
	Cr	0.081	0.09	1.4	<0.06 ± 0.017	5.0
	As	—	0.05	<0.002	<0.52 ± 0.50	5.0
	Cd	0.03	0.01	—	<0.04 ± 0.038	1.0

^{a}TCLP: toxicity characteristic leaching procedure.
^{b}MEPT: multiple extraction procedure tests.
—, data not available.

dry abrasive blasting.[79] In practice, copper slag, which is qualified and relatively inexpensive, is mixed with high-density, high-hardness crystalline material such as secular hematite to form abrasives that can cut ductile materials such as metals. The crystalline copper slag is formed by cooling molten slag in thin slag layers (1–5 cm) at a moderate rate which induces the formation of small crystals rather than glass or large crystals. Annealing of the slag at around 100 °C is carried out after air cooling for 24 h. The abrasive materials, with a particle size of 0.075–0.8 mm, can provide a smooth surface for the target material.[80] Glass–ceramic based fine abrasives can be produced by quenching a mixture of copper slag, soda lime glass, silica sand and alumina. The glass–ceramic abrasives can be separated and recycled by magnetic attraction owing to the presence of sufficient ferrite.[81]

The application of spent copper slag from abrasive cleaning in land reclamation has been studied. Its physical, geotechnical and geochemical characteristics make the spent copper slag suitable for use for land reclamation. However, the potential environmental consequences of the massive use of spent copper slag in reclaimed sites have to be further studied.[76]

2.3.2.4.4 Tile and Pavement.   Copper slag mixed with clay and sand can be used to produce unglazed floor tiles. The addition of copper slag is limited to 40 wt% owing to the bloating effect caused by SO_2 emission from the oxidation of sulfides. A promising sample contained 40 wt% copper slag and after firing at 1025 °C for 1 h, it obtained a flexural strength of 57 MN m^{-2}, a water absorption of 2 wt%, a hardness of 750 HV and a good acid resistance owing to the presence of alkali oxides.[82] Copper slag has been used in the fabrication of ceramic roofing tiles at 10 wt%. The application of copper slag as additive changes the quality and the pore size distribution of the tiles. This has a positive effect on the tiles response to the oxalic acid action of lichens which causes ceramic matrix deterioration and consequently ageing.[83]

Research on the feasibility of using copper slag as a fine aggregate in bituminous pavements has been carried out. With up to 30 wt% copper slag addition, the obtained material had improved volumetric as well as mechanical properties owing to good interlocking between the aggregates provided by copper slag.[84]

2.3.2.4.5 Glass Ceramics.   Glass ceramics can be produced from a mixture of fayalite based slag (22 wt%), blast furnace slag (68 wt%) and a small amount of quartz sand (10 wt%). A glass ceramic was obtained with a melting temperature, a glass transition temperature and a crystallisation temperature of 1300 °C, 700 °C and 860 °C, respectively.[85] Glass ceramics can be produced from fayalite based smelter slag after iron removal and addition of CaO and Al_2O_3. The obtained CaO–MgO–Al_2O_3–SiO_2 based glass ceramic exhibits higher mechanical strength as well as better wear and corrosion resistance than traditional construction materials. 10 wt% TiO_2 is added as a nucleating agent to improve grain growth and lower the cryst-allisation temperature.[86]

2.3.2.4.6 Wastewater Treatment. Copper smelter slag can also be used as a reducing agent for Cr(VI) removal in aqueous systems owing to its high Fe(II) content. The reduction efficiency depends on the acid content of the solution, the particle size of the slag and the reacting temperature. Toxic Cr(VI) can be completely reduced rapidly with a high acid dosage (stoi-chiometry of 2), a low slag particle size (less than 105 μm) and moderate temperature (80 °C). After the reduction, the dissolved heavy metals can be totally precipitated from the supernatant by adjusting the pH to 9.0 using a NaOH solution. The feasibility of the process needs further investigation.[32]

Besides the utilisation mentioned above, copper slag also has potential applications in solar power capture, aquatic food production and catalytic destructive distillation of biomass.[24] However, despite the various promising applications, the sales of fayalite based copper slag are limited. This situation can be attributed to the high grinding costs of cementitious applications, the high maturity of the aggregate market, widely available alternative materials, long shipping distances, and so on. The utilisation of such materials still requires encouragement from the technical–scientific–entrepreneurial community.[24]

2.3.2.5 Combined Schemes

Comprehensive research has been done into metal recovery and utilisation of fayalite slag. However, in order to achieve full valorisation of the slag, combination of the two schemes is worthy of consideration. Attempts have been made by several researchers but results are far from implementation.

Metal values, such as iron and copper, can be recovered by high temperature direct reduction with the addition of coke and fluxing agents. A Fe–Cu–C alloy containing precious metals is obtained. Meanwhile, a liquid slag phase will form, which has the potential to be used for cement production owing to its similar composition to steelmaking slags.[57] The flow sheet obtained is demonstrated in Figure 2.8.

The flotation waste of copper smelter slag (FWCS) is also a potential iron source for Portland cement clinker production owing to its sufficiently high content of iron in the form of fayalite and magnetite. Cement replacement tests have been performed on an industrial scale. The content of the slag in the clinker is only 2.5–6%. The products obtained had similar chemical composition and mechanical performance (40–50 MPa compressive strength) as the currently produced cement made with iron ore. Although the release of heavy metals from the FWCS can be of environmental concern, leaching of heavy metals is no longer a problem when the FWCS is used in the production of cement clinker because there is so little. Meanwhile, the FWCS is readily available at a low cost for use as cement raw material since no mining and

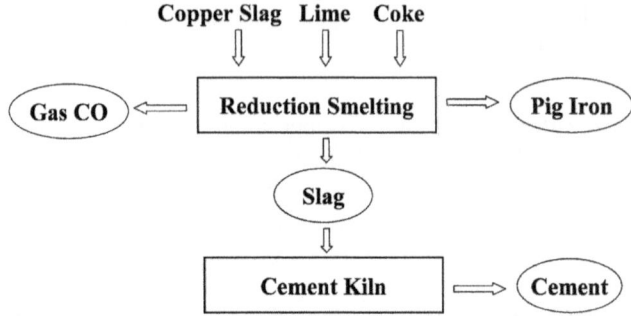

Figure 2.8   Flow sheet for a high temperature direct reduction process.[57]

material processing (size reduction) is needed and the use of FWCS can reduce the operating cost of cement making.[87]

2.4   Conclusions

Extractive metallurgical processes have been utilised and optimised in past centuries. Many of these methods can be applied to metallurgical slags, which often have similar or even higher metal concentrations than current ores. In addition, the combination of metal extraction from slag with utilisation of the residues from the extraction process can provide a more beneficial flow sheet, both from the economic and environmental point of view.

This was illustrated by a case study on fayalite slags from copper production. Rather than being treated as a waste, fayalite slags can be utilised as feed material for metal extractions and/or substitute of natural resources in different applications. The overall metal recovery of copper, nickel, cobalt and zinc can be up to 90% and the recovery of iron can be higher than 80% using different procedures. Owing to their chemical and physical properties, fayalite slags reveal promising potentials in many applications, such as cement, aggregate, abrasive tool, tile and pavement, glass ceramics, wastewater treatment, and so on.

Although many studies have been done on the metal recovery and utilisation of fayalite slag, most have mainly been carried out on a laboratory scale and have not yet been applied in industry. When compared with the utilisation of ferrous slags, the valorisation of fayalite slags is still in its infancy. The important parameters, efficiency and economic effects of different procedures need to be further studied and optimised. Moreover, the combination of metal recovery and the utilisation of the processing residues is worthy of further study in order to achieve a zero-waste process.

References

1. B. A. Wills, *Will's Mineral Processing Technology*, Elsevier, Oxford, 7th edition, 2006.
2. P. Hayes, *Process Principles in Minerals and Materials Production*, Hayes Publishing, Brisbane, Australia, 3rd edition, 2003.

3. USGS, *Mineral Commodity Summaries 2012*, Virginia, US, 2012.
4. F. Habashi, *Handbook of Extractive Metallurgy*, Wiley-VCH, 1997.
5. C. Feneau, *Non-ferrous Metals, from Ag to Zn*, Umicore, Brussels, 2002.
6. T. E. Norgate, S. Jahanshahi and W. J. Rankin, *J. Clean Prod.*, 2007, **15**, 838.
7. T. Norgate and S. Jahanshahi, *Miner. Eng.*, 2010, **23**, 65.
8. M. E. Schlesinger, M. J. King, K. C. Sole and W. G. Davenport, *Extractive Metallurgy of Copper* Elsevier, Oxford, 5th edition, 2011.
9. K. Nakajima, O. Takeda, T. Miki and T. Nagasaka, *Mater. Trans.*, 2009, **50**, 453.
10. J. Geiseler, *Waste Manage.*, 1996, **16**, 59.
11. P. Hewlett, *Lea's Chemistry of Cement and Concrete*, Elsevier, Oxford, 2003.
12. Environment-Agency, *BF Slag: a technical report on manufacturing of BF slag and material status in UK*, Waste and Resources Action Programme, Environment Agency, Oxon, 2007.
13. I. Z. Yildirim and M. Prezzi, *Adv. Civ. Eng.*, 2011, 13.
14. G. Power, M. Gräfe and C. Klauber, *Hydrometallurgy*, 2011, **108**, 33.
15. M. Gräfe, G. Power and C. Klauber, *Hydrometallurgy*, 2011, **108**, 60.
16. C. Morrison, R. Hooper and K. Lardner, *Cem. Concr. Res.*, 2003, **33**, 2085.
17. C. Weeks, R. J. Hand and J. H. Sharp, *Cem. Concr. Comp.*, 2008, **30**, 970.
18. W. J. Bruckard, M. Somerville and F. Hao, *Miner. Eng.*, 2004, **17**, 495.
19. V. Petkov, *Degradation Mechanisms of Copper Anode Furnace Referactory Linings*, PhD Dissertation, Katholieke Universiteit Leuven, 2007.
20. B. Gorai, R. Jana and K. Premchand, *Resour., Conervs. Recycl.*, 2003, **39**, 299.
21. K. Zhao, X.-L. Cheng, Y.-H. Qi, J.-J. Gao and X.-F. Shi, *Zhongguo Youse Yejin (China Nonferrous Metallurgy)*, 2012, **B**, 56.
22. H. Y. Cao, N. X. Fu, C. G. Wang, L. Zhang, F. S. Xia, Z. T. Sui and N. X. Feng, *Multipurp. Util. Miner. Resour.*, 2009, 2.
23. D. Busolic, F. Parada, R. Parra, J. Palacios, M. Hino, F. Cox, A. Sánchez and M. Sánchez, Recovery of iron from copper flash smelting slags, In: *VIII International conference on molten slags, fluxes and salts*, eds. M. Sánchez, R. Parra, G. Riveros and C. Díaz, GECAMIN Ltd, Santiago, Chile, 2009, 621–628.
24. M. Sudbury, Smelter slag – seeking market opportunities & carbon credits and in a changing world. In: *VIII International conference on molten slags, fluxes and salts*, eds. M. Sánchez, R. Parra, G. Riveros and C. Díaz, GECAMIN Ltd, Santiago, Chile, 2009, 523–534.
25. International Copper Study Group, *The World Copper Factbook 2010*, 2010.
26. C. Atzeni, L. Massidda and U. Sanna, *Cem. Concr. Res.*, 1996, **26**, 1381.
27. H. Shen and F. Forssberg, *Waste Manage.*, 2003, **23**, 933.
28. C. Shi and J. Qian, *Resour., Conserv. Recycl.*, 2000, **29**, 195.

29. A. K. Biswas and W. G. Davenport, *Extractive Metallurgy of Copper*, Elsevier Science, Oxford, 3rd edition, 1994.
30. A. N. Banza, E. Gock and K. Kongolo, *Hydrometallurgy*, 2002, **67**, 63.
31. Y. Zhang, R. L. Man, W. D. Ni and H. Wang, *Hydrometallurgy*, 2010, **103**, 25.
32. B. Kiyak, A. Özer, H. S. Altundogan, M. Erdem and F. Tümen, *Waste Manage.*, 1999, **19**, 333.
33. H. S. Altundogan, M. Boyrazli and F. Tumen, *Miner. Eng.*, 2004, **17**, 465.
34. A. V. Bese, *Ultrason. Sonochem.*, 2007, **14**, 790.
35. A. V. Bese, O. N. Ata, C. Çelik and S. Çolak, *Chem. Eng. Process.*, 2003, **42**, 291.
36. O. Pavéz, F. Rojas, J. Palacios and A. Nazer, Pozzolanic activity of copper slag, In: *VI International Conference on Clean Technologies for the Mining Industry*, ed. M. Sanchez, F. Vergara, S. H. Castro and F. Parada. University of Concepcion, Concepcion, Chile, 2004.
37. B. Mobasher, M. Asce, R. Devaguptapu and A. M. Arino, Effect of copper slag on the hydratation of blended cementitious mixtures. In: *Materials Engineering Conference, Materials for the New Millenium*, ed. K. P. Chong. ASCE, New York, USA, 77–86.
38. O. Herreros, R. Quiroz, E. Manzano, C. Bou and J. Viñals, *Hydrometallurgy*, 1998, **49**, 87.
39. Y. J. Li, V. G. Papangelakis and I. Perederiy, *Hydrometallurgy*, 2009, **97**, 185.
40. K. Maweja, T. Mukongo, R. Mbaya and E. A. Mochubele, *J. Hazard. Mater.*, 2010, **183**, 294.
41. D. M. Mihailova, *J. Univ. Chem. Technol. Metall.*, 2010, **45**, 317.
42. N. Cardona, P. Coursol, J. Vargas and R. Parra, *Can. Metall. Q.*, 2011, **50**, 330.
43. I. Imris, A. Sánchez and G. Achurra, Copper losses to slags obtained from the El Teniente process. In: *VII International Conference on Molten Slags, Fluxes and Salts*, ed. P. C. Pistorius. The South African Institute of Mining and Metallurgy, Johannesburg, South Africa, 2004, 177–182.
44. F. Parada, M. Sanchez, A. Ulloa, J. C. Carrasco, J. Palacios and A. Reghezza, *Miner. Proc. Extractive Metall.*, 2010, **119**, 171.
45. US-EPA, *Innovative Methods of Managing Environmental Releases at Mine Sites*, San Manuel, Arizona, 1994.
46. A. Sarrafi, B. Rahmati, H. R. Hassani and H. H. A. Shirazi, *Miner. Eng.*, 2004, **17**, 457.
47. M. C. Fuerstenau, K. N. Han, *Principles of Mineral Processing*, Society for Mining, Metallurgy and Exploration, Littleton, Colorado, 2003.
48. F. Carranza, R. Romero, A. Mazuelos, N. Iglesias and O. Forcat, *Hydrometallurgy*, 2009, **97**, 39.
49. B. Das, B. K. Mishra, S. Angadi, S. K. Pradhan, S. Prakash and J. Mohanty, *Waste Manage. Res.*, 2010, **28**, 561.
50. E. Tabosa and J. Rubio, *Miner. Eng.*, 2010, **23**, 1198.

51. K. L. C. Gonçalves, V. L. L. Andrade and A. E. C. Peres, *Miner. Eng.*, 2003, **16**, 1213.
52. S. Demetrio, J. Ahumada, M. A. Duran, E. Mast, U. Rojas, J. Sanhueza, P. Reyes and E. Morales, *JOM*, 2000, **52**, 20.
53. P. K. Ackerman, G. H. Harris, R. R. Klimpel and F. F. Aplan, *Int. J. Miner. Process.*, 2000, **58**, 1.
54. C. M. Acuna and M. Sherrington, *Mater. Sci. Forum*, 2005, **475–479**, 2745.
55. J. W. Donaldson, S. N. Sharma and N. J. Themelis, *Pyrometallurgical recovery of copper from slag material*, U.S. Patent 4,032,327, June 28, 1977.
56. R. Matusewicz and E. Mounsey, *JOM*, 1998, **50**, 53.
57. C. González, R. Parra, A. Klenovcanova, I. Imris and M. Sánchez, *Scand. J. Metall.*, 2005, **34**, 143.
58. J. C. Agarwal, P. R. Amman and J. J. Kim, *Pyrometallurgical recovery of iron from iron silicate slags*, Kennecott Copper Corporation, U.S. Patent No. 4,001,011, 1977.
59. C. Arslan and F. Arslan, *Hydrometallurgy*, 2002, **67**, 1.
60. H. S. Altundogan and F. Tümen, *Hydrometallurgy*, 1997, **44**, 261.
61. D. Andrews, A. Raychaudhuri and C. Frias, *J. Power Sources*, 2000, **88**, 124.
62. Z. Yang, M. Rui-lin, N. Wang-dong and W. Hui, *Hydrometallurgy*, 2010, **103**, 25.
63. M. Baghalha, V. G. Papangelakis and W. Curlook, *Hydrometallurgy*, 2007, **85**, 42.
64. M. Penpolcharoen, *Cem. Concr. Res.*, 2005, **35**, 1050.
65. K. S. Al-Jabri, M. Hisada, S. K. Al-Oraimi and A. H. Al-Saidy, *Cem. Concr. Comp.*, 2009, **31**, 483.
66. M. F. M. Zain, M. N. Islam, S. S. Radin and S. G. Yap, *Cem. Concr. Comp.*, 2004, **26**, 845.
67. R. Tixier, R. Devaguptapu and B. Mobasher, *Cem. Concr. Res.*, 1997, **27**, 1569.
68. W. A. Moura, J. P. Gonçalves and M. B. L. Lima, *J. Mater. Sci.*, 2007, **42**, 2226.
69. K. Al-Jabri, R. Taha, A. Alhashmi and A. Alharthy, *Constr. Build. Mater.*, 2006, **20**, 322.
70. T. Y. Tu, Y. Y. Chen and C. L. Hwang, *Cem. Concr. Res.*, 2006, **36**, 943.
71. S. Biswas and A. Satapathy, *Waste Manage. Res.*, 2010, **28**, 615.
72. C. Shi, C. Meyer and A. Behnood, *Res. Cons. Recycl.*, 2008, **52**, 1115.
73. M. Khanzadi and A. Behnood, *Constr. Build. Mater.*, 2009, **23**, 2183.
74. K. S. Al-Jabri, A. H. Al-Saidy and R. Taha, *Constr. Build. Mater.*, 2011, **25**, 933.
75. N. Piatak, *Appl. Geochem.*, 2004, **19**, 1039.
76. T. T. Lim and J. Chu, *Waste Manage. Res.*, 2006, **24**, 67.
77. H. Alter, *Resour., Conserv. Recycl.*, 2005, **43**, 353.
78. P. Shanmuganathan, P. Lakshmipathiraj, S. Srikanth, A. Nachiappan and A. Sumathy, *Resour., Conserv. Recycl.*, 2008, **52**, 601.

79. K. Kambham, S. Sangameswaran, S. Datar and B. Kura, *J. Cleaner. Prod.*, 2007, **15**, 465.
80. M. S. Chopra and Mehlman, S. F., *Abrasive formulation for waterjet cutting and method employing same*, Minerals Research and Recovery, U.S. Patent No. 5,637,030, 1997.
81. R. D. Hale, *Copper slag reclamation and recycling method*, U.S. Patent No. 2006/0207289 A1, 2006.
82. V. K. Marghussian and A. Maghsoodipoor, *Ceram. Int.*, 1999, **25**, 617.
83. M. Radeka, J. Ranogajec, J. Kiurski, S. Markov and R. Marinkovic-Neducin, *J. Eur. Ceram. Soc.*, 2007, **27**, 1763.
84. N. K. S. Pundhir, C. Kamaraj and P. K. Nanda, *J. Sci. Ind. Res.*, 2005, **64**, 997.
85. Z. J. Wang, W. Ni, Y. Jia, L. P. Zhu and X. Y. Huang, *J. Non-Cryst. Solids*, 2010, **356**, 1554.
86. J. H. Li, W. Ni and M. Ma, *Cons. Util. Miner. Res.*, 2006, 6.
87. I. Alp, H. Deveci and H. Sungun, *J. Hazard. Mater.*, 2008, **159**, 390.

CHAPTER 3

Ionometallurgy: Processing of Metals using Ionic Liquids

ANDREW P. ABBOTT* AND GERO FRISCH

Department of Chemistry, University of Leicester, Leicester, LE1 7RH, United Kingdom
*Email: apa1@le.ac.uk

Metallurgy is the foundation of the greatest technological advances made by mankind, where for every major metal, significant progress has followed the discovery of extraction and processing methods. Metals are predominantly found in nature as oxides, sulfides, carbonates or silicates and must be reduced primarily from the liquid state, that is high temperature salt melts or aqueous solutions, into a metallic form. In addition to naturally occurring ores there is an increasing demand for metallurgical processing of finished metal items and recycling waste from end of life goods, sludges, slags and tailings. The recent discovery of ionic liquids, that is salts which are molten at or around ambient temperature, has generated new metallurgical methods, which combine low temperature processing with the advantages of working in ionic media. These techniques constitute the new field of ionometallurgy.[1]

3.1 Metal Extraction and Recovery

Traditionally, metallurgical processes centre on three main methods: pyrometallurgy, electrometallurgy and hydrometallurgy. The principles underlying all three methods are the same: the metallic species is taken into a liquid state and then reduced. The method chosen depends upon the abundance of the

RSC Green Chemistry No. 22
Element Recovery and Sustainability
Edited by Andrew J. Hunt
© The Royal Society of Chemistry 2013
Published by the Royal Society of Chemistry, www.rsc.org

metal in the ore, its chemical form and the value of the metal. Metals can be split into four main categories:

- rock-forming metals, such as Al, Fe, Mg, Ti;
- major industrial metals, such as Cr, Cu, Ni, Zn, Sn and Pb;
- rare earth metals, such as La, Ce, Pr, Nd, Sm, Eu, Gd, Dy and Lu;
- precious metals, such as Ru, Rh, Ag, Pd, Re, Os, Ir, Pt and Au.

The abundance of the rock forming metals is about 10^3 to 10^4 times larger than the industrial metals, 10^5 to 10^6 times more than the rare earth elements and 10^9 to 10^{10} times larger than the precious metals. Extraction of rock forming metals tends to be by using pyrometallurgy or electrometallurgy and the technologies are relatively mature. Metal concentrations in workable ores can be as low as 1 to 10 ppm, so the large volume of waste created, in both liquid and solid form, is a significant issue in metal processing. Finished metallic objects tend to be processed using low temperature aqueous solutions, for example pickling, polishing, anodising or plating and this can also result in large volumes of dilute waste. Commonly used processing steps for metal extraction and recovery are shown in Figure 3.1.

The dissolution/melting step is dependent on the chemical state of the metal in the source material. If the source material contains the metal in an ionic form it can be introduced into the solution by acidic or basic digestion methods in either salt melts or aqueous solution. In the case of very insoluble salts, the anionic component can sometimes be removed by oxidation, for example digestion of sulfides by oxidising acids. If the sample is already in the metallic form, like it is in the recycling of metal scrap, it needs to be dissolved either by electrolytic or by chemical oxidation.

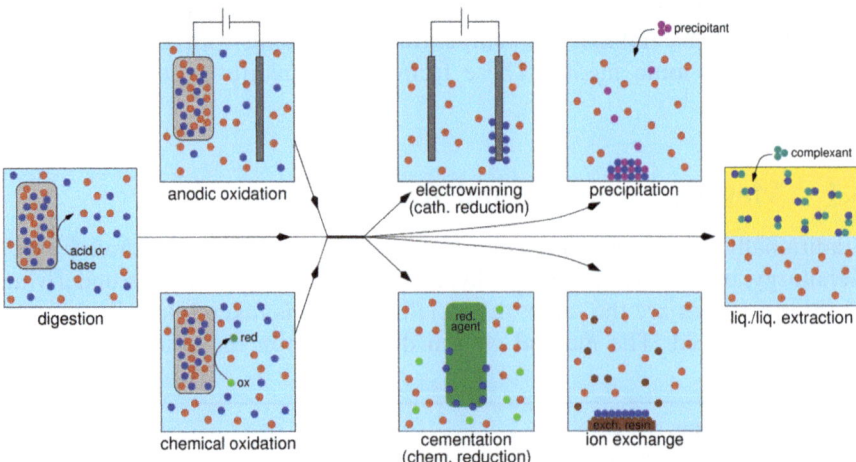

Figure 3.1   Flow chart showing steps involved in metallurgical processing of an ore, metal or alloy.

Electrolytic dissolution is a generic and efficient method of dissolving metals but is only feasible if the raw material can be cast into suitable anodes. This is often not the case for scrap sources like printed circuit boards. Digestion and chemical oxidation in aqueous solution is usually achieved using mineral acids, most commonly sulfuric acid or *aqua regia*. These have two advantages: the protons act as good oxygen acceptors for metal oxides and HNO_3 and H_2SO_4 are good oxidising agents for metals or sulfides. In some cases the medium chosen for extraction (the lixiviant) can be selective for a given fraction of the sample. For example, for silver purification HNO_3 is first added to extract the base metals but it does not dissolve silver. Once the solution has been removed concentrated HNO_3 or *aqua regia* is added to digest the silver.

Once in solution, a variety of extraction techniques can be used to convert the metal ions into either a metal or an insoluble salt. This process is shown schematically in Figure 3.1. Electrowinning uses an electrode to supply the electrons to reduce the metal ions in solution, in other words the opposite of the anodic dissolution stage. The metal crystallises on the electrode surface and is mechanically removed at the end of the process. Anodic oxidation and electrowinning can often be carried out in the same apparatus and this is called electrorefining. This is a common purification technique, for example for raw copper. Cementation uses a chemical reducing agent, for example a sacrificial metal like zinc, to exchange with the metal of interest in solution. Gold, for instance, can be cemented from ammoniacal thiosulfate solutions using copper as a sacrificial metal.[2] Ion exchange uses an ionic polymeric resin which in the case of metal extraction usually has acidic moieties. It exchanges the metal ions for protons and binds the metal ions to the polymer support. The precipitation of metal ions from solution requires the addition of a counter-ion which binds to the cation to render it insoluble. An alternative technology is to use a chelating agent in an immiscible solvent to extract the metal ions from a dilute aqueous solution into a more concentrated (usually non-polar) environment.[3] The chelating agents therefore contain a non-polar and frequently fluorinated moiety to enhance solubility in low dielectric constant media.[4]

3.2 Speciation and Phase Control

Metal ions in solution are often dilute and this can make extraction inefficient. The concentration of water molecules in an aqueous solution is in excess of $50 \, mol \, dm^{-3}$. Given that the concentration of any ion or neutral ligand is usually at least an order of magnitude smaller it is evident to see why aquo-species dominate in solution. In order to give an ionic solute the desired chemical properties its speciation must be adapted, for example by complexation:

$$\left[M(H_2O)_y\right]^{x+} + zL^- \rightleftharpoons \left[M(L)_z\right]^{x-z} + yH_2O$$

In order to compete against the high activity of water, that is to shift the equilibrium to the product side, ligands must either form a strong bond to the metal or be present in very high concentrations. Strong binding ligands are usually toxic and the use of cyanide in gold extraction processes is probably the best known example. In weaker, less toxic ligands the required high concentration is commonly achieved by using concentrated acids or bases, for example concentrated HCl for high chloride concentration. Both methods have clear disadvantages regarding handling, safety and environmental impact.

These issues can be circumvented in an ionometallurgical process: the desired ligand can be tailored into the ionic formulation of the solvent whilst keeping the concentration of water at an insignificant level. The effect that concentration of ionic and molecular species has upon speciation and reactivity can be seen in Figure 3.2. The difference in the reactivity of $CuCl_2$ in a deep eutectic solvent (DES) of choline chloride:ethylene glycol (1:2 ratio), an ionic solvent with properties similar to ionic liquids, compared to water can clearly be seen. In the former, $[CuCl_4]^{2-}$ is the dominant species whereas in water $[Cu(H_2O)_6]^{2+}$ is present. In the chloride-based ionic solvent, transfer in $Cu^{2+} + e^- \rightarrow Cu^+$ can be distinguished from $Cu^+ + e^- \rightarrow Cu$, whereas in water a complex redox response is observed owing to the instability of Cu^+.

Superimposed on this voltammogram is the effect of adding LiCl to the aqueous solution. The addition of 12.5% LiCl (molar fraction) clearly pushed the species equilibrium from $[Cu(H_2O)_6]^{2+}$ towards $[CuCl_4]^{2-}$ and the addition of 25% LiCl produces an electrochemical response which is almost indistinguishable from that of the DES. This is confirmed by the UV–vis spectra of solutions shown in Figure 3.2(b). This clearly reinforces the difficulty of defining an ionic liquid as simply a system which consists only of ions.

The generality of this approach can be seen from Figure 3.3 which shows solutions of various transition metal ions in water, the chloride-based DES ethaline and aqueous solutions of concentrated HCl. It can be seen that the aqueous solutions are different in colour from the solutions containing chloride ions. The colour of the metal ions in concentrated HCl solutions shows that chloride ions dominate ligation over water in cases where the concentration of chloride is very high or the concentration of water is low.

3.3   Current Issues

Hydrometallurgy understandably dominates ambient temperature metal processing: water was a convenient and cheap solvent when we could afford to ignore the consequences of direct waste water discharge into the environment. Today more responsible conduct is required and waste water is purified by energetically and chemically rather demanding methods. The main issue in hydrometallurgy is that dissolution of metal compounds usually involves the use of either toxic additives, for example CN^-, or concentrated mineral acids and bases, principally *aqua regia*, HNO_3, H_2SO_4 and caustic soda. These are needed to attain the requisite concentrations of metals in solution either by oxidation, oxide protonation or metal complexation. This results in large volumes of acidic and basic waste containing low but significant levels of metals. A comparison of metallurgical methods is shown in Table 3.1.

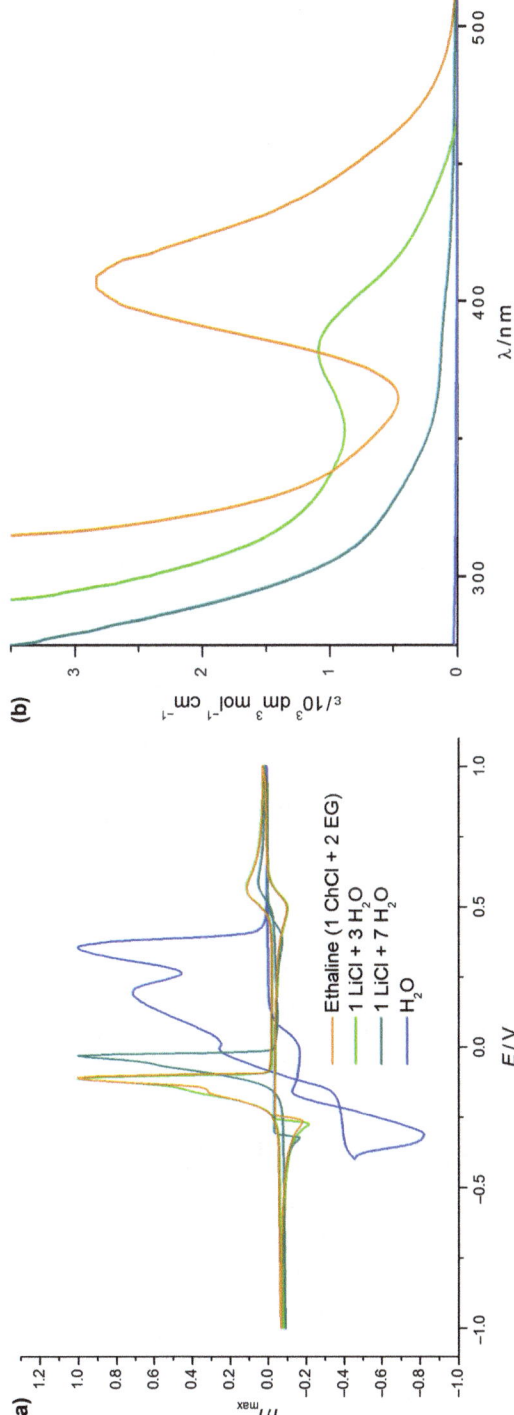

Figure 3.2 (a) Normalised cyclic voltammograms of CuCl$_2$ in choline chloride : 2 ethylene glycol (ethaline), in an aqueous solution and in an aqueous solution containing a molar fraction of 0.125 and 0.25 of LiCl, respectively. All voltammograms were measure using a Pt working electrode and a silver wire pseudo reference electrode. (b) UV-Vis spectra of the solutions shown in (a). In the above diagrams, I is electric current, E is electrode potential, ε is molar absorptivity and λ is wavelength.

(a)   (b)   (c)

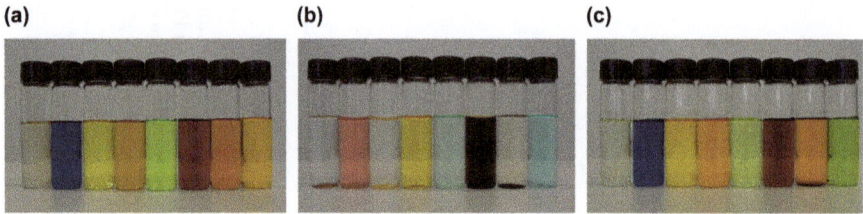

Figure 3.3 In each photograph, solutions from left to right, are $MnCl_2$, $CoCl_2$, $FeCl_2$, $FeCl_3$, $NiCl_2$, $PdCl_2$, $PtCl_2$, $CuCl_2$ in (a) ethaline (b) water and (c) concentrated HCl.

Table 3.1 Comparison of process traits.

	Advantages	*Disadvantages[6,7]*
Hydrometallurgy	Low temperature Useful for small scale Can be selective	Large volumes of aqueous waste streams Limited potential windows
Pyrometallurgy	Avoids oxo-chemistry Fast reaction rates Ability to reduce reactive metals Wide potential windows Inexpensive Large scale possible	High temperature High plant costs Less suitable for small scale Significant atmospheric pollution

One specific issue associated with the extraction particularly of rare earth and precious metals is the similar chemistry of the related and generally co-located metals. The large demand for these metals in electronic, magnetic and catalytic applications has pushed the extraction of ever leaner ore sources. Modern techniques involve processes such as *in situ* leaching. Ores such as copper carbonate and uranium oxide are leached using acid or carbonate solutions.[5] This can clearly lead to pollution of ground water, co-dissolution of heavy metals and destruction of delicate ecosystems.

Separation of mixed metal solutions can be achieved either at the digestion step or any of the recovery steps owing to differences in redox potential, speciation (charge or ligation) or solubility of the various metal-containing species.

3.4 Ionometallurgy: A Potential Solution

Ionometallurgy is the study of metal processing using ionic liquids. This includes not only solvents which consist solely of ions but any solution where the chemistry of solutes is dominated by solvent ions. It seeks to gain the advantages of low temperature processing methods while predominantly avoiding aqueous chemistry. It has the potential to benefit from the advantages of classic molten salt methods, for example wide potential windows, whilst avoiding the issues associated with high temperature. The desire to replace aqueous solutions is not an academic curiosity, but is driven by a need for facile, efficient methods of separating complex metal mixtures. Properties of ionic solvents can be varied for the application, for example they can be made immiscible with water or tuned to be polar and hydroscopic. The flexible architecture of anion and cation means that the chemical and physical properties of the ionic phase can be tailored to the specific digestion, extraction or recovery technique. High selectivity can be obtained by judicious choice of the anionic component in particular. This can be achieved with much less aggressive constituents or additives than in aqueous solution, where chemistry is dominated by a large excess of water molecules.

3.5 What is an Ionic Liquid?

Ionic liquids are officially defined as systems composed solely of ions which are liquid below 100 °C.[8] This is a rather arbitrary classification which is easy to define but impossible to achieve and does not describe useful systems since molecular species are omnipresent. Even very hydrophobic ionic liquids can support up to 1.4 wt% water.[9] A more useful definition could be based upon the physical properties of ionic liquids whose conductivity differs from dilute salt solutions in molecular solvents because mass transport is limited by hole size and mobility.[10] Alternatively, an ionic liquid could be defined as a solvent in which the chemistry of solutes is dominated by the ionic components of the solvent. As for metals, this is usually the anion.

Traditionally, ionic liquids are classified as either having one discrete anion and one cation or a series of complex anions formed from the interaction between a Lewis basic anion and a Lewis or Brønsted acidic constituent. The majority of cations are based on aromatic or aliphatic ammonium salts and the most studied systems involve either ethyl- or butyl-methyl imidazolium (C_2mim or C_4mim) as these tend to have the lowest viscosity and the highest conductivity.[11] Table 3.2 lists some of the anionic and cationic systems studied.

3.5.1 Properties of Ionic Liquids

The eco-toxicity of many of these liquids has been determined and reported with very varied results.[12] In general the toxicity of the ionic liquid increases as

Table 3.2 Common anions and cations used in ionic liquids.

Cations	Alkyl methyl imidazolium ($C_n$mim), Alkylpyridinium (C_nPy), Alkylpyrollidinium, Alkylpiperidinium Ammonium Phosphonium Cholinium $HOC_2H_4N^+$ (CH_3) (Ch) Beatine $HOOCC_2H_4N^+(CH_3)$ (Hbet)
Anion (discrete)	BF_4^-, PF_6^-, $CF_3SO_2^-$, $[(CF_3SO_2)_2N]^-$, $C_2H_5SO_4^-$, Cl^-, $(CN)_2N^-$
Anion (complex)	$AlCl_4^-$, $Zn_2Cl_5^-$, $[Cl^- \cdot 2urea]$

the alkyl chain length on the cation increases.[13] A range of ionic liquids with low eco-toxicity are now available and some have been applied to metal extraction. The versatility of the ionic liquid design has meant that benign cations have been produced and these include functionalised imidazoles to aid biodegradation,[14] lactams,[15] amino acids[16] and choline.[17] Our group has focussed on the last of these and taken them to the 10^2 to 10^3 kg scale for metal processing.[18]

The physical properties of ionic liquids span a wide range, for example density $1.00-2.40\,\mathrm{g\,cm^{-3}}$, viscosity $5-20\,000\,\mathrm{cP}$, mp -40 to $100\,^{\circ}\mathrm{C}$ and the properties change with temperature more than most molecular liquids.[19] However, the viscosity of ionic liquids is considerably higher than that of most molecular liquids and this will be a significant issue for industrial scale-up. As a consequence, although most texts state that most ionic liquids have relatively high conductivities, they do in general fall in the range $0.1-50\,\mathrm{mS\,cm^{-1}}$. This is considerably lower than aqueous acidic solutions which are in the range $100-500\,\mathrm{mS\,cm^{-1}}$ and is particularly important for the electrowinning step of the recovery.

The other important parameter is naturally one of cost which is harder to quantify as process scale is the controlling variable. It is however easy to speculate that the cost of most aqueous lixiviants will be in the range $1-10\,€$ per kg, whereas the most optimistic projections for ionic liquids would put the cost in the region $10-100\,€$ per kg. This clearly affects the areas into which this technology will reach. This also dictates that a recycling stage is essential and this must include at least 10 cycles for the process to be viable.

Ionic liquids are not a metallurgical panacea and they are not inherently green, however, through judicious design of the appropriate lixiviant they have the potential to introduce improved selectivity for some difficult to separate systems. This therefore offers the ability to decrease environmental emissions and energy consumption through process intensification. The most probable application will come where ionic liquids decrease the number of processing steps and the volume of lixiviant or where they are used primarily for high value metals. Potential candidates include platinum group metals, rare earth metals and trans-uranic elements.

The requirements of a medium in which to carry out metallurgical processing are:

- chemical and electrochemical inertness
- high solubility for solutes
- facile tuning of solvent properties
- immiscibility with water for biphasic extraction
- high temperature stability
- insensitivity to water.

With over 10^{18} ionic liquids available[20] and 10^2 available on a commercial scale it is pertinent to investigate properties systematically in order to predict suitable solvents for given applications. The cation tends to dominate the physical properties of an ionic liquid, whereas the anion controls the chemical properties. By selecting the appropriate anion–cation combination, a liquid with suitable chemical and physical properties can be produced. This does not simply allow the ultimate solvent to be created since physical, toxicological and financial constraints have to be considered. It is, however possible to identify classes of ionic liquids which may be suitable for a particular application.

3.6 Previous Studies in Ionometallurgy

3.6.1 General Observations

The practicality of using ionic liquids for metal processing on a large scale can be easily assessed. Given that metal solubility will usually not exceed 10 wt% in any solvent a 10–100 fold excess will be needed for processing. The cost of the liquid will have to be a fraction of the value of the metal to be economical since physical losses alone are likely to be in the range of 2–10 wt%. Given that the cost of most viable ionic liquids will be in the range 10–100 € per kg the recovery of 1 kg of metal may result in the loss of more than 1 kg of ionic liquid, at which point there are only a limited number of metals which can be recovered in this manner. Absorption losses during ore extraction will tend to be larger than those experienced using solvent extraction, however mutual solubility in the aqueous phase can also be significant. The high cost and viscosity of these liquids means that they are better suited to small volume, high value applications. The prospect of concentrating metals from large volumes of dilute aqueous solution into small volumes of ionic liquids is clearly beneficial. This is where most of the prior studies have focussed.

3.6.2 Systems Studied

While ionic liquids were first shown to be useful for metal processing more than 15 years ago they have only been applied to show proof of principle for the fundamental aspects, for example extraction, digestion cementation and so on. Numerous, often unrelated systems have been studied and these are covered in detail in other reviews.[21–24] Some typical examples are summarised in Table 3.3. In most cases, ionic liquids have been applied to biphasic extraction processes as a direct replacement for non-polar, water-immiscible phases but other aspects of the digestion and recovery processes have been probed.

The first systems studied were based on a chloroaluminate anion simply because these were the only ones commonly used at the time. Dai *et al.*[25] were the first to study dissolution in ionic liquids using imidazolium chloroaluminate to investigate the dissolution of UO_3. Ideally the digestion medium should not contain metal as it could lead to contamination of the extracted metal. Most of the complexing agents investigated are classically used in organic biphasic extraction and the results are relatively similar in most cases. The miscibility of ionic liquids with aqueous solutions is often worse than that with most organic

Table 3.3 Summary of metal digestion and extraction studies using ionic liquids.

Metal species	Ionic liquid	Complexing agent/oxidant	Reference
		Digestion	
UO_3	C_4mim Al_2Cl_7		18
Zn	C_4mim PF_6		37
V_2O_5, Cr_2O_3, CrO_3, MnO, FeO, Fe_2O_3, Fe_3O_4, CoO, Cu_2O, ZnO	ChCl • 2urea ChCl • 2ethylene glycol ChCl • oxalic acid		38
Eu_2O_3, yttrium oxide	([Hbet][Tf$_2$N])	Hbet$^+$	39
Uranium oxide	C_4mim Tf$_2$N		40
Au	C_4mim HSO_4	$Fe_2(SO_4)_3$	41, 42
Au	ChCl • 2ethylene glycol	I_2	
		Extraction	
Cd(II) Pb(II) Ni(II) Zn(II)	C_4mim PF_6	2-Pyridinealdoxime	43
Mn(II), Cd(II), Co(II), Ni(II), Zn(II)	C_4mim Tf$_2$N	Thenoyltrifluoroacetone	44
Ag(I) Pb(II)	Various PF_6 and Tf$_2$N	Dithizone	45
Ce(IV) Th(IV) La(III)	C_8mim PF_6	None	46
Ni(II), Cu(II), Pb(II)	C_4mim PF_6	2-Aminothiophenol	47
Ca(II), Sr(II), Ba(II)	C_4mim Tf$_2$N	tetraocty-3-oxapentane diamide	48
Y(III), Eu(III), Zn(II)	[$C_n$mim][Tf$_2$N] $n = 4$, 8, 12	N,N-Dioctyldiglycol amic acid	49
U(VI), Sr(II), Cs(I)	C_4mim BF_4, PF_6 and Tf$_2$N	Dicyclohexyl-18-crown-6	50
Zn(II), Cd(II), Cu(II) Fe(III)	N_{1888} Cl and BF_4	None	51

solvents which does not make sense either from an environmental or a financial point of view.

One aspect of ionic liquids which has gained considerable attention is the ability to engineer the architecture of the anion or the cation to make it specific to the solution species with which it is designed to interact. As a concept, Davis[26] is credited with task-specific ionic liquids (TSIL) where the liquid plays a dual role as both solvent and complexing agent. Many studies have chosen to functionalise the cation, primarily because the chemistry is relatively easy. It could be as simple as taking a ligand with an amine functionality and quaternising it. The simplest functionalised cation is beatine $HOOCC_2H_4N^+(CH_3)$ where the carboxylate functionality is an effective complexant for a variety of metals. Nockemann et al.[27] used the bistriflamide salt to extract a variety of metals from aqueous solutions. They showed a useful temperature induced phase behaviour which improved extraction efficiency although ionic liquid loss in the aqueous phase is a significant issue.[28] Other classical complexants such as monoaza-crown ether fragments[29] or urea- and thiourea functionalities have been covalently attached to imidazolium cations and used for biphasic extraction.[30,31] As many complexing agents are also anionic, their incorporation into a liquid formulation requires the choice of a suitable cation which delivers the appropriate physical properties to the liquid. In general the cations control the physical properties of the liquids whereas the anions tend to complex the metal and therefore control the chemistry.

Probably the most studied area is that of lanthanides and actinides as the facile separation of uranium from fission products is one of the most pressing issues in metal processing. A review by Binnemans[32] covers the chemistry of the *f*-block elements in ionic liquids. The extraction of lanthanides from aqueous solutions was investigated by Jensen *et al.*[33] using 2-thenoyltrifluoroacetone (Htta) and the metals were found in the form $Nd(tta)_4^-$ and $Eu(tta)_4^-$ as opposed to $M(tta)_3(H_2O)_n$ ($n = 2$ or 3), which was found in non-polar molecular solvents. Similar studies have used hexafluoroacetylacetonate[34] and salicylate anions[35] to extract a variety of metals from aqueous solutions. A comprehensive review of TSILs was written by Davis.[36]

3.7 Current Research and Future Direction

3.7.1 Oxidising Metals Electrolytically

One of the most important requirements is the ability to oxidise and reduce metals in the operating medium. The most efficient method for doing this is often anodic oxidation. While relatively common in aqueous solutions it has barely been reported in ionic liquids. Oxidic surface passivation is a well-known issue in aqueous electrodissolution processes. This can be addressed by the addition of acids or bases which solubilise the oxides but these create further problems concerning the environmental impact and by lowering the potential window of the solvent. The high solubility of many oxides in ionic liquids[26] suggests that anodic oxidation could become a key area for application of ionometallurgy.

We have previously demonstrated electropolishing of a variety of nickel based alloys in DES but not for the purposes of metal digestion. Figure 3.4 shows a nickel alloy which was anodically polarised for *ca* 8 h at a constant voltage of 4 V. This electrodissolution was carried out at room temperature. Figure 3.4(b) shows the solution after digestion and it is clear that two layers have formed. Nickel is relatively insoluble in the ionic liquid because it forms a complex with ethylene glycol and this means that the nickel is easy to recover by simple filtration. The filtrate only contains nickel which allows facile separation of the major component of the alloy from trace elements.

3.7.2 Oxidising Metals Chemically

In aqueous solutions a variety of organic and inorganic oxidants can be used to oxidise metals, including H_2SO_4, HNO_3, H_2O_2, chlorate, perchlorate and hypochlorite. Many of these processes involve the transfer not only of an

Figure 3.4   (a) Nickel alloy rod (left) before and (right) after anodic oxidation in a DES. (b) shows the DES after digestion.

electron but also a proton or OH^- ion. This is clearly not an issue in aqueous solutions but in aprotic ionic liquids common oxidising agents have not been studied. In metal digestion it is important to have a pure electron transfer oxidant that is not metal based. Oxidants such as H_2O_2 would become involved in the complexation and oxidants such as Fe(III) salts would complicate the extraction of metals at the end of the process. These restrictions rule out the majority of commonly used oxidants. Trihalide anions have been studied in ionic liquids with some success.

We have used I_2 as an oxidation catalyst for the digestion of metals in DESs. This was found to be ideal as there is a simple electron transfer where there are no complex side products and iodine does not oxidise the ionic liquid, which bromine, for example, would. I_2 partitions very effectively into DESs owing to the high concentration of Cl^- which produces I_2Cl^-. The high solubility of I_2 in the DES can be seen clearly in Figure 3.5 where organic solutions, each containing 0.5 M I_2, were placed in contact with a DES consisting of choline chloride : ethylene glycol (ethaline) (1 : 2 ratio respectively). All of the organic solvents are effectively decolourised and the iodine partitions into the ionic liquid. The redox potential of the I_2/I^- couple is more positive than would be expected from aqueous electrochemistry and it is above the oxidation potential of most metals, including gold. This has led to the practical application of stripping gold from microfossil scanning electon microcopy (SEM) samples.[52] Iodine can therefore act as an effective oxidant for most metals. Experiments have also attempted to use bromine, which oxidises the solvent constituents in DESs but may be suitable for use with more redox-stable ions. The trichloride

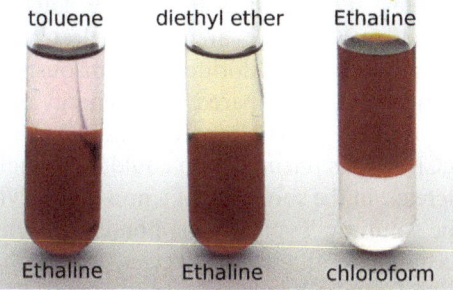

Figure 3.5 Partition of iodine between the DES ethaline and a variety of non-polar organic solvents.

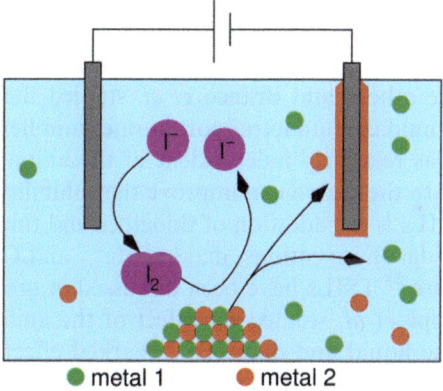

Figure 3.6 Diagram of the recovery of alloys using iodine as the oxidising agent in a solution of ethaline.

anion, Cl_3^- has even been reported to be a breakdown product of ionic liquids and may exist in media with high ionic strength.[53]

This has led to its use as an oxidation catalyst for the digestion of metals and particularly alloys. Ionic liquids effect a change in the relative redox potential of metals owing to differences in solution speciation. This can enable systems to be designed in which all the metals in an alloy can be digested using iodine and one of the metals can then be retrieved by electrowining from the solution. The use of iodine effectively avoids having to establish electrical contact between the anode and the metal. Anodic reoxidation of iodine in solution is very fast. This has been successfully achieved for a variety of alloys and is shown schematically in Figure 3.6.

3.7.3 Solubility

In aqueous solutions approximate rules exist for the solubility of metal salts, for example most hydroxides, sulfides and carbonates tend to be insoluble, whereas halides, nitrates and acetates tend to be soluble with only a few exceptions. Many of these general rules are pH and electrolyte dependent owing to speciation, for example silver chloride is notoriously insoluble in water owing to the high lattice energy compared to the energy of solvation, however in concentrated HCl or chloride-based ionic liquids, AgCl is soluble owing to the formation of $AgCl_2^-$. This phenomenon could be thought of as a salting-in effect whereby an increase in the solubility of a non-electrolyte is observed when a salt is added to the solution. Typically the non-electrolyte interacts with the electrolyte to produce an overall charged species.

No analogous consideration has been made for ionic liquids although the AgCl example is one that is commonly found. Because ionic liquids typically contain >4 M anions many salts which are insoluble in aqueous solutions are soluble in ionic liquids. In contrast to aqueous solutions, salts of Na^+, K^+, Cs^+ and most group 2 salts show relatively poor solubility in ionic liquids.[54]

Surprisingly, relatively few systematic studies of metal salt solubility in ionic liquids have been carried out. The solubilities of metals have been previously reviewed.[55,56] On the other hand Branco *et al.* studied the solubility of LiCl, $HgCl_2$ and $LaCl_3$ in imidazolium tetrafluoroborates and hexafluorophosphates and found that it was relatively independent of the anion or cation.[57] Functional groups added to the cation can improve the solubility of metal salts. An example of these TSILs is the addition of thioether and thiourea groups to the side chain of the imidazolium cations, making Hg^{2+} and Cd^{2+} ions soluble in the ionic liquids phase.[58] TSILs have been discussed in greater detail above.

A study by Chiappe *et al.* studied the effect of the anion of the salt being dissolved in the ionic liquid and observed a marked effect. The highest solubility occurred when the anion of the metals salt was the same as that of the ionic liquid.[59]

3.7.4 Precipitating Agents

Although the use of precipitating agents in ionic liquids is not a topic which has been systematically studied, qualitative observations have been made albeit from the opposite perspective. Studies have attempted to construct suitable homogeneous catalysts for use in ionic liquids and it has been noted that uncharged complexes tend to exhibit very low solubility.[53]

From the general rule that "like dissolves like" it could be concluded that most ionic species should be soluble in ionic liquids whereas neutral molecules should not. This is a rather simplistic model since the rule tends to apply to small monophilic molecular solvents. Taking a simple ionic liquid such a C_6mim $C_2H_5SO_4^-$, it will naturally have the ability to facilitate charge–charge interactions and the cation's aromatic ring will have a high degree of polarisability. The C_1 proton on the imidazolium ring will be relatively acidic, whereas the anion will be Lewis basic. Removing the charge–charge interactions will clearly be the main effect on solubility which is the reason that the more covalent oxides are less soluble in charged media. An alternative, although as yet unproven way, would be to add an amphiphilic ligand, for example a fluorous surfactant which specifically interacts with the metal forming a hydrophobic complex. An approach that has been applied successfully is to use an anti-solvent. It has previously been shown that iron, nickel and chromium salts could be precipitated from a choline chloride : ethylene glycol DES by the addition of an equal volume of water. After filtration the water could be evaporated from the solution.

3.7.5 Redox Potentials

An understanding of redox potentials is essential in metallurgy as it governs the ease of oxidation for dissolution and reduction for deposition or cementation. It is possible to alter the redox potentials of metals significantly through judicious choice of the ionic liquid, although this has not been quantified in a systematic way. The difference in reactivity of metals in ionic liquids depends primarily upon the complexation of the metal in solution, that is the anionic component of the solvent. An example of this is that iodine can oxidise gold in some ionic liquids but it will not oxidise it in water. Complexation of the metal ion with the anion of the ionic liquid tends to result in anionic complexes.

Comparison of redox potentials from one ionic liquid to another are complicated owing to the lack of reliable reference electrodes for use in ionic liquids; this is part of the reason for a lack of authoritative data. Most studies have used either an electrode with a reference electrode that is unreliable owing to unknown liquid junction potentials or a quasi-reference electrode (QRE). Silver or platinum wires are the most commonly used QREs.[60] The wires are frequently put directly in the analyte solutions which may affect the reference potential by reaction with the analyte or species that adsorb to the surface of the electrode. Reference redox couples such as ferrocene/ferrocinium or cobaltocene/cobaltocinium are also commonly added at the end of an experiment to determine a reference potential.[61,62] The instability of the $Ag|Ag^+$ couple has been studied by Torriero and MacFarlane[63] and Rogers *et al.*[64] who suggested a standard approach for reference potentials in ionic liquids which is similar to that adopted in non-aqueous solvents. This avoids QREs except where they can be calibrated against a standard reference redox system. The $Ag|0.01$ M AgOTf reference electrode is recommended, along with the $Fc|Fc^+$ and $[Co(Cp)_2]^+|[Co(Cp)_2]$ reference couples.

The first hydrogen electrode for use in an ionic liquid was recently reported. It was found that the $Ag|Ag^+$ couple exhibited ideal Nernstian behaviour in ionic liquids with anions such as Cl^- whereas anions such as SCN^- deviated significantly from ideality. It was shown that other couples such as the $Cu^{2+/+}$ and H^+/H_2 were also Nernstian up to 1 mol kg^{-1} in DESs.[65]

A recent paper introducing the concept of ionometallurgy published the first electrochemical series in an ionic liquid.[1] The redox potentials in ionic liquids were compared to standard aqueous redox potentials and it was shown that significant scatter existed between the two sets of data, as shown in Figure 3.7.

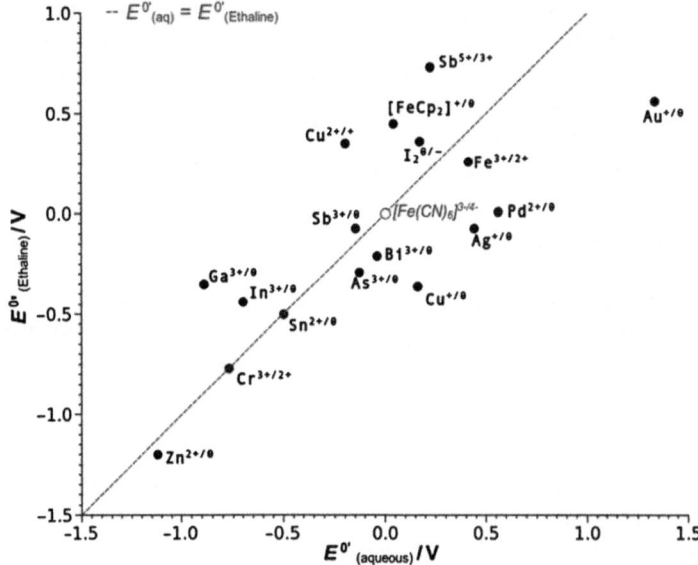

Figure 3.7   Plot showing the correlation between redox potentials in water and the chloride-based DES ethaline.

3.8   Issues to be Addressed

3.8.1   Material handling

To date most studies in ionic liquids have used analogous handling procedures to hydrometallurgy. One of the main differences between these media and aqueous solutions is the viscosity and hence the mass transport. The intimate contact of a solid particle and a viscous liquid phase generates difficulties associated with slow extraction kinetics and material loss resulting from entrapment or physical loss, particularly in processes like filtering. Since evaporation of the solvent is not an option, excess ionic liquid has to be extracted with a molecular solvent or left on the tailings of the digestion process. If these are released to the environment, the residual ionic liquid will have to have negligible environmental toxicity and be of low cost since mechanical losses could be in excess of 10 wt%. Filtration is the most common separation technique used and so one technical issue that will have to be addressed is how to filter viscous solutions containing suspensions of fine particles. In the only scale-up study to date, the particulates were <1 μm dust particles suspended in an ionic liquid with a viscosity of 20 cP. A double bag filter was used but visible amounts of particles were still left in the ionic liquid which became entrapped during the electrowinning stage of upstream processing. The physical losses of ionic liquid on the filtrate were in the order of 10 wt%.[5]

3.8.2 Material Recycling

The high cost and environmental incompatibility of many ionic liquids makes the repeated recycling of these systems essential. Physical losses, as described above, are an issue but so is the removal of extraneous waste like inert metal ions, water and various organic materials. There will always be waste water streams used either up-stream or downsteam of the ionic liquid stages so removal of ionic liquids from water is an issue that is relatively unaddressed. Biphasic extraction usually leads to the transfer of an ion from the ionic liquid to the aqueous phase and the metal complex is taken up by the ionic liquid. Methodologies for large scale cleaning of ionic liquids are not given in the open literature although some are clearly known from the handful of processes which are already carried out on a scale in excess of 1 tonne.

3.8.3 Fundamental Data

Little is known about speciation, reactivity, redox properties or solubility of metals in ionic liquids. In aqueous systems hydrometallurgical processing is greatly aided by Pourbaix diagrams,[66] which show thermodynamically preferred metal species in a variety of aqueous solutions as a function of pH and electrode potential. These do not exist in ionic liquids, partially because of the lack of reference potentials but mainly due to the lack of a suitable theory to quantify Lewis or Brønsted acidities. The conventional concept of acidity resulting from H_3O^+ does not work owing to the high ionic strength and the number of different ionic species present. A review of Brønsted acidity of ionic liquids[67] suggested that the position of acid base equilibria in a wide range of liquids correlates more closely with gas phase proton affinities (of the conjugate base) than with aqueous pK_a values because of the high aqueous solvation energies. Definitive pH measurement could, in principle, be attempted by potentiometric methods. This has been shown in the construction of a standard hydrogen electrode in an ionic liquid but data has still to be collected for useful acid systems.[62]

3.8.4 Environmental Impact

Some work on environmental impact was carried out by Matthijs and co-workers who studied choline chloride and ethylene glycol DESs.[68] They found that the heavy metals present in the rinse solutions constituted the main environmental impact but some electrolytic decomposition was also identified. The advantages of DESs are that the liquids themselves can be designed to be non-hazardous and biodegradable. Mechanical losses have been quantified but much of the metal loss was treated in a standard method using ion exchange resins to trap metal residues.[69] This area still requires extensive research before ionometallurgy can become successful on a large scale.

3.9   Conclusions

Ionometallurgy is a method of controlling the dissolution and recovery of metal species from solution by dominating the environment surrounding the metal cation with ionic species. It circumvents issues where metal separation and solubility are limited by aquo-controlled speciation. The main areas of application appear to be for precious metals and rare earth metals primarily because of the cost of the liquids but also due to the potential simplification of process chemistry resulting from tailored speciation in solution. Although the topic is in its infancy, ionometallurgy is probably most tantalising for its ability to recycle metals from complex composite materials. Significantly, more work needs to be carried out to characterise the most suitable systems for each complex material but some evidence exists already in the open literature that shows unprecedented tunability of ionic media over aqueous solutions. Ultimately, however it is most probable that a hybrid system of molecular and ionic components will be the most likely to deliver the optimum separation capability.

References

1. A. P. Abbott, G. Frisch, S. J. Gurman, A. R. Hillman, J. Hartley, F. Holyoak and K. S. Ryder, *Chem. Commun.*, 2011, **47**, 10031.
2. E. Guerra and D. B. Dreisinger, *Hydrometallurgy*, 1999, **51**, 155.
3. H. Freiser, *Acc. Chem. Res.*, 1984, **17**, 126.
4. H. Mehdi, K. Binnemans, K. Van Hecke, L. Van Meervelt and P. Nockemann, *Chem. Commun.*, 2010, 234.
5. L. E. Murr, *Miner. Sci. Eng.*, 1980, **12**, 121.
6. D. A. Wright and P. Welbourn, *Environmental Toxicology*, Cambridge University Press, Cambridge, 2002.
7. V. T. McLemore, Society of Mining, Metallurgy and Exploration, Littleton, CO, USA, *Basics of Metal Mining Influenced Water*, 2008, **1**.
8. K. Seddon, *Nat. Mater.*, 2003, **2**, 363.
9. P. Bonhôte, A.-P. Dias, N. Papageorgiou, K. Kalyanasundaram and M. Grätzel, *Inorg. Chem.*, 1996, **35**, 1168.
10. A. P. Abbott, R. C. Harris and K. S. Ryder, *J. Phys. Chem. B*, 2007, **111**, 4910.
11. H. Tokuda, K. Hayamizu, K. Ishii, M. Abu Bin Hasan Susan and M. Watanabe, *J. Phys. Chem. B*, 2005, **109**, 6103.
12. F. Stock, J. Hoffmann, J. Ranke, Stoermann, B. Ondruschka and B. Jastorff, *Green Chem.*, 2004, **6**, 286.
13. J. Ranke, K. Moelter, F. Stock, U. Bottin-Werner, J. Poczubutt, J. Hoffmann, B. Ondruschka, J. Filser and B. Jastorff, *Ecotoxicol. Environ. Saf.*, 2004, **58**, 396.
14. N. Gathergood, P. J. Scammells and M. T. Garcia, *Green Chem.*, 2006, **8**, 156.
15. Z. Du, Z. Li, S. Guo, J. Zhang, L. Zhu and Y. Deng, *J. Phys. Chem. B*, 2005, **109**, 19542.

16. K. Fukumoto, M. Yoshizawa and H. Ohno, *J. Am. Chem. Soc.*, 2005, **127**, 2398.

17. A. P. Abbott, G. Capper, D. L. Davies, R. Rasheed and V. Tambyrajah, *Chem. Commun.*, 2003, **7**, 70.

18. A. P. Abbott, J. Collins, I. Dalrymple, R. C. Harris, R. Mistry, F. Qiu, J. Scheirer and W. R. Wise, *Aust. J. Chem.*, 2009, **62**, 341.

19. P. Wasserscheid and T. Welton, *Ionic Liquids in Synthesis*, Wiley VCH, 2nd edition, 2008.

20. K. R. Seddon, Ionic Liquids: Designer Solvents?, in *The International George Papatheodorou Symposium: Proceedings*, S. Boghosian, V. Dracopoulos, C. G. Kontoyannis and G. A. Voyiatzis, Institute of Chemical Engineering and High Temperature Chemical Processes, Patras, 1999, pp. 131–135.

21. M. L. Dietz, *Sep. Sci. Technol.*, 2006, **41**, 2047.

22. M. L. Dietz, *Ion Exch. Solvent Extr.*, 2010, **19**, 617.

23. R. G. Reddy, *Metall. Mater. Trans. B*, 2003, **34B**, 137.

24. A. P. Abbott, G. Frisch, J. Hartley and K. S. Ryder, *Green Chem.*, 2011, **13**, 471.

25. S. Dai, Y. S. Shin, L. M. Toth and C. E. Barnes, *Inorg. Chem.*, 1997, **36**, 4900.

26. J. H. Davis Jr., *Chem. Lett.*, 2004, **33**, 1072.

27. P. Nockemann, B. Thijs, S. Pittois, J. Thoen, C. Glorieux, K. Van Hecke, L. Van, L. Meervelt, B. Kirchner and K. Binnemans, *J. Phys. Chem. B*, 2006, **110**, 20978.

28. P. Nockemann, B. Thijs, T. N. Parac-Vogt, K. Van Hecke, L. Van, L. Meervelt, B. Tinant, I. Hartenbach, T. Schleid, V. T. Ngan, M. T. Nguyen and K. Binnemans, *Inorg. Chem.*, 2008, **47**, 9989.

29. H. Luo, S. Dai, P. V. Bonnesen and A. C. Buchanan III, *J. Alloys Compd.*, 2006, **418**, 195.

30. A. E. Visser, R. P. Swatloski, W. M. Reichert, R. Mayton, S. Sheff, A. Wierzbicki, J. H. Davis, Jr. and R. D. Rogers, *Environ. Sci. Technol.*, 2002, **36**, 2523.

31. A. E. Visser, R. P. Swatloski, W. M. Reichert, J. H. Davis, R. D. Rogers, R. Mayton, S. Sheff and A. Wierzbicki, *Chem. Commun.*, 2001, **37**, 135.

32. K. Binnemans, *Chem. Rev.*, 2007, **107**, 2592.

33. M. P. Jensen, J. Neuefeind, J. V. Beitz, S. Skanthakumar and L. Soderholm, *J. Am. Chem. Soc.*, 2003, **125**, 15466.

34. H. Mehdi, K. Binnemans, K. Van Hecke, L. Van Meervelt and P. Nockemann, *Chem. Commun.*, 2010, **46**, 234.

35. V. M. Egorov, D. I. Djigailo, D. S. Momotenko, D. V. Chernyshov, I. I. Torocheshnikova, S. V. Smirnova and I. V. Pletnev, *Talanta*, 2010, **80**, 1177.

36. J. H. Davis Jr., *Chem. Lett.*, 2004, **33**, 1072.

37. H-L. Huang, H. P. Wang, E. M. Eyring and J. E. Chang, *Environ. Chem.*, 2009, **6**, 268.

38. A. P. Abbott, G. Capper, D. L. Davies, K. J. McKenzie and S. U. Obi, *J. Chem. Eng. Data*, 2006, **51**, 1280.

39. P. Nockemann, B. Thijs, K. Lunstroot, P. V. Kyra, N. Tatjana, C. Gorller-Walrand, K. Binnemans, K. Van Hecke, L. Van Meervelt, S. Nikitenko, J. Daniels, C. Hennig and R. Van Deun, *Chem. Eur. J.*, 2009, **15**, 1449.

40. K. Servaes, C. Hennig, I. Billard, C. Gaillard, K. Binnemans, C. Gorller-Walrand and R. Van Deun, *Eur. J. Inorg. Chem.*, 2007, **32**, 5120.

41. J. A. Whitehead, G. A. Lawrance and A. McCluskey, *Green Chem*, 2004, **6**, 313.

42. J. A. Whitehead, J. Zhang, N. Pereira, A. McCluskey and G. A. Lawrance, *Hydrometallurgy*, 2007, **88**, 1.

43. K. Fujinaga, K. Kozaka, Y. Watanabe, Y. Komatsu and J. Noro, *J. Ion Exch.*, 2007, **18**, 374.

44. K. Kidani, N. Hirayama and H. Imura, *Anal. Sci.*, 2008, **24**, 1251.

45. U. Domanska and A. Rekawek, *J. Solution Chem.*, 2009, **38**, 739.

46. Y. Zuo, Y. Liu, J. Chen and D. Q. Li, *Ind. Eng. Chem. Res.*, 2008, **47**, 2349.

47. R. Lertlapwasin, N. Bhawawet, A. Imyim and S. Fuangswasdi, *Sep. Purif. Technol.*, 2010, **72**, 70.

48. A. N. Turanov, V. K. Karandashev and V. E. Baulin, *Solvent Extr. Ion Exch.*, 2010, **28**, 367.

49. F. Kubota, Y. Shimobori, Y. Baba, Y. Koyanagi, K. Shimojo, N. Kamiya and M. Goto, *J. Chem. Eng. Jpn*, 2011, **44**, 307.

50. V. V. Yakshin, N. A. Tsarenko, A. M. Koshcheev, I. G. Tananaev and B. F. Myasoedov, *Radiochemistry*, 2012, **54**, 54.

51. A. P. de los Rios, F. J. Hernandez-Fernandez, F. J. Alguacil, L. J. Lozano, A. Ginesta, I. Garcia-Diaz, S. Sanchez-Segado, F. A. Lopez and C. Godinez, *Sep. Purif. Technol.*, 2012, **97**, 150.

52. D. Jones, J. Hartley, G. Frisch, M. Purnell and L. Darras, *Palaeontol. Electron*, 2012, **15**, 2.4T.

53. K. Haerens, E. Matthijs, K. Binnemans and B. Van der Bruggen, *Green Chem.*, 2009, **11**, 1357.

54. I. M. Alnashef, *Adv. Mater. Res.*, 2011, **233–235**, 276.

55. V. A. Cocalia, A. E. Visser, R. D. Rogers and J. D. Holbrey, In *Ionic Liquids in Synthesis*, P. Wasserscheid and T. Welton, Wiley-VCH, Weinheim, 2nd edition, 2008, Vol. 1, 89.

56. C. L. Hussey, *Pure. Appl. Chem.*, 1988, **60**, 1763.

57. L. C. Branco, J. N. Rosa, J. J. Moura Ramos and C. A. M. Alfonso, *Chem. Eur. J.*, 2002, **8**, 3671.

58. A. E. Visser, R. P. Swatloski, W. M. Reichert, R. Mayton, S. Sheff, A. Wierzbicki, J. H. Davis, Jr. and R. D. Rogers, *Environ. Sci. Tech*, 2002, **36**, 2523.

59. C. Chiappe, M. Malvaldi, B. Melai, S. Fantini, U. Bardi and S. Caporali, *Green Chem.*, 2010, **12**, 77.

60. G. A. Snook, A. S. Best, A. G. Pandolfo and A. F. Hollenkamp, *Electrochem. Commun.*, 2006, **8**, 1405.

61. J. Zhang and A. M. Bond, *Anal. Chem.*, 2003, **75**, 2694.

62. A. A. J. Torriero and A. M. Bond, in *Electroanalytical Chemistry Research Trends*, ed. K. Hayashi, Nova Science Publishers, New York, 2009.
63. A. A. J. Torriero and D. R. MacFarlane, *Electrochemistry in Ionic liquids in Ionic Liquids* UnCOIL eds, K. R. Seddon and N. V. Plechkova, Wiley, 2012.
64. E. I. Rogers, D. S. Silvester, S. E. W. Jones, L. Aldous, C. Hardacre, A. J. Russell, S. G. Davies and R. G. Compton, *J. Phys. Chem. C*, 2007, **111**, 13957.
65. A. P. Abbott, G. Frisch, H. Garrett and J. Hartley, *Chem. Commun.*, 2011, **47**, 11876.
66. M. Pourbaix, *Atlas of Electrochemical Equilibria in Aqueous Solutions*, 2nd English edition, National Association of Corrosion Engineers, Houston, Texas, 1974.
67. K. E. Johnson, R. M. Pagni and J. Bartmess, *Monatsh. Chem.*, 2007, **138**, 1077.
68. K. Haerens, E. Matthijs, A. Chmielarz and B. Van der Bruggen, *J. Environ. Manage.*, 2009, **90**, 3245.
69. K. Haerens, E. Matthijs, K Binnemans and B. Van der Bruggen, *Green Chem.*, 2009, **11**, 1357.

CHAPTER 4

Biosorption of Elements

PEI PEI GAN AND SAM FONG YAU LI*

Department of Chemistry, National University of Singapore, 3 Science Drive 3, Singapore, Republic of Singapore 117543
*Email: chmlifys@nus.edu.sg

4.1 Science of Biosorption

4.1.1 Mechanisms of Biosorption

Biosorption can be used to describe any system where interaction between a sorbate (atom, molecule or ion) and a biosorbent (solid surface of a biological matrix) results in an accumulation of the sorbate at the biosorbent interface and therefore a reduction of the sorbate concentration in solution.[1] There are two schools of thought regarding the nature of a biosorbent, one refers to both living and non-living biomass and the other refers to non-living biomass only.[1–4] Since many biosorption studies have been done on both living and dead biomass, in this chapter biosorption is regarded as a physicochemical process that involves both living and dead organisms as well as their excreted components.

The metal-binding mechanisms responsible for metal uptake in biosorption involve chemisorption (by ion exchange, complexation, coordination and chelation), physical adsorption and micro-precipitation. There are also possible oxidation/reduction reactions taking place on the biosorbent.[5] In the case of living organisms, the mechanisms could be further complicated by the change of microenvironment around living cells caused by metabolic activities such as respiration, nutrient uptake and metabolite release. Various plausible

RSC Green Chemistry No. 22
Element Recovery and Sustainability
Edited by Andrew J. Hunt
© The Royal Society of Chemistry 2013
Published by the Royal Society of Chemistry, www.rsc.org

mechanisms of biosorption of metal ions are depicted in Figure 4.1.[6] Owing to the complexity of the biomass, the biosorption process might be a result of the interplay of several different mechanisms and the identification of each single step is difficult to achieve.

Amongst various plausible mechanisms of biosorption, ion exchange appears to be the principal mechanism because it explains many of the observations made during heavy metal uptake experiments.[7,8] In biosorption, ion exchange could be considered as the replacement of an ion previously existing in the biosorbent with another ion in solution. Several functional groups that are available in the structural components of many biomass material are known to be potential ion exchange sites, such as carboxyl, sulfate, phosphate and amine, making them potential biosorbents for sequestering metal ions.[9] As an example, ion exchange could occur between light metal ions such as K^+, Na^+, Ca^{2+} and Mg^{2+} that are originally bound to the acid functional groups of algal biomass and heavy metals at the binding sites, resulting in a decrease of heavy metal concentration. Carboxylic groups (pK_a in the range 3.5–5.5) that are abundantly available in the algal and fungal cell-wall constituents are directly related to the adsorption capacity of the algae. The binding mechanisms involved during this ion exchange process may range from physical (electrostatic or van der Waals forces) to chemical processes (ionic and covalent),

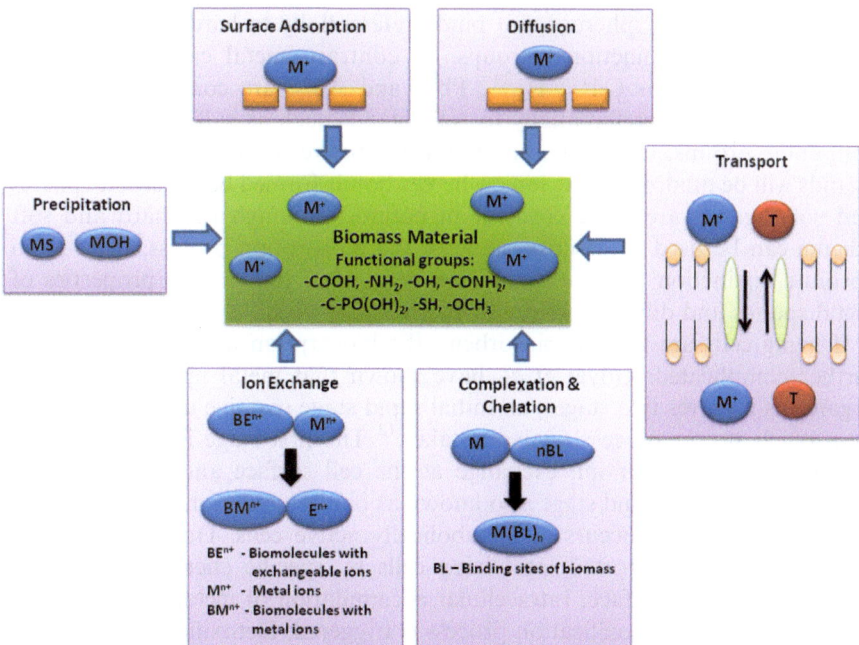

Figure 4.1 Plausible mechanisms of biosorption (reproduced with permission from Sud *et al.*[6]).

depending on the solution chemistry and available binding sites on the biosorbent. Since most of the heavy metals precipitate at pH >5.5, the bound metal species can act as loci for subsequent deposition and contribute to the metal uptake. Hence, micro-precipitation arising from the hydrolysis product might take place simultaneously in addition to ion exchange at higher pH.

Complexation could happen between any cations and molecules or anions containing free electron pairs (ligands). Chelation refers to complex formation with multidentate ligands. If the interacting ligand is immediately adjacent to the metal cation it is termed an inner-sphere complex, which is largely covalent in nature. On the other hand, if ions of an opposite charge are attracted and approach each other within a critical distance (usually separated by one or more water molecules), it is termed an outer-sphere complex which is largely electrostatic in nature.[10] It has been shown in several studies that heavier metal ions usually have higher binding affinities to the multidentate ligands where chelation of metal ions occurs. The underlying reason was related to the stereochemical effect, wherein a larger ion might fit better to a binding site with two distant functional groups.[11] Since most biosorption processes appear to be reversible, the nature of a bond formed during the chelation process was believed most likely to be electrostatic in nature (owing to the formation of an outer-sphere complex).

In some cases, the nature of complexation in biosorption could be predicted by hard and soft acids and bases (HSAB) theory. According to HSAB theory, metal cations with low polarisability such as Cr^{3+}, Fe^{3+} and Co^{3+} are considered as "hard spheres" that bind preferentially to hard ligands such as oxygen-containing functional groups. In contrast, metal cations with high polarisability such as Cd^{2+}, Hg^{2+}, Pb^{2+} and Au^{3+} are considered as "soft spheres" that bind preferentially to soft ligands such as sulfur and nitrogen-containing groups. Generally, bonds formed between hard spheres and hard ligands will be predominantly ionic whereas bonds formed between soft spheres and soft ligands are more covalent in nature. Although this hard and soft scheme can be used to describe some of the biosorption process, there is no absolute distinction that we can follow because of the varying properties of metal species and different nature of biomass.[1]

If living cells are used as biosorbent, the biosorption mechanism could be further complicated. Goyal *et al.* have shown that metal uptake by micro-organisms involves two stages, an initial rapid stage (passive uptake) followed by a much slower process (active uptake).[12] The first stage is thought to be physical adsorption or ion exchange at the cell surface and is metabolism independent. The second stage also known as bioaccumulation, is metabolism-dependent and only occurs in metabolically active cells. During this stage, metal ions are transported into living cells by specific chemical functional groups on the cell surface. Intracellular accumulation of metals could lead to cell death unless a detoxification process is triggered. Detoxification of metals could be achieved by binding to specific intracellular compounds (metallo-thioneins), internal compartmentalisation (vacuoles or polyphosphate bodies) and efflux of metals back into solution by active transport.[13,14]

4.1.2 Modelling Biosorption for Evaluation of Sorption Performance

The capacity of a particular biosorbent is assessed by considering the amount of sorbate being attracted and immobilised by the biosorbent at equilibrium. Basic evaluation of the sorption performance can be done by using a biosorption isotherm derived from equilibrium batch contact experiments performed under controlled environmental conditions (pH, temperature, ionic strength *etc.*) by varying the initial sorbate concentrations (C_i). During the equilibrium sorption studies, a certain amount of biosorbent (M) is allowed to react with the sorbate in an aqueous solution with volume V until an equilibrium state is established (the residual sorbate concentration in the solution become time invariant). A biosorption isotherm can be derived by plotting sorbate uptake (q_e, derived from Equation 4.1) against the final sorbate concentration at equilibrium (C_f):

$$q_e = (C_i - C_f)V/M \qquad (4.1)$$

This sorption isotherm can be used to compare the performance of different biosorbents towards the same sorbate or the affinities of different sorbates towards the same biosorbent. However, the comparison can only be done at the same equilibrium concentration, usually one at low C_f and another at high C_f (See Figure 4.2).[8]

Many sorption isotherms have been employed to describe the biosorption process. Two-parameter models such as those of Langmuir (See Equation 4.2) and Freundlich (See Equation 4.3) appear to be the most widely used in

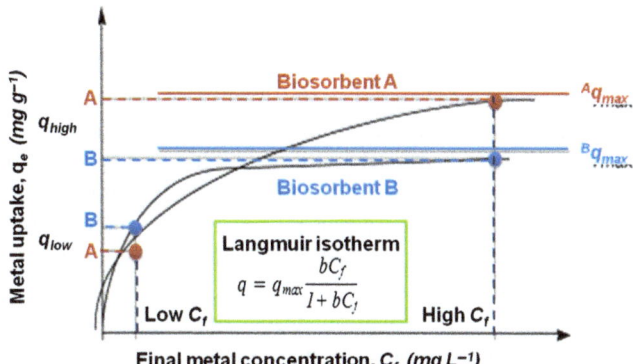

Figure 4.2 Biosorption equilibrium isotherm plots of metal uptake (q_e) against the residual (metal) concentration in the solution. When different biosorbents are being compared, biosorption performance in terms of metal uptake at q_{high} and q_{low} has to be judged at the same equilibrium (final) metal concentration. Comparisons of q_{max} are also useful (reproduced with permission from Kratochvil and Volesky[8]).

literature.[15,16] The extensive use of these two models might be due to their well-established and interpretable physical meaning:

$$q = q_{max}\frac{bC_f}{1 + bC_f} \tag{4.2}$$

$$q = KC_f^{1/n} \tag{4.3}$$

These models were originally developed to describe the gas–solid phase adsorption of activated carbon. The Langmuir isotherm is primarily theoretical with some underlying assumptions: (1) all binding sites have the same affinity for the sorbate; (2) adsorption is limited to the formation of a monolayer and (3) the sorbates are assumed to be incapable of interacting with neighbouring sorbate molcules.[10] For biosorbents that might consist of cell walls or other biological components that have multiple binding sites, these assumptions are not applicable as the affinities of different functional groups to the metal of interest might vary greatly. However, the Langmuir isotherm provides two interpretable constants that are useful for evaluating the biosorbent performance, that is, q_{max} and the coefficient b that correspond to the loading capacity (the highest possible sorbate uptake) and affinity to sorbate, respectively. A biosorbent with good sorption performance should have both high capacity and affinity. As shown in Figure 4.2, biosorbent A has a higher loading capacity than biosorbent B but biosorbent B has a higher affinity to the metal of interest at lower residual metal concentration. Thus, to judge which biosorbent is "better", it is always necessary to consider the intended purpose of different applications.

The Freundlich isotherm was originally developed empirically but was later interpreted as sorption to heterogeneous surface or surface supporting sites with varied affinities. The Freundlich model assumes that stronger binding sites are occupied first and the binding strength then decreases with increasing degree of site occupation.[17] In Equation 4.3, K is related to the maximum binding capacity whereas n is related to the affinity or binding strength.[7] Apart from the Langmuir and Freundlich models, there are other two-parameter models that have been frequently used to characterise the biosorption isotherms, such as the Temkin isotherm and the Dubinin–Radushkevich model.[18–21] In some studies, three-parameter models such as the Redlich–Peterson and the Sips models are also able to fit the experimental data accurately.[22,23]

Besides equilibrium batch sorption tests, the evaluation of biosorbent performance has been supplemented by process-oriented studies using dynamic continuous flow tests. Models for dynamic sorption studies include bed-depth service time (BDST), Thomas, Bohart–Adams and Yan models. In particular, the BDST model can be used to estimate the service time for a given bed height and specific sorbate concentrations at the bed inlet and outlet while the Thomas model can be used to predict the breakthrough curve and the maximum sorbate uptake by the biosorbent.[24,25]

However, it should be noted that the theoretical characteristics of a particular isotherm model are often too simplistic for actual biosorption processes. Moreover, the parameters obtained from these models are only applicable to the environmental conditions under which the experiment was conducted by assuming that all the external sorption system parameters remain constant over time.

4.2 Biosorbent Materials

4.2.1 Potential Candidates

Various types of biomass have been tested as potential biosorbents for metal removal. This includes microorganisms such as yeast, bacteria and algae as well as some waste materials of plant origin. A common rationale for selecting the potential biosorbent material is based on its ease of availability and cost of production. In this section, the discussion will focus on the potential of yeast, micro- and macroalgae, agro-based waste and vermicompost as biosorbents, taking into account their ease of availability in terms of industrial wastes and the massive growth in biomass with low production costs.

4.2.1.1 Yeast as Biosorbent

Amongst yeasts of different genera (*i.e. Saccharomyces*, *Candida* and *Pichia*), *Saccharomyces* yeast appears to be the most frequently studied for biosorption probably because it is the most important commercial yeast. Brewer's *Saccharomyces* yeast biomass is the second major by-product from the brewing industry and it is generally sold as inexpensive animal feed after heat inactivation.[26] The surplus yeast produced from fermentation industries can easily be obtained at a low price.

Yeasts are eukaryotic cells with a cell size of about 2.5–10 μm wide by 4.5–21 μm long, which are considerably larger than bacteria. The cell wall of yeast consists of a large number of complex organic components such as glucan (28.8%), mannan (31%), protein (13%), lipid (8.5%), chitin and chitosan (2%).[27,28] Several functional groups such as carboxyl, amino, amide, hydroxyl, sulfhydryl and phosphate groups that are available in the cell wall structure possess the ability to interact with heavy metals and provide to yeast the possibility of binding different elements. In particular, the outer mannan–protein layer appears to be more important than the inner glucan–chitin layer in heavy metal cations accumulation.[28] The biosorption capability of *Saccharomyces cerevisiae* cells has been shown in several studies in which they are able to remove a wide variety of metals from synthetic and real effluents.[29–35]

Compared with other microorganisms there is an additional advantage of using yeast as biosorbent owing to its flocculation characteristics. Yeast flocculation is a non-reproductive and reversible process of cell aggregation into multicellular masses (flocs), which enables them to settle rapidly in suspension media.[36] The flocculation characteristic of yeast is very useful in practical applications in which they can be easily separated from the treated effluent after biosorption without the need of cell immobilisation or solid–liquid separation. Since this is an intrinsic property of yeast cells, this cell separation process is also simple, green and efficient with low operational and maintenance costs. Furthermore, it is noteworthy that the flocculent yeast cells are able to retain their flocculation properties even after heat inactivation of biomass (at 45 °C), which is highly desirable for the separation process after effluent treatment.[30]

In addition, it has been reported that the flocculent yeast cells seem to have a higher ability to accumulate Cu^{2+} than non-flocculent yeast cells. The difference in metal uptake ability between the flocculent and non-flocculent yeast cells might be attributed to the presence of additional metal-binding sites on the cell walls of flocculent strains. These additional metal-binding sites are used to fix Ca^{2+} ions that regulate the interaction between specific zymolectin and the surface mannose residue (receptors) of the neighbouring cells during the flocculation process.[37,38] It was found that the flocculent strains are able to flocculate in the presence of Zn^{2+}, Cu^{2+}, Cd^{2+}, Ni^{2+} and Cr^{2+} but not in the presence of Pb^{2+}. The inhibition process exerted by Pb^{2+} is most likely related to its inability to induce the correct conformation required by zymolectins during its interaction with the mannose receptor on the neighbouring flocculent cell surface.[30,38] However, this inhibition effect could be alleviated by small amounts of calcium which are commonly found in industrial effluents and hence it will not affect the utilisation of this flocculation property in actual applications.[39]

In addition the metal uptake capability of yeasts is also affected by the changes in the composition and structure of the yeast cell wall owing to variation in the culture age, growth and nutritional conditions. Specifically, younger biomass was found to perform better in removing uranium than older biomass.[40,41] Yeast cultured in cysteine-supplemented media shows higher metal uptake than in a non-supplemented medium.[12] This behaviour is most likely related to the increased formation of sulfhydryl groups in metal-binding protein such as metallothioneins which are cysteine-rich. Figure 4.3 shows the culture of a brewing flocculent strain of *Saccharomyces cerevisiae* NCYC 1364 and the Ca^{2+} dependent cell flocculation mechanism.[42,43]

4.2.1.2   Macro- and Microalgae as Biosorbent

The term algae considered in this section refers to a large and diverse assemblage of aquatic photosynthetic organisms that are included in the plant kingdom. Macroalgae, commonly known as "seaweeds", are multicellular plants growing in salt or fresh water, whose sizes can reach up to 60 m in length.[44] Macroalgae are classified into three broad categories based mainly on their contents of phytopigments: (i) brown algae (Phaeophyceae); (ii) red algae (Rhodophyceae) and (iii) green algae (Chlorophyceae).[45] Microalgae are unicellular photoautrophic organisms that adapt to a wide range of conditions including soil, fresh and salt water, as well as industrial and domestic effluent

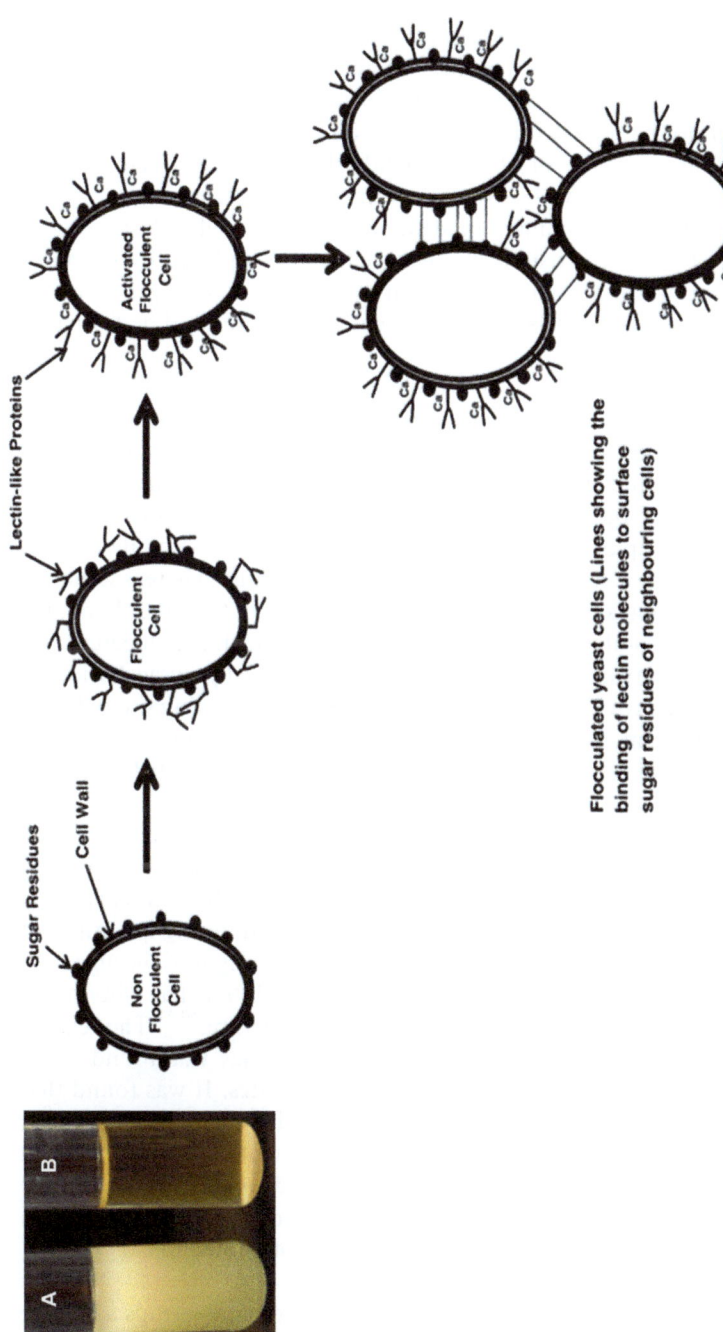

Figure 4.3 The lectin theory of yeast flocculation. Sugar residues are present on the cell walls of yeast cells. Flocculent cells bear lectin-like surface proteins, activated in the presence of calcium ions. These interact with sugar receptors on the cell walls of neighbouring cells and form flocs. The photo shows the culture of a non-flocculent strain (A) and a brewing flocculent strain (B) of *Saccharomyces cerevisiae* (reproduced with permission from Soares and Soares[42] and Singh *et al.*[43]).

dumping sites. The three most important classes of micro-algae in terms of abundance are: (i) diatoms (Bacillariophyceae); (ii) green algae (Chlorophyceae) and (iii) golden algae (Chrysophyceae).[46] However, it should be noted that cyanobacteria that are commonly known as "blue-green algae" are actually one of the groups of bacteria and do not truly belong to the algae group, so it will not be included in our discussion here.

The biosorption capability of algae is mainly determined by their cell wall properties in which the cell wall serves as the first cellular structure to come in contact with metal ions. A summary of algal divisions with cell wall structure is given in Table 4.1.[7] Other divisions of algae such as Crytophyta, Pyrrhophyta and Chrysophyta are either naked (without cell wall) or only protected by cellulosic "thecal" plates or scales, hence none of them show good performance in sequestration of heavy metals.[45,47]

The algal cell walls of Phaeophyta, Rhodophyta and many Chlorophyta are composed of two layers including an inner and an outer layer. The inner fibrillar skeleton layer in brown algae consists of cellulose as the major component while it can be replaced by either xylan, or xylan/mannan in the red and green algae. The outer amorphous embedding layer in brown algae is predominantly made up of cellulose with the presence of alginate or alginic acid and a smaller amount of sulfated polysaccharides (fucoidan), whereas it mainly consists of cellulose and sulfated galactans in red algae. On the other hand, the cell wall of green micro-algae is mainly made of cellulose with a high percentage of glycoproteins. The availability of alginic acid and sulfated polysaccharides in brown and red algae provides a platform of possible metal binding sites where a biosorption mechanism like electrostatic attraction and complexation could take place. The typical cell wall structure of brown algae and related structural components are depicted in Figure 4.4.[7,48,49]

Alginate is the main component in the outer layer of brown algal cell wall composed of a linear copolymer with homopolymeric blocks of 1,4-linked β-D-mannuronic (M) and α-L-guluronic (G) acid residues that are covalently linked in a non-regular sequence. The varying ratios and differing quantities of G and M residues differ according to the age, season and origin of the algae. As shown in Figure 4.4, polymannuronic acid (MM) exhibits a flat ribbon-like chain whereas polyguluronic acid exhibits a rod-like structure.[50,51] The relative conformation and proportion of these two block residues was found to be responsible for the difference in metal affinities of alginates. It was found that the affinities of alginates for divalent cations such as Pb^{2+}, Cu^{2+}, Cd^{2+}, Zn^{2+}, Ca^{2+}, increased with increasing guluronic acid (G) content.[51] The higher metal affinities of divalent metals for polyguluronic acid residues were believed to be related to their zigzag structures which can accommodate them (*e.g.* Ca^{2+}) more easily. Metal uptake was also observed for but not limited to Co^{2+}, Ba^{2+}, Sr^{2+}, Ni^{2+}, Mn^{2+} and Mg^{2+}.

Carboxylic groups which are the most abundant functional groups available on the alginate polymer play an important role in imparting metal binding ability to brown algae. It has been shown that many divalent metal ions have their maximal biosorption capability at pHs close to the pK_a of carboxylic

Table 4.1 Three algal divisions and significant characteristics (reproduced with permission from Davis *et al.*[7]).

Division	Common name	Pigments	Storage product	Cell wall	Flagella
Chlorophyta	Green algae	Chlorophyll a, b; α-, β-,γ-carotenes and several xanthophylls	Starch (amylase and amylopectin) (oil in some)	Cellulose in many (β-, 1,4-glucopyroside), hydroxyproline glucosides; xylans and mannans; or wall absent; calcified in some	Present
Phaeophyta	Brown algae	Chlorophyll a, c; β-carotene and fucoxanthin and several other xanthophylls	Laminaran (β-1, 3-glucopyranoside, predominantly); mannitol	Cellulose, alginic acid and sulfated muco-polysaccharides (fucoidan)	Present
Rhodopyta	Red algae	Chlorophyll a (d in some Florideophyceae); R- and C-phycocyanin, allophycocyanin; R- and B-phycoerythrin. α- and β carotene and several xanthophylls	Floridean starch (amylopectin-like)	Cellulose, xylans, several sulfated polysaccharides (galactans) calcification in some; alginate in corallinaceae	Absent

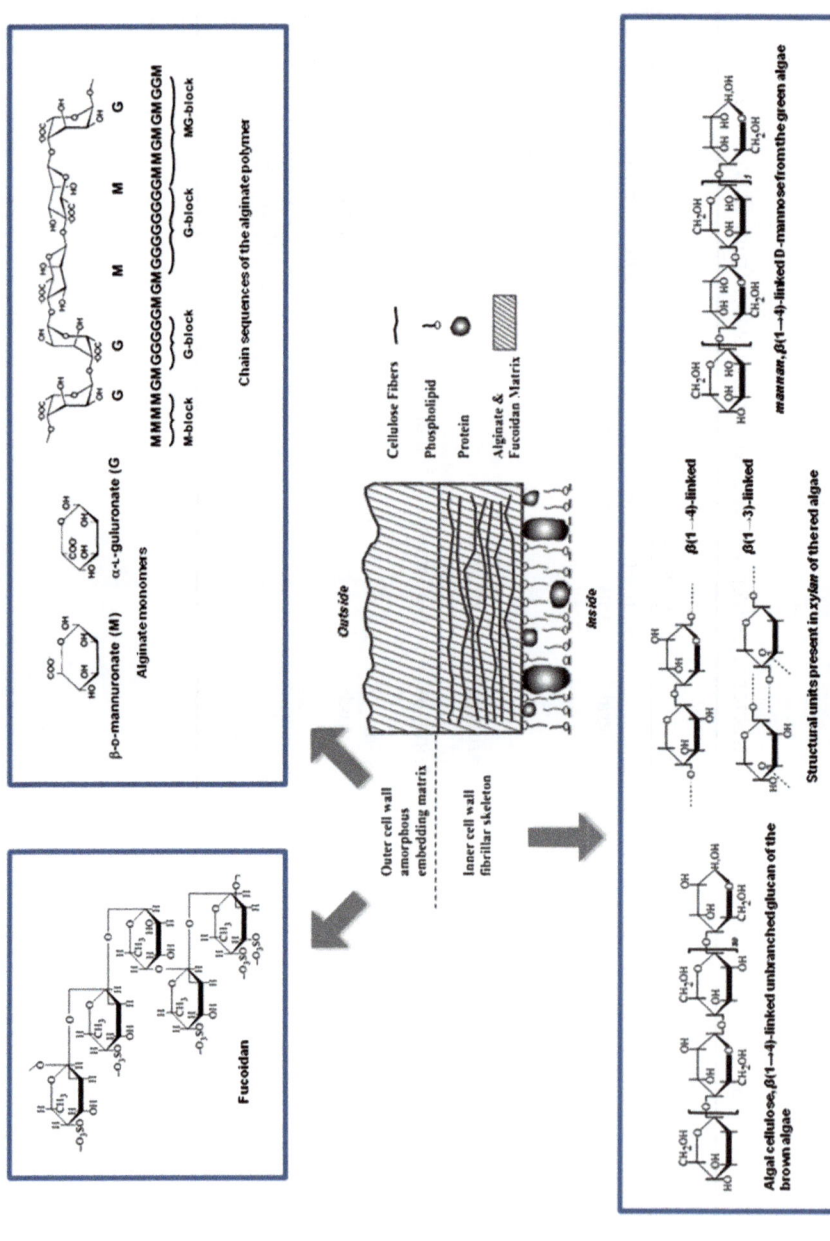

Figure 4.4 Typical cell wall structure of brown algae and related structural components of outer and inner cell wall layers (reproduced with permissions from Davis *et al.*[7] and Draget and Taylor[49]).

groups (≈ 5).[52–54] Although carboxyl groups have been proposed to be the main metal-binding sites, the role of other functional groups such as sulfated esters in cellular polysaccharides including fucoidans, galactans and xylans as well as polyuronides present in galacturonic, glucuronic, guluronic and mannuronic acids in the algal biomass could also serve as active metal-binding sites for metal biosorption.[55]

According to the existing literature, it seems that various species of brown macroalgae have been extensively studied for their metal biosorption ability whereas studies focusing on the metal biosorption of green and red macroalgae have been assessed to a lesser extent. The underlying reason might be related to the difference in their cell wall composition which presents a direct connection with the metal-binding sites, in which functional groups such as alginic acid and sulfated polysaccharides in brown macroalgae show great potential in metal biosorption. The biosorption potential of red macroalgae, namely *Coralline mediterranea*, *Galaxaura oblongata*, *Jania rubens* and *Pterocladia capillacea* to remove Co^{2+}, Cd^{2+}, Cr^{3+} and Pb^{2+} ions from aqueous solution have been examined by Ibrahim *et al.*[56] Their study demonstrated the high capacity of metal biosorption for red seaweed (the highest removal efficiency was obtained by *G.oblongata* biomass with a mean biosorption efficiency of 84%), high-lighting its potential for treating heavy metal ions from industrial effluents. In addition, the ability of red (*Corallina officinalis* L., *Porphyra columbina*) and green (*Codium fragile*) marine macroalgae to remove Cd^{2+} ions from aqueous solutions was examined as well. The results obtained show that species of red macroalgae possess better sorption capacities than green macroalgae in sequestering Cd^{2+} which might be attributed to the presence of sulfated galactans in red macroalgae.[57]

In addition to macroalgae, recent research interest has been diverted toward the investigation of microalgae as new biosorbent material. Some species of microalgae exhibit high metal tolerance when living cells are used.[58–60] The much smaller cell size of microalgae provides a larger specific surface area where the contact between active metal-binding sites and surrounding metals species in solution is greatly enhanced. The biosorption ability of microalgae is promoted by the negatively charged functional groups on the cellular surface where metal cations are adsorbed and subsequently transported across the cell membrane into the cytoplasm and bound to some intracellular components. Cell walls of microalgae mainly consist of polysaccharides, proteins and lipids where several negatively charged functional groups such as carboxyl, hydroxyl, phosphate, amino and sulfhydryl are available for metal binding. To investigate the metal uptake mechanism by microalgae, several studies have been performed and it seems that adsorption onto the cell surface via ion exchange appears to be the main mechanism.[51,58,61–64] In contrast, the mechanism of metabolic uptake was suggested in only a few reports.[65,66]

Although ion exchange is generally regarded as the dominant mechanism, other mechanisms such as complexation and microprecipitation could act simultaneously to varying degrees depending on the species of microalgae and other environmental factors. When the extracellular concentration of metal ions is considerably higher than that of intracellular ions, metal ions being adsorbed on the cell surface can penetrate into the cell via several mechanisms. One possible mechanism is molecular mimicry in which metal ions compete to bind to multivalent ion carriers or enter the cell by active transport after binding to low molecular weight thiols (*e.g.* cysteine). Alternatively, metal ions can enter the cell by endocytosis after binding to chelating proteins (*e.g.* metallothioneins) or by the disruption of the cell wall by natural or artificial forces.[67] When the metal ions have penetrated into the cells they are compartmentalised into different subcellular organelles. This intracellular accumulation process is an active metabolic process driven by energy generated by living cells and the extent of metal accumulation might vary between different species of microalgae.[67]

Once metal ions are inside the cell, heavy metal ions could cause toxicity by binding to sulfhydryl groups in proteins causing disruption of the protein structure or displacement of an essential element. Intracellular metal detoxification in microalgae is commonly achieved by the formation of metal-binding peptides or proteins, that is, phytochelatins (PCs) or metallothioneins (MTs). These organometallic complexes can further partition inside vacuoles to facilitate appropriate control of cytoplasmic concentration of metal ions and thus prevent or neutralise the potential toxic effects.[68,69] MTs and PCs are the two best-characterised heavy metal-binding ligands in plant cells. MTs are cysteine-rich polypeptides encoded by a family of genes, whereas PCs are a family of enzymatically synthesised cysteine-rich peptides (also known as class III metallothioneins or MtIII). The general structure of PCs has been determined to be (γ-Glu-Cys)$_n$-Gly with the chain length, n, ranging from 2 to 11 between different species of microalgae.[70]

Studies have shown that glutathione, which is the substrate in MtIII synthesis, appears to be the primary peptide involved in heavy metals binding.[71,72] MtIII are enzymatically synthesised by phytochelatin synthase, which catalyses the transpeptidation of the γ-Glu-Cys moiety of glutathione (γ-Glu-Cys-Gly) onto a second glutathione molecule to form MtIII$_2$ ($n = 2$) or onto another MtIII molecule to produce an $n + 1$ oligomer.[73] In the structure of MtIII, Cys appears to be the chelating core as it is able to chelate metallic ions via the thiol group (-SH). It has been shown that higher metal tolerance was exhibited by microalgae with longer chain lengths (n) owing to the availability of more chelating sites for metal binding.[69,74]

The biosynthesis of MtIII can be induced by heavy metals such as Cd^{2+}, Ag^{2+}, Bi^{3+}, Pb^{2+}, Zn^{2+}, Cu^{2+}, Hg^{2+} and Au^{2+} both *in vivo* and *in vitro*. Cd^{2+} appears to be the most potent activator, followed generally by Pb^{2+}, Zn^{2+} and Cu^{2+} and other heavy metals.[68] Pérez-Rama *et al.* claimed that *Tetraselmis suecica* appeared to be one of the microalgal species exhibiting the highest tolerance to Cd^{2+} because of its ability to synthesise longer MtIII.[69] Furthermore, it was proposed that MtIII synthesis could be related to the degree of heavy metal pollution in an aquatic environment in which increased MtIII production was found at sites with higher metal concentrations.[75] The heavy metal detoxification mechanism mediated by MtIII allows some species and ecotypes of algae to survive in the presence of toxic metals at concentrations that are lethal to other species or populations. The general scheme of heavy metal detoxification mechanism mediated by MtIII in microalgae is illustrated in Figure 4.5.[68]

In addition, the content of polyphosphate in microalgal cells also plays a role in sequestering heavy metal in which the divalent cations associated with polyphosphate could be trafficked into vacuoles. However, it is worth noting that when the uptake of metals exceeds the cellular tolerance, the excess metals inside the microalgae cell will be expulsed by efflux of metals back into solution via active transport, for example Cu exclusion has been reported to be a resistance mechanism of heavy metal tolerance adopted by the green microalga, *Chlorella vulgaris*.[76] In this case, the use of living microalgal cells is restricted if the effluent to be treated contains metal concentrations that exceed the tolerance of microalgae.

In fact the excellent ability of microalgae in metal biosorption and bioaccumulation has been proven in many bioremediation processes.[77–80] In particular, inactivated microalgal biomass has been commercialised as adsorbent (algaSORB®) to remove metal pollutants from industrial effluents whereas living microalgae have been used for the removal of heavy metals by immobilising the cells in a bioreactor (BIOALGA).[81,82] Nevertheless, it should be noted that the yield of microalgal biomass in the original volume of sea or fresh water is relatively poor compared to macroalgae. In addition, large-scale culturing or harvesting of algae biomass from the sea or fresh water can sometimes be a difficult task owing to the associated costs of transport and labour involved.

4.2.1.3 Plant-based Agro waste as Biosorbent

Most plant-based agro waste materials are mainly composed of cellulose, lignin and hemicellulose with the presence of other components including extractives, pectin, lipids, proteins, simple sugars, starches, water, hydrocarbons, ash and many more compounds in varying composition. The biosorption capability of plant-based agro waste is attributed to the availability of numerous reactive functional groups such as acetamido groups, carbonyl, phenolic, structural polysaccharides, amido, amino, sulfhydryl, carboxyl groups, alcohols and esters in the biomass molecules. Metal ions adsorb mainly to carboxylic (primarily present in hemicelluloses, pectin and lignin), phenolic (lignin and extractives), carbonyl groups (lignin) and to some extent hydroxylic (cellulose, hemicelluloses, lignin, extractives and pectin). These functional groups are able to immobilise the metal ions via mechanisms such as chemisorption, complexation, adsorption on surface and pores, ion exchange, chelation, adsorption by physical forces, entrapment in inter and intra-fibrillar capillaries and spaces in the structural polysaccharides network as a result of the concentration gradient and diffusion through cell wall and membrane.[6,83]

A wide range of agricultural residues including rice and wheat-based biomass, coconut-based biomass, oil palm-based biomass, sawdust of various plants, tree barks, tea wastes, maize corn cob, sugarcane bagasse, fruit peels

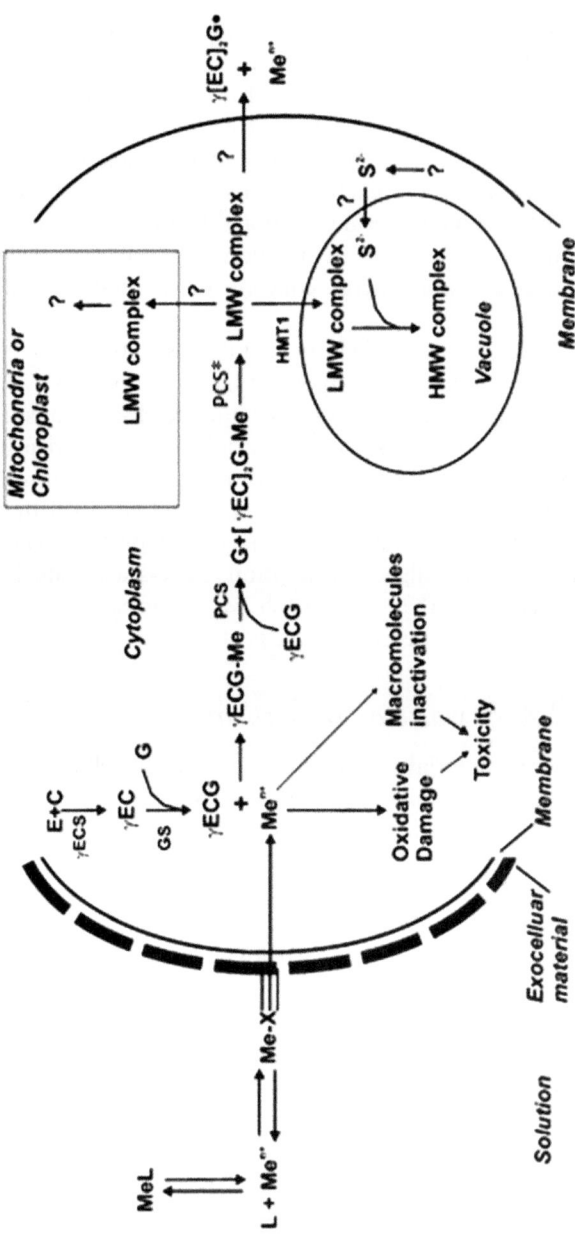

Figure 4.5 General scheme of a heavy metal detoxification mechanism mediated by MtIII in microalgae. Abbreviations: (MeL) metal complex in solution, (Men$^+$) free heavy metal ion, (X) biotic exocellular ligand, (E) glutamic acid, (C) cysteine, (G) glycine, (γEC) gamma glutamylcysteine, (γECG) glutathione, ([γEC]$_2$G$\bullet$) metallothionein $n = 2$, (LMW) low molecular weight, (HMW) high molecular weight, (γECS) gamma glutamylcysteine synthetase, (GS) glutathione synthetase, (PCS*) phytochelatin synthase, (HMT1) vacuolar ABC transporter. *When this step is repeated $\bullet$MtIII with a longer chain can be synthesized, MtIII is disassociated when released to the medium (reproduced with permission from Perales-Vela *et al.*[68]).

and various parts of many other agricultural crops have been demonstrated to be promising biosorbent material for the removal of metal ions either in their natural form or after some physical or chemical modifications.[84–92] It has been demonstrated in many studies that plant-based agro waste particularly, that containing high levels of lignocellulosic materials show promising performance in metal biosorption.[85,93,94] Lignocellulosic substances are mainly composed of high-molecular weight compounds including cellulose, hemicelluloses and lignins which contribute most of the mass. The structures of cellulose, hemicelluloses and lignin are given in Figure 4.6.[95] In addition, lignocellulosics

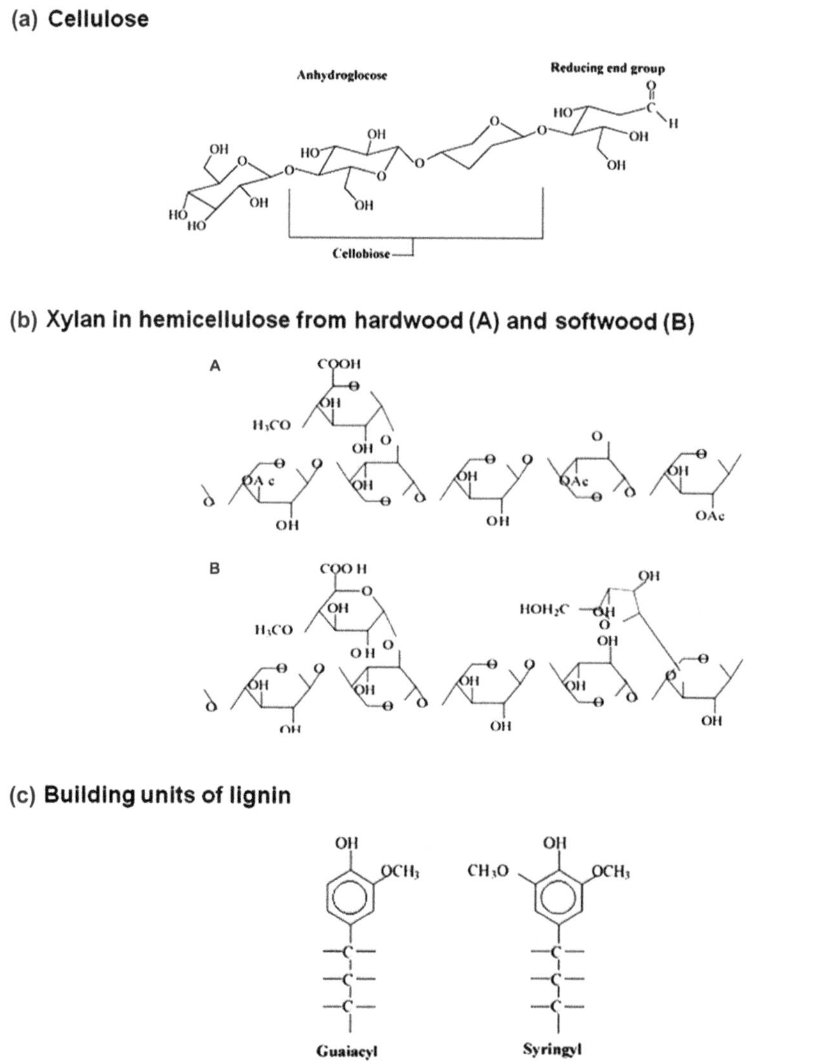

(a) Cellulose

(b) Xylan in hemicellulose from hardwood (A) and softwood (B)

(c) Building units of lignin

Figure 4.6 Structure of cellulose, hemicelluloses and lignin (reproduced with permission from Demirbas[95]).

also contain small quantities of low-molecular weight extractives such as resin (terpenes, lignans and other aromatics), fats, waxes, fatty acids, alcohols, terpentines, tannins and flavonoids. The main components of lignocellulosic materials are probably the most abundant polymeric renewable resources in the world, readily available in large quantities as residual plant parts after harvests.[85] The abundant availability of these lignocellulosic materials at low or no cost presents a significant advantage for their large-scale application as biosorbent for the removal of heavy metals in water treatment.

Lignins are polymeric aromatic macromolecules covalently linked to xylans in hard woods and to galactoglucomannans in soft woods. The chemical structure of lignin basically consists of phenylpropane units primarily syringyl, guaiacyl and *p*-hydroxyphenol which are bonded together by a set of linkages to form a very complex three-dimensional structure. Their function is to provide structural strength, sealing the water conducting system that links roots with leaves and protect plants against degradation. This structure, with apparent infinite molecular weight, imparts to lignin a strong resistance to chemical reactions and a high surface area, providing it with great potential to be used as an effective biosorbent. In fact, lignin is known to be effective in binding many heavy metal ions in water owing to the availability of a variety of reactive polar functional groups such as hydroxyl, methoxyl and carbonyl that have the ability to complex with heavy metals by donating an electron pair from these groups.[96–99] Harman *et al.*[93] have shown that hard wood bark was capable of binding the chromite to a level more than 3% by its weight and the binding between the bark and the metal ion is sufficiently strong to pass toxicity characteristic leaching procedure (TCLP) tests. They proposed that materials retaining cellular structures with high lignin contents were highly effective in biosorption. The role of lignin in Cr^{3+} biosorption from aqueous solution was also demonstrated in the study of Garcia-Reyes and Rangel-Mendez in which they proposed that lignin was the major constituent responsible for the removal of Cr^{3+}.[100]

Apart from lignin, sugar-type macromolecules such as cellulose and hemicelluloses appear to be the larger carbohydrate categories that have significant value in plant-based agro waste materials. Cellulose, as the most abundant biomass fraction in the plant-based biomass, is located predominantly in the secondary cell wall of plant. It is a giant straight chain molecule that is formed by joining units of anhydroglucose by β-(1, 4)-glycosidic linkages in which cellobiose is established as the repeating unit. Intramolecular and intermolecular hydrogen bonds could be formed between OH groups within the same cellulose chain and the surrounding cellulose chains. These chains tend to be arranged parallel in a crystalline supra-molecular structure that forms a microfibril oriented in the cell wall structure. Cellulose is insoluble in most solvents and has a high resistance to acid and enzymatic hydrolysis. Chemical modification of cellulose is a promising technique for modifying its physical and chemical properties to improve its adsorption properties with regard to removal of various pollutants (or recovery of elements).[101] The study of Rezić shows that natural cellulose samples (cotton, flax and hemp) are suitable choices of biosorbents which can also act as indicators of environmental pollution.[102]

In contrast to cellulose, hemicelluloses consist of different monosaccharide units that are mainly derived from chains of pentose. The backbone of the chains of hemicellulose can be a homopolymer (generally consisting of single sugar repeat unit) or heteropolymer (mixture of different sugars). The functions of hemicelluloses are either to interconnect the cellulose microfibrils by hydrogen bonds or to regulate as "scaffolds" the space between cellulose fibrils. The polymer chains of hemicelluloses have short branches and are amorphous in nature. The amorphous morphology of hemicellulose allows it to be partially soluble or swellable in water. It can also be largely soluble in alkaline solution and hence more easily hydrolysed. In addition, the solubility of hemicellulose in common solvents also makes it easy to modify so that its biosorption capability could be further enhanced by suitable modifications.[101,103] The application of hemicellulose in biosorption has been demonstrated. A novel and high-loading biosorbent based on xylan-type hemicelluloses was developed for heavy metal ion removal and recovery (Pd^{2+}, Cd^{2+}, and Zn^{2+}).[104]

Amongst those lignocellulosic materials that have been studied as potential biosorbents, the use of oil palm-based biomass is worth highlighting owing to the efficient use of land, water, nitrogen and energy resources with relatively low pesticide consumption during the growth of this biomass. De Vries *et al.*[105] used a set of nine sustainability indicators focused on resource use efficiency, soil quality, net energy production and greenhouse gas emissions to evaluate the production-ecological sustainability of several major crops. Their study reveals that oil palm (South East Asia), sugarcane (Brazil) and sweet sorghum (China) appeared to be the most sustainable crops and oil palm was the most sustainable with respect to maintenance of soil quality.[105] This makes oil palm biomass and its associated waste a comparatively superior and sustainable material compared to many current large-scale agricultural crops. The biosorption performance of various parts of oil palm biomass such as trunks, fronds, leaves, empty fruit bunches and shells have been extensively studied by different researchers and promising results were obtained.[85] This further implies the great potential of oil-palm based biomass in the production of biosorbent in which different parts of the biomass or waste generated from oil palm industry can be effectively utilised.

The use of plant-based agro waste as biosorbent is also more affordable for many developing countries because it requires little processing, is abundant in nature and is a waste material from agro industry. Moreover, compared to microorganisms which require specialised facilities and elaborate work to maintain their growth, the collection and storage of plant-based agro waste requires much less effort. Nevertheless, the feasibility of using a particular plant-based agro waste is restricted by its geographical distribution. A low-cost material could become expensive if it is not locally available and could still incur costs of transport and storage. Hence, it is important to conduct more studies to assess the performance of different agro plants while considering their respective growth location. If a commercial biosorption process were to be developed using a particular agro waste, it is possible that the waste might soon become a commodity with rising costs. In this regard, proper control of their availability and price is of another great concern.

4.2.1.4   Vermicompost as Biosorbent

Vermicompost is a product of biodegradation and stabilisation of organic materials by interaction between earthworms and microorganisms. It usually contains about 30% humified organic matter with a characteristic black colour caused by the presence of humic substances.[106,107] There are several different sources of vermicompost, such as municipal solid waste, pig manure, cattle manure and decomposed pods of green gram which promote plant growth.[108] The process of vermicomposting allows the transformation of some organic species such as proteins, carbohydrates, fats, nucleic acids into a more stable product (vermicompost). Vermicompost has wide particle size distribution, large surface area and exhibits high capacity for cationic exchange and for water adsorption.[109] The high cationic exchange capacity and large surface area of vermicompost are desirable characteristics of biosorbent, indicating its potential applicability in the adsorption of different meal ions.

In addition, vermicompost is also an excellent soil conditioner with a finely divided peat-like structure, which possesses high porosity, aeration, drainage, water-holding capacity and microbial activity. The nutrient content in vermicompost, such as nitrates, phosphates, exchangeable calcium and potassium, is usually much richer and bioavailable to plants than the organic substrate from which it is produced.[106,110] In fact, vermicompost contains plant-growth regulating materials that are produced by the action of microbes like fungi, bacteria, actinomycetes and earthworms. This includes plant-growth regulators like auxins, gibberellins, cytokinins and other plant-growth regulating materials, such as humic acid-like components, which are responsible for promoting plant growth and greater yields of many crops.[106] Humic acid is naturally oxidised, giving it a negative charge that is able to complex strongly with polyvalent metals. Humic substances comprise a number of organic compounds with complex molecular structures including aromatic rings, carbonyl groups, phenolic and hydroxyl groups.[109] The presence of humic and fulvic substances in the context of vermicompost allows it to interact with metal ions owing to the presence of numerous reactive functional groups. The processes involved are mineral dissolution, formation of metal complexes that are soluble in water, formation of metal–ligand complexes in water, adsorption/desorption processes (solid phase) of metallic ions, adsorption onto external mineral surfaces and adsorption onto interlaminar mineral surfaces.[111]

The high metal removal capacity of vermicompost for Pb, Ni, V and Cr was demonstrated by Urdaneta *et al.* Their results showed that the mean removal efficiency for Pb, Ni and Cr was around 95% in the multi-elemental synthetic sample.[111] Carboxylic acid in the molecular structure of vermicompost appears to be an important adsorption site for heavy metals. One process involving the exchange of protons from weak organic acids in vermicompost with Pb^{2+} ions dissolved in aqueous solutions was proposed.[112] Apart from aqueous samples, vermicompost was also found to be able to reduce the uptake of lead by a white bean plant growing in a soil irrigated by an acidic lead solution. The results illustrated that the presence of vermicompost effectively limited the uptake and translocation of the metallic pollutant, showing its efficiency as a resource for controlling the mobility of pollutants such as lead.[107]

The application of vermicompost in the adsorption of Cd^{2+} was demonstrated in the work of Pereira and Arruda.[109] In addition, Jordão *et al.*[108] have also shown the efficient removal of Cu, Cr, Ni, Zn and Cd from electroplating wastes and synthetic solutions using vermicompost from cattle manure.

In view of the ability of vermicompost to treat both water and soil contaminated with high levels of metals, its great potential as a biosorbent material should not be overlooked. Moreover, vermicompost residues obtained by the metal retention process are enriched with organic material and micronutrients and hence may be applied as a fertiliser to agricultural lands. Nevertheless, very limited studies have been performed by using vermicompost for biosorption of elements. Hence more work is necessary to further investigate its potential in biosorption.

4.2.2 Living versus Dead Biomass

Both living and dead biomass can be used as sorbent material for biosorption of elements. The use of each one has its own merits and demerits. The advantages of living biomass include self-renewing properties, the ability to take up metal via active transport into the cell and the removal of metals by binding to some metabolic products that are excreted by living cells.[113] The application of living biomass could be of value in a system where additional benefit will result from metabolic activity such as biodegradation of organic substances that are highly important in sewage treatment, biofilm reactors for pollutants and anaerobic digestion. However, the use of living biomass is limited by the possible toxic effects of metals as well as the cost of nutrients and facilities required to support cell growth.[1,113]

Owing to the involvement of metabolic activities in living biomass, living cells are quite sensitive to the chemical composition of the effluent being treated. Hence, it is not suitable to treat wastewaters with too high concentrations of metals or with the presence of other toxic impurities. In addition, it is also not appropriate to use living biomass at temperatures and pHs that are intolerable to the growth of living cells. In fact, biosorption studies using living biomass are often performed in a culture medium which contains a supplement of glucose, amino acids, ammonium and phosphate. These substances might interfere with overall metal removal and give misleading results.[114–116] Culture age is another factor that may affect metal removal performance of living biomass, for example older cultures of *Chlorella vulgaris* were found to have a higher Cu^{2+} adsorption capacity than the exponentially growing ones.[61] Another limitation of using living biomass is the difficulty of recovering metals that are bound intracellularly or that complex with extracellular metabolites that are being actively retained in solution.[81] In this regard, it would be difficult to reuse the living biomass for a subsequent sorption process. Thus, it seems that the use of living biomass has greatly increased the complexity of the overall biosorption process because the influence of metabolic processes on sorption is often unpredictable.

Conversely, the limitations of using living biomass can be avoided if dead biomass is used. The advantages of using dead biomass for industrial applications has been emphasised by some authors. It has been reported that the metal sorption capacity of dead biomass is comparable or even greater than living biomass.[63,113,117–119] Unlike living biomass, the metal removal process involved in dead biomass entails only a passive process in which metal cations are predominantly adsorbed onto the functional groups on the cell surface.[13] Compared to living biomass, enriched media are not required for dead biomass and they can be stored easily without additional costs for cell maintenance. In addition, dead cells are immune to the toxic effects of heavy metals and are less affected by extreme pH and temperature conditions. Thus, their metal sorption capability can be enhanced by chemical and physical pretreatment, providing greater flexibility than living biomass. Since the biosorption mechanism of dead biomass is mostly reversible, it allows the regeneration of spent biosorbent in multiple sorption/desorption cycles as well as possible recovery of sorbed metal.[13,42,79,94]

However, the use of dead biomass still suffers from some disadvantages. Dead biomass is not applicable to treatments that require biological alteration in the valency of metals or degradation of organometallic species. It is not possible to improve biosorption performance by mutant isolation or genetic engineering of the cell structure. One thing to mention is that the metal sorption capability of dead cells is also affected by the method of inactivation. It has been reported that inactivation by heat might destroy some of the cellular structures and decrease the number of active metal binding sites.[60,79,120,121] However, a contrasting effect of heat inactivation has been observed in another study in which a higher capacity of dead cells actually resulted upon thermal processing.[63] Similarly, both positive and negative effects on metal sorption capability have been observed for cells that undergo chemical inactivation.[122,123]

In fact, whether living or dead biomass is more suitable for biosorption is still under debate. In the early research on biosorption of heavy metals, the majority of studies were actually performed on living biomass. However, attention shifted to dead biomasses when it was realised that they could have the same or even higher uptake capacity of metal ions, while overcoming the limitations of living biomass. Therefore, the use of dead or pretreated biomass dominated biosorption research during the 1980s and 1990s.[124,125] However, it was later realised in some pilot plants that techniques used for immobilisation and pretreatment of dead biomass could be expensive in real applications. The facilities required for the process of regeneration and reuse of dead biomass could be quite complex and have a high cost. Hence, a renewed interest of using living cells together with other physicochemical processes has been observed in recent research.[115,126,127]

4.3 Metal Recovery and Post-treatment of Spent Biosorbent

Safe post-treatment of spent biosorbent is important because metal ions that are being adsorbed are simply removed from one phase (*e.g.* industrial effluent or wastewater) and transferred to another phase (biosorbent) in the process of biosorption. The metal-loaded biosorbent itself could be another source of pollutant if it is not properly treated in the environment. Strategies for managing spent biosorbent include metal desorption from the biomass, biomass dissolution and biomass incineration.[42] The most feasible post-treatment strategy is the regeneration of biosorbent by metal desorption from the biomass which allows repeated use of the regenerated biosorbent and also offers the possibility of reclaiming the desorbed metals. Metals sorbed on biomass can be desorbed by a suitable eluent and a majority of the work carried out has focused on testing the efficiency of various desorbing agents. Some commonly tested desorbing agents include diluted mineral acids (HCl, HNO_3 and H_2SO_4) and organic acids (CH_3COOH), bases (Na_2SO_4, Na_2CO_3, $CaCO_3$, NaOH and KOH), inorganic salts (KCl and $CaCl_2$) and metal chelators (*e.g.* EDTA).[42]

In most cases, heavy metal cations can be effectively desorbed from the biosorbent using dilute inorganic acids. During the desorption process, metal cations are released into the desorbing solution after being displaced by protons at the binding sites.[81] The high desorption capacity of HCl has been widely demonstrated.[128–131] However, it was also reported that HCl could decrease the metal sorption capability of the regenerated biosorbent when applied in successive cycles. This is most likely due to damage of metal binding sites and possible hydrolysis of polysaccharides on the cell wall surface.[132] Contradictory results were obtained in the study of Al-Rub *et al.*[133] in which immobilised *Chlorella vulgaris* cells could be reused in up to three cycles using 0.1 M HCl as the desorbing agent with an improved efficiency of Ni^{2+} removal being observed after the first cycle. The authors attributed this improvement to the release of more active sites after some contaminants bound previously to the algal cell were removed during the desorption process.

On the other hand, NaOH appears to be effective for desorbing metal anions such as arsenate from biosorbents.[134] The use of dilute acid solutions containing $CaCl_2$ and NH_4Cl as desorbing agents could be advantageous because they can function as electrolytes in the electrolytic process of metal recovery.[135] As a strong metal chelator, the desorbing ability of EDTA has been evaluated in several studies.[136,137] However, the use of EDTA as a desorbing agent is not recommended because EDTA is not biodegradable in the environment. In addition, it is also not suitable for desorption of alginate immobilised biomass because it dissolves alginate.[81] Keeping the above facts in mind, it seems that the performance of different desorbing agents varies between different biosorbents. The type and strength of a desorbing agent are determined by how the deposited metal has been deposited. Nevertheless, choice of desorbing agents should be made based on the cost and desorption efficiency as well as the ability to restore the biosorbent without causing significant loss of performance. To utilise the material fully, the exhausted biosorbent could be converted into energy fuels such as methane, hydrogen and bioethanol by a fermentation process. Alternatively, they can also be transformed to thermal energy by combustion or composted as fertilisers if sufficient removal of adsorbed metals from the biosorbents is confirmed.[138]

In cases where the regeneration of biosorbent is not feasible, destruction of biomass by heat dissolution with strong acids or incineration followed by acid digestion of the ash appear to be effective alternatives. These treatments allow the generation of a concentrated heavy metal solution which could be used for subsequent elemental recovery. Meanwhile the remaining ashes of biomass could be immobilised in an inert support or land filled.[42] Metal recovery from the highly concentrated solution obtained from the sorption–desorption process could be achieved by applying co-precipitation, flocculation or electrowinning. In general, recovered solid metal with high purity could be obtained from the concentrate by electrowinning. Selective electrodeposition of Cu, Pb, Cd and Zn in purities above 99 mol% from a highly concentrated chloride solution was demonstrated.[139] Nevertheless, the consumption of electrical energy and costs involved in the electrowinning process are a consideration. In contrast, recovery of Al, Cd, Co, Cu, Fe, Mn, Ni and Zn from acid mine drainage by precipitation agents such as sulfide and hydroxides has been shown by Tabak *et al.*[140] However, it should be noted that sludge containing metal sulfides is toxic and additional precautions are necessary to ensure safe transport and storage. Ideally the metals recovered can be sent back and reused in industry so that it can slow down the depletion of metal resources without losing the financial benefits.

4.4   Further Consideration of Practical Applications

4.4.1   Availability of Reactors for Biosorption

For practical purposes, the availability of reactors for biosorption is important in order to evaluate the feasibility of scaling up the process from a laboratory scale procedure to commercialisation. The process of biosorption for metal removal is basically a solid–liquid contact process. The technological configuration is very similar to that used for an ion-exchange process or activated carbon applications.[141] Based on the mode of contact between metal solution and biosorbent, there are two major types of treatment systems that have been widely studied for biosorption treatment. One is batch systems (including batch stirred-tank reactors) and the other is continuous flow systems (including continuous-flow stirred-tank reactor, fixed packed-bed reactor, fluidised-bed reactor and fixed biofilm column *etc.*).[81]

Batch reactors were one of the earliest used for evaluation of metal-biosorbent systems because a large number of theoretical studies have been performed on the batch process. In a batch reactor, biomass is contacted with a metal-containing solution for a limited retention time by means of air bubbling or magnetic stirring. The separation of metal-loaded biosorbent from the liquid phase is usually done by centrifugation or filtration. The removal and recovery of Cd using free suspensions of dead *Streptomyces* biomass in a stirred tank reactor has been developed by Butter *et al.*[135] The solids were separated from the aqueous phase by flotation or sedimentation followed by immobilisation in a filtration unit. The biomass was regenerated by an elution step and the Cd was recovered from the eluate (concentration around 50 times the original Cd solution) using a rotating cathode cell, resulting in Cd powder and Cd-depleted electrolyte which is recycled as eluant in the next desorption process. This system shows the commercial possibilities of batch reactors to achieve very effective Cd removal while producing clean water and recovering solid metal.[135] A batch multistage system for the removal of Cu^{2+} and Ni^{2+} ions by *Rhizopus arrhizus* using three batch stirred reactors operating in series has been investigated by Sağ and Kutsal *et al.*[142] In addition, for cases where living biomass is to be used as a biosorbent, a batch reactor seems to be a more suitable choice. However, the separation of biosorbent by centrifugation and filtration from the liquid appears to be impractical for handling large volumes of effluent owing to their high operational and maintenance costs.[81]

On the other hand, a continuous treatment system is necessary for the treatment of large volumes of metal-containing effluents. Several types of reactors can be used in continuous systems. A continuous-flow stirred tank reactor (CFST) uses a vessel similar to that used for batch systems with the supply of a continuous flow of metal solution. There are two basic operating modes for CFST. One uses a continuous feed of fresh biosorbent which is then continuously harvested from the reactor in the effluent. The other alternative uses a batch of biosorbent which is retained in the system using a special arrangement of the reactor. The metal solution continuously flows through the reactor until the batch of the biosorbent becomes saturated with metal and then withdrawn as a suspension and recovered by a solid–liquid separation. The biosorption of Pb^{2+}, Ni^{2+} and Cu^{2+} in single component, binary and ternary systems was studied using *Rhizopus arrhizus* in a CFST by Sag *et al.*[141]

A fix packed-bed column appears to be the most convenient configuration where a continuous operation is required. The packed bed column reactor basically consists of a column filled with biosorbent in either free or immobilised form. Influent solution is fed into the packed-bed column by either flowing upward or downward. Breakthrough curves showing the relationship between metal concentration in effluent and breakthrough time or volume is useful for the evaluation of column performance.

A large number of studies have been performed to evaluate the performance of biosorbent in the removal of heavy metals using column studies. Some examples of recent studies include dynamic biosorption of pine cone shell for the removal of Cu^{2+} using a packed-bed column as well as the removal of Cd^{2+} and Pb^{2+} in a fixed-bed column by protonated alga and acid-treated olive stone in multiple sorption–regeneration cycles.[21,143,144] In fact, the performance of biosorbent in column studies is different in theoretical studies that are conducted in batch experiments. Hence, it is necessary to optimise the operational parameters for packed-bed reactors such as flow rate, bed height, pH, influent metal concentration, biomass loading and size of biosorbent particles before putting the reactors to actual practice.

Volesky *et al.*[9] suggested that the size of biosorbent particles should be about 0.7–1.5 mm which is similar to the size of general ion-exchange resins manufactured for the same purpose. In this size range, the sorbent particles should be hard enough to withstand the application pressures, appear to be porous and "transparent2 to sorbate species and feature high and fast sorption uptake even after repeated regeneration cycles.[9]

4.4.2 Immobilisation of Biomass

Most of the time, biomass cannot be used directly for continuous applications because it is either too soft to withstand the pressure in the reactor or has too small a particle size which makes its separation from the treated effluents a difficult task. Thus, the direct use of native biomass is not desirable as it can clog the column and is vulnerable to shear stress in a stirred-tank reactor. Immobilisation of the biomass within a suitable matrix can overcome these problems by offering ideal size, mechanical strength, rigidity and porous characteristics to the biological material.[145] Several immobilisation techniques have been used for the granulation of biosorbent materials, which can be primarily divided into passive and active modes of immobilisation.

Passive immobilisation refers to the natural tendency of many micro-organisms to attach to surfaces and grow on them. Both natural and synthetic materials can be used as the carriers for passive immobilisation.[13,146] One example of a natural carrier is the Loofa sponge which has been previously suggested as an immobilisation matrix for algal, fungal and yeast cells. Loofa sponge is a highly porous and strong biomatrix, made of an open network of fibrous support from the dry fruit of *Luffa cylindrica*. Study of the removal and recovery of Ni^{2+} from aqueous solution by a loofa-sponge immobilised biomass of *Chlorella sorokiniana* has been conducted by Akhtar *et al.*[145] The biosorption capacity of immobilised biomass was observed to be 25% higher than the free biomass, showing the promising potential of loofa-sponge immobilised biomass in heavy metal remediation.[145] On the other hand, synthetic materials can also be used as passive immobilisation carriers. A nonwoven short-fibre polyester material (named 7607) was successfully used as an immobilising support for the natural adhesion of plant cells in a laboratory-scale bioreactor.[147] However, it should be noted that passive immobilisation processes are easily reversible and the leached cells could cause undesirable contamination of effluents.

Alternatively there are some commonly used active immobilisation techniques such as flocculant agents, chemical attachment and gel entrapment. The advantage of using flocculant agents is to avoid the tedious and expensive centrifugation step in the separation of biomass from liquid medium. Chitosan has been the most widely used flocculant agent among those commonly used. The presence of positively charged amino groups on chitosan allows it to adsorb negatively charged particles. In addition, the biodegraded chitosan can also be used for nutritional purposes. The disadvantage of using chitosan is mainly due to its poor stability. Chemical attachment is more suitable for immobilising dead rather than living biomass because the chemical interaction between cross-linking agents and the cell surface can cause damage to the cells and reduce viability. Commonly used cross-linking agents usually have specific functional groups capable of cross-linking the cell wall of the biomass. For example, formaldehyde can enable cross-linking between adjacent hydroxyl groups of sugars on the cell wall, whereas glutaraldehyde cross-links with the amino groups of cell wall. Cross-linked biomass can have either improved or reduced sorption capacity depending on the cross-linking procedure.[81,124,146]

Gel entrapment has also been widely used for the active immobilisation of biomass. It can be performed using synthetic polymers (acrylamide, photocross-linkable resins, polyurethanes), proteins (gelatine, collagen or egg white) or natural polysaccharides (agars, carrageenans or alginates). Among these, gel entrapment in natural polysaccharide matrixes is the most widely used immobilisation technique owing to its inherent low toxicity compared to synthetic polymer. In particular, the most widely used polysaccharide gel for entrapping living cells is alginate. Commercial alginates are extracted from brown algae as water soluble sodium salts. The most common cation used to form alginate gels is Ca^{2+}. A major advantage of alginate gel entrapment comes from the permeability, non-toxicity and transparency of the formed matrix. However, the use of alginate beads is limited by its vulnerability to extreme pH conditions. Loss of alginate occurs at a pH of below 3 and above 9 whereas breakage of alginate beads occurs at pH 11. Hence, alginate-immobilised biosorbent is not applicable for treating effluents with high acidity or alkalinity. It is also not suitable to be acid or alkali washed during the desorption process. In addition, alginate is not resistant to orthophosphate or Na^+ which limits its application in effluents containing high levels of Na^+ or orthophosphate.[81,146] On the other hand, the use of synthetic polymers for entrapment is limited by their high cost and toxicity to living cells. However, the use of synthetic polymer for immobilisation of dead biomass is worth considering. The binding of Au^{3+}, Cu^{2+} and Ag^{1+} to 10 biogenic materials, including sphagnum peat, a top soil, peat, compost peat, organic peat, peat replacer, dead *Chlorella vulgaris* and a cultured cell material from the plant *Datura innoxia* immobilised within a polysilicate matrix has been investigated in a flowing system.[148]

4.5 Current Challenges and Future Direction

Over the past few decades a huge number of studies have been conducted based on the use of different living or dead biomass for the biosorption of elements. The great potential of biosorption relative to conventional methods has been widely claimed in different studies, mainly attributed to its advantages, including low operating cost, minimisation of chemical and biological sludge generation, effective removal of metals at low concentration, reduced consumption of chemicals, ease of regeneration of biosorbents and possibility of metal recovery. Despite the great efforts that have been put into developing possible biosorption systems, the commercial exploitation of biosorption is progressing painfully slowly and most research is still limited to laboratory-scale studies. The underlying reasons that limit the commercial success of biosorption are possibly the lack of understanding of the biosorption mechanism with regard to the vast variety of living and dead biomass used in different studies and the lack of specificity and robustness of biosorbent owing to its inherent complex biological structure.

Most of the time the use of a biosorbent is considered to be a cost-effective alternative to synthetic ion-exchange resins. As mentioned by Volesky, the price of ion-exchange resins that are hydrocarbon derivatives is subject to the stability of crude oil. Biomass (agricultural residues) as a renewable resource seems to be more sustainable for long term usage in comparison to crude oil, which in recent years has had a dramatic impact on global markets.[149] However, unlike biosorbent, ion-exchange resins can be specially synthesised with functional groups required for the removal of certain target elements and the ion-exchange process has been well-established with predictable performance in large-scale applications. Additionally, the pre-processing of biomass (include immobilisation, chemical and physical pretreatment) as well as the storage and safe disposal of spent biosorbent could significantly increase the overall cost. Hence, the cost-effectiveness of biosorbent has to be carefully assessed by taking into account all the relevant aspects. In general, waste material appears to be an attractive choice of biosorbent, especially in developing countries that produce large amounts of agricultural waste. The development of biomass pre-processing plants could not only help to solve the disposal problem but also provide new economic opportunities for these developing countries.

Undoubtedly, the successful implementation of biosorption for pollution control or metal recovery is still full of challenges. Nevertheless, it is worth noting that the potential is always there, whether biosorption is to be used as a replacement strategy for existing methods or as an adjunct technology in a hybrid process. The promising biosorption performance exhibited by different types of biomass is encouraging and worth further exploration. Along with the continuous efforts to discover new potential biosorbents, future research could be focused more on study of the mechanism based on some selected representative types of biomass such as flocculent yeast, algae and lignocellulosic materials. It is believed that the application of various advanced techniques along with proper process modelling will allow a better understanding of the biosorption mechanism especially at the molecular level. Meanwhile, a more well-defined relationship between biosorption performance with those important influencing factors such as pH, ionic strength, temperature and biomass loading could be established.

A large number of biosorption studies have been performed based on single-element solutions owing to the ease of correlating the experimental data with theoretical models. However, real industrial effluents always contain many elements coexisting with interfering ionic species. Hence, studies based on multi-element solutions should be conducted in which the development of multi-element binding models and the competing adsorption between different elements are the focus. It should also be noted that instead of using synthetic wastewater, it is necessary to test the efficiency of biosorbent in real wastewater as the properties of industrial effluents generated from different sources might differ greatly from each other. Preliminary assessment of the commercial feasibility of certain biosorption systems could be achieved by pilot-scale studies in which reactors similar to large-scale applications are used. In addition, advanced biotechnology also provides the possibility of producing genetically modified microorganisms equipped with higher metal sorption capability and resistance to extreme physicochemical conditions. Hence, the feasibility of applying biosorption is laid on an interdisciplinary basis which requires the contributions of expertise from different fields such as chemistry, biotechnology and process engineering.

4.6 Conclusions

The launch of a new technology for water remediation is never an easy task. This is especially true for the biosorption of elements (for clean-up and recovery) which exploits biomass with highly complex biological structures. Therefore, much more effort is required to evaluate the commercial potential of a particular biosorbent from the initial stage of pre-processing to the final stage of biomass post-treatment. Although the commercialisation of biosorbent has not yet met with success, the potentially huge market, possible environmental and economic benefits (for metal recovery and bioremediation) for some developing countries that are rich in biomass resources suggests that continuing biosorption-related work in future research is still worthwhile.

Acknowledgement

We acknowledge financial support from the National University of Singapore, National Research Foundation and Economic Development Board (SPORE, COY-15-EWI-RCFSA/N197-1), and Ministry of Education (R-143-000-441-112 and R-143-000-519-112).

References

1. G. M. Gadd, *J. Chem. Technol. Biotechnol.*, 2009, **84**, 13.
2. F. Veglio and F. Beolchini, *Hydrometallurgy*, 1997, **44**, 301.
3. J. Wang and C. Chen, *Biotechnol. Adv.*, 2006, **24**, 427.
4. J. Wang and C. Chen, *Biotechnol. Adv.*, 2009, **27**, 195.
5. D. Kratochvil, P. Pimentel and B. Volesky, *Environ. Sci. Technol.*, 1998, **32**, 2693.
6. D. Sud, G. Mahajan and M. P. Kaur, *Bioresour. Technol.*, 2008, **99**, 6017.
7. T. A. Davis, B. Volesky and A. Mucci, *Water Res.*, 2003, **37**, 4311.
8. D. Kratochvil and B. Volesky, *Trends Biotechnol.*, 1998, **16**, 291.

9. B. Volesky, *Hydrometallurgy*, 2001, **59**, 203.
10. W. Stumm and J. J. Morgan, *Aquatic Chemistry*, Wiley Interscience, New York, 1996.
11. A. Haug, B. Larsen and O. Smidsrod, *Acta Chem. Scand.*, 1966, **20**, 183.
12. N. Goyal, S. C. Jain and U. C. Banerjee, *Adv. Environ. Res.*, 2003, **7**, 311.
13. C. M. Monteiro, P. M. L. Castro and F. X. Malcata, *Biotechnol. Prog.*, 2012, **28**, 299.
14. G. M. Gadd, *Microbiology*, 2010, **156**, 609.
15. T. Akar, S. Celik, A. G. Ari and S. T. Akar, *J. Chem. Technol. Biotechnol.*, 2012, **88**, 680.
16. G. Yuvaraja, M. V. Subbaiah and A. Krishnaiah, *Ind. Eng. Chem. Res.*, 2012, **51**, 11218.
17. H. Freundlich, Ueber die adsorption in loesungen, *Z Phys Chem*, 1907.
18. N. Saravanan, C. Ahmed Basha, V. Manivasagan, N. G. R. Babu and T. Kannadasan, *Eur. J. Sci. Res.*, 2012, **81**, 231.
19. C. Gérente, Y. Andrès, G. McKay and P. Le Cloirec, *Chem. Eng. J.*, 2010, **158**, 593.
20. Ç. Kirbiyik, M. Kiliç, Ö. Çepelioğullar and A. E. Pütün, *Water Sci. Technol.*, 2012, **66**, 231.
21. G. Değirmen, M. Kiliç, Ö. Çepelioğullar and A. E. Pütün, *Water Sci. Technol.*, 2012, **66**, 564.
22. B. Preetha and T. Viruthagiri, *Sep. Purif. Technol.*, 2007, **57**, 126.
23. S. Al-Asheh, F. Banat, R. Al-Omari and Z. Duvnjak, *Chemosphere*, 2000, **41**, 659.
24. H. Muhamad, H. Doan and A. Lohi, *Chem. Eng. J.*, 2010, **158**, 369.
25. U. Kumar and M. Bandyopadhyay, *J. Hazard. Mater.*, 2006, **129**, 253.
26. P. Binod, R. Sindhu, R. R. Singhania, S. Vikram, L. Devi, S. Nagalakshmi, N. Kurien, R. K. Sukumaran and A. Pandey, *Bioresour. Technol.*, **101**, 4767.
27. M. T. Madigan, J. M. Martinko and J. Parker, *Brock biology of microorganisms*, Prentice Hall, Upper Saddle River, NJ, 8th edition, 1997.
28. D. Brady, A. D. Stoll, L. Starke and J. R. Duncan, *Biotechnol. Bioeng.*, 1994, **44**, 297.
29. C. Chen and J. Wang, *J. Hazard. Mater.*, 2008, **151**, 65.
30. M. D. Machado, M. S. F. Santos, C. Gouveia, H. M. V. M. Soares and E. V. Soares, *Bioresour. Technol.*, 2008, **99**, 2107.
31. M. D. Machado, E. V. Soares and H. M. V. M. Soares, *Environ. Sci. Pollut. Res.*, 2010, **18**, 1279.
32. M. D. Machado, E. V. Soares and H. M. V. M. Soares, *J. Hazard. Mater.*, 2010, **180**, 347.
33. M. D. Machado, E. V. Soares and H. M. V. M. Soares, *J. Chem. Technol. Biotechnol.*, 2010, **85**, 1353.
34. K. Parvathi and R. Nagendran, *Sep. Sci. Technol.*, 2007, **42**, 625.
35. A. Stoll and J. R. Duncan, *Environ. Pollut.*, 1997, **97**, 247.
36. X. Q. Zhao and F. W. Bai, *Biotechnol. Adv.*, 2009, **27**, 849.

37. B. L. A. Miki, N. H. Poon, A. P. James and V. L. Seligy, *J. Bacteriol.*, 1982, **150**, 878.
38. E. V. Soares, G. De Coninck, F. Duarte and H. M. V. M. Soares, *Biotechnol. Lett.*, 2002, **24**, 663.
39. C. Gouveia and E. V. Soares, *J. Inst. Brew.*, 2004, **110**, 141.
40. B. Volesky and H. A. May-Phillips, *Appl. Microbiol. Biotechnol.*, 1995, **42**, 797.
41. P. Simmons, J. M. Tobin and I. Singleton, *J. Ind. Microbiol.*, 1995, **14**, 240.
42. E. V. Soares and H. M. V. M. Soares, *Environ. Sci. Pollut. Res.*, 2012, **19**, 1066–1083.
43. R. S. Singh, R. Bhari and H. P. Kaur, *Biotechnol. Adv.*, 2011, **29**, 726.
44. D. J. McHugh, *A Guide to the Seaweed Industry*, International Government Publication Edition, Food and Agriculture Organization of the United Nations, Rome, 2003.
45. H. C. Bold and M. J. Wynne, *Introduction to the Algae: Structure and Reproduction*, Prentice-Hall of India Pvt. Ltd, New Delhi, 1978.
46. A. S. Carlsson, *Micro- and Macro-algae: Utility for Industrial Applications: Outputs from the EPOBIO Project*, CPL Press, University of York, United Kingdom, 2007.
47. R. E. Lee, *Phycology*, Cambridge University Press, United Kingdom, 1999.
48. S. Schiewer and B. Volesky, *Biosorption by Marine Algae*, Kluwer Academic Publishers, Dordrecht, The Netherlands, 2000.
49. K. I. Draget and C. Taylor, *Food Hydrocolloids*, 2011, **25**, 251.
50. E. D. T. Atkins, I. A. Nieduszynski, W. Mackie, K. D. Parker and E. E. Smolko, *Biopolymers*, 1973, **12**, 1879.
51. A. Haug and O. Smidsrød, *Acta Chem. Scand.*, 1965, **19**, 341.
52. T. A. Davis, B. Volesky and R. H. S. F. Vieira, *Water Res.*, 2000, **34**, 4270.
53. M. M. Figueira, B. Volesky and V. S. T. Ciminelli, *Biotechnol. Bioeng.*, 1997, **54**, 344.
54. E. Fourest and B. Volesky, *Appl. Biochem. Biotechnol.*, 1997, **67**, 215.
55. R. H. Crist, K. Oberholser, J. McGarrity, D. R. Crist, J. K. Johnson and J. M. Brittsan, *Environ. Sci. Technol.*, 1992, **26**, 496–502.
56. W. M. Ibrahim, *J. Hazard. Mater.*, 1827, **2011**, 192.
57. M. C. Basso and A. L. Cukierman, *Int. J. Environ. Pollut.*, 2008, **34**, 340.
58. H. Doshi, A. Ray and I. L. Kothari, *Biotechnol. Bioeng.*, 2007, **96**, 1051.
59. C. Monteiro, P. Castro and F. Xavier Malcata, *Environ. Chem. Lett.*, 2011, **9**, 169.
60. C. Monteiro, A. Marques, P. Castro and F. Xavier Malcata, *Biodegradation*, 2009, **20**, 629.
61. S. K. Mehta, A. Singh and J. P. Gaur, *J. Environ. Sci. Health, Part A*, 2002, **37**, 399.
62. A. C. A. da Costa and F. P. de Franca, *Aquacult. Int.*, 1998, **6**, 57.
63. J. Kaduková and E. Virčíková, *Environ. Int.*, 2005, **31**, 227.

64. N. La Rocca, C. Andreoli, G. Giacometti, N. Rascio and I. Moro, *Photosynthetica*, 2009, **47**, 471.
65. M. Pérez-Rama, J. Abalde Alonso, C. Herrero López and E. Torres Vaamonde, *Bioresour. Technol.*, 2002, **84**, 265.
66. K. Wilde, J. Stauber, S. Markich, N. Franklin and P. Brown, *Arch. Environ. Contam. Toxicol.*, 2006, **51**, 174.
67. K. Arunakumara and X. Zhang, *J. Ocean Univ. China (Engl. Ed.)*, 2008, **7**, 60.
68. H. V. Perales-Vela, J. M. Peña-Castro and R. O. Cañizares-Villanueva, *Chemosphere*, 2006, **64**, 1.
69. M. Pérez-Rama, C. H. López, J. A. Alonso and E. T. Vaamonde, *Environ. Toxicol. Chem.*, 2001, **20**, 2061.
70. C. Cobbett and P. Goldsbrough, *Annu. Rev. Plant Biol.*, 2002, **53**, 159.
71. G. Howe and S. Merchant, *Plant Physiol.*, 1992, **98**, 127.
72. O. K. Vatamaniuk, S. Mari, Y. P. Lu and P. A. Rea, *Proc. Natl. Acad. Sci. U. S. A.*, 1999, **96**, 7110.
73. E. Grill, S. Löffler, E. L. Winnacker and M. H. Zenk, *Proc. Natl. Acad. Sci.*, 1989, **86**, 6838.
74. E. Torres, A. Cid, P. Fidalgo, C. Herrero and J. Abalde, *Aquat. Toxicol.*, 1997, **39**, 231.
75. B. A. Ahner, N. M. Price and F. M. M. Morel, *Proc. Natl. Acad. Sci. U. S. A.*, 1994, **91**, 8433.
76. P. L. Foster, *Nature*, 1977, **269**, 322.
77. N. la Rocca, C. Andreoli, G. M. Giacometti, N. Rascio and I. Moro, *Photosynthetica*, 2009, **47**, 471.
78. A. Fraile, S. Penche, F. González, M. L. Blázquez, J. A. Muñoz and A. Ballester, *Chem. Ecol.*, 2005, **21**, 61.
79. R. Vannela and S. K. Verma, *Biotechnol. Prog.*, 2006, **22**, 1282.
80. A. A. Hamdy, *Curr. Microbiol.*, 2000, **41**, 232.
81. S. K. Mehta and J. P. Gaur, *Crit. Rev. Biotechnol.*, 2005, **25**, 113.
82. L. Travieso, A. Pellón, F. Benítez, E. Sánchez, R. Borja, N. O'Farrill and P. Weiland, *Biochem. Eng. J.*, 2002, **12**, 87.
83. B. Pejic, M. Vukcevic, M. Kostic and P. Skundric, *J. Hazard. Mater.*, 2009, **164**, 146.
84. U. Farooq, J. A. Kozinski, M. A. Khan and M. Athar, *Bioresour. Technol.*, 2010, **101**, 5043.
85. T. Ahmad, M. Rafatullah, A. Ghazali, O. Sulaiman and R. Hashim, *J. Environ. Sci. Health, Part C: Environ. Carcinog. Ecotoxicol. Rev.*, 2011, **29**, 177.
86. A. Bhatnagar, V. J. P. Vilar, C. M. S. Botelho and R. A. R. Boaventura, *Adv. Colloid Interface Sci.*, 2010, **160**, 1–15.
87. S. Akmal, J. Jaya Malathi, Y. Vijaya, S. R. Popuri and M. Venkata Subbaiah, *Desalin. Water Treat.*, 2012, **47**, 59.
88. M. G. Yu and Y. X. Chen, *Chin. J. Appl. Ecol.*, 2010, **21**, 505.
89. A. I. Adeogun, A. E. Ofudje, A. I. Mopelola and O. K. Sarafadeen, *Res. J. Appl. Sci.*, 2011, **6**, 302.

90. R. Leyva-Ramos, L. E. Landin-Rodriguez, S. Leyva-Ramos and N. A. Medellin-Castillo, *Chem. Eng. J.*, 2012, **180**, 113.

91. I. Alomá, M. A. Martín-Lara, I. L. Rodríguez, G. Blázquez and M. Calero, *J. Taiwan Inst. Chem. Eng.*, 2012, **43**, 275.

92. C. Liu, H. H. Ngo, W. Guo and K.-L. Tung, *Bioresour. Technol.*, 2012, **119**, 349.

93. G. Harman, R. Patrick and T. Spittler, *Ind. Biotechnol.*, 2007, **3**, 366.

94. M. Betancur, P. R. Bonelli, J. A. Velásquez and A. L. Cukierman, *Bioresour. Technol.*, 2009, **100**, 1130.

95. A. Demirbas, *J. Hazard. Mater.*, 2008, **157**, 220.

96. A. B. Albadarin, A. H. Al-Muhtaseb, N. A. Al-laqtah, G. M. Walker, S. J. Allen and M. N. M. Ahmad, *Chem. Eng. J.*, 2011, **169**, 20.

97. D. L. R. Crist, R. H. Crist and J. R. Martin, *J. Chem. Technol. Biotechnol.*, 2003, **78**, 199.

98. Q. F. Lü, Z. K. Huang, B. Liu and X. Cheng, *Bioresour. Technol.*, 2012, **104**, 111.

99. M. B. Šćiban, M. T. Klašnja and M. G. Antov, *Ecol. Eng.*, 2011, **37**, 2092.

100. R. B. Garcia-Reyes and J. R. Rangel-Mendez, *J. Chem. Technol. Biotechnol.*, 2009, **84**, 1533.

101. A. Demirbas, *J. Hazard. Mater.*, 2009, **167**, 1.

102. I. Rezić, *Microchem. J.*, 2013, **107**, 63.

103. C. Schädel, A. Richter, A. Blöchl and G. Hoch, *Physiol. Plant.*, 2010, **139**, 241.

104. X.-W. Peng, L.-X. Zhong, J.-L. Ren and R.-C. Sun, *J. Agric. Food Chem.*, 2012, **60**, 3909.

105. S. C. de Vries, G. W. J. van de Ven, M. K. van Ittersum and K. E. Giller, *Biomass Bioenergy*, 2010, **34**, 588.

106. R. Singh, R. R. Sharma, S. Kumar, R. K. Gupta and R. T. Patil, *Bioresour. Technol.*, 2008, **99**, 8507.

107. A. Carrasquero Durán, I. Flores, C. Perozo and Z. Pernalete, *Int. J. Environ. Sci. Technol*, 2006, **3**, 203.

108. C. P. Jordão, M. d. G. Pereira, R. Einloft, M. B. Santana, C. R. Bellato and J. W. Vargas de Mello, *J. Environ. Sci. Health, Part A*, 2002, **37**, 875.

109. M. G. Pereira and M. A. Z. Arruda, *J. Braz. Chem. Soc.*, 2003, **14**, 39.

110. P. Garg, A. Gupta and S. Satya, *Bioresour. Technol.*, 2006, **97**, 391.

111. C. Urdaneta, L. M. M. Parra, S. Matute, M. A. Garaboto, H. Barros and C. Vázquez, *Spectrochim. Acta, Part B*, 2008, **63**, 1455.

112. A. Carrasquero and I. Flores, *Chem. Edu.*, 2004, **9**, 1.

113. S. Schiewer and B. Volesky, *Biosorption Process for Heavy Metal Removal*, D. R. Lovely, ASM Press, Washington, 2000.

114. J. Miranda, G. Krishnakumar and R. Gonsalves, *J. Appl. Phycol.*, 2012, **24**, 1439.

115. J. Miranda, G. Krishnakumar and R. Gonsalves, *Ann. Microbiol.*, 2012, 1.

116. M. J. Melgar, J. Alonso and M. A. García, *Sci. Total Environ.*, 2007, **385**, 12.

117. P. K. Jjemba, *Interaction of Metals and Metalloids with Microorganisms in the Environment*, Science Publishers, New Hampshire, 2004, ch. 12.
118. D. Özer, A. Özer and G. Dursun, *J. Chem. Technol. Biotechnol.*, 2000, **75**, 410.
119. R. Flouty and G. Estephane, *J. Environ. Manage.*, 2012, **111**, 106.
120. C. Monteiro, P. Castro and F. Xavier Malcata, *Environ. Chem. Lett.*, 2011, **9**, 169.
121. P. A. Terry and W. Stone, *Chemosphere*, 2002, **47**, 249.
122. J. M. Tobin, D. G. Cooper and R. J. Neufeld, *Enzyme Microb. Technol.*, 1990, **12**, 591.
123. S. K. Mehta and J. P. Gaur, *Ecol. Eng.*, 2001, **18**, 1.
124. B. Volesky and Z. R. Holan, *Biotechnol. Prog.*, 1995, **11**, 235.
125. A. Malik, *Environ. Int.*, 2004, **30**, 261.
126. M. d. C. Vargas-García, M. J. López, F. Suárez-Estrella and J. Moreno, *Sci. Total Environ.*, 2012, **431**, 62.
127. M. N. Sepehr, M. Zarrabi and A. Amrane, *J. Taiwan Inst. Chem. Eng.*, 2012, **43**, 420–427.
128. J. P. Chen and L. Yang, *Ind. Eng. Chem. Res.*, 2005, **44**, 9931.
129. J.-Z. Chen, X.-C. Tao, J. Xu, T. Zhang and Z.-L. Liu, *Process Biochem.*, 2005, **40**, 3675.
130. B. Southichak, K. Nakano, M. Nomura, N. Chiba and O. Nishimura, *Water Res.*, 2006, **40**, 2295.
131. K. Santhy and P. Selvapathy, *Sep. Sci. Technol.*, 2004, **39**, 3331.
132. K. H. Chu, M. A. Hashim, S. M. Phang and V. B. Samuel, *Water Sci. Technol.*, 1997, **35**, 115.
133. F. A. A. Al-Rub, M. H. El-Naas, F. Benyahia and I. Ashour, *Process Biochem.*, 2004, **39**, 1767.
134. Y. Wu, Y. Wen, J. Zhou, Q. Dai and Y. Wu, *Environ. Sci. Pollut. Res.*, 2012, **19**, 3371.
135. T. J. Butter, L. M. Evison, I. C. Hancock, F. S. Holland, K. A. Matis, A. Philipson, A. I. Sheikh and A. I. Zouboulis, *Water Res.*, 1998, **32**, 400.
136. R. Gong, Y. Ding, H. Liu, Q. Chen and Z. Liu, *Chemosphere*, 2005, **58**, 125.
137. L. Deng, Y. Su, H. Su, X. Wang and X. Zhu, *J. Hazard. Mater.*, 2007, **143**, 220.
138. T. Kikuchi and S. Tanaka, *Crit. Rev. Environ. Sci. Technol.*, 2012, **42**, 1007.
139. L. Doulakas, K. Novy, S. Stucki and C. Comninellis, *Electrochim. Acta*, 2000, **46**, 349.
140. H. H. Tabak, R. Scharp, J. Burckle, F. K. Kawahara and R. Govind, *Biodegradation*, 2003, **14**, 423.
141. Y. Sag, A. Yalcuk and T. Kutsal, *Process Biochem.*, 2000, **35**, 787.
142. Y. Sağ and T. Kutsal, *Process Biochem.*, 1996, **31**, 561.
143. P. Lodeiro, R. Herrero and M. E. Sastre de Vicente, *J. Hazard. Mater.*, 2006, **137**, 1649.

144. M. A. Martín-Lara, G. Blázquez, A. Ronda, I. L. Rodríguez and M. Calero, *J. Ind. Eng. Chem.*, 2012, **18**, 1006.
145. N. Akhtar, J. Iqbal and M. Iqbal, *J. Hazard. Mater.*, 2004, **108**, 85.
146. I. Moreno-Garrido, *Bioresour. Technol.*, 2008, **99**, 3949.
147. J. Archambault, B. Volesky and W. G. W. Kurz, *Biotechnol. Bioeng.*, 1990, **35**, 702.
148. P. C. Stark and G. D. Rayson, *Adv. Environ. Res.*, 2000, **4**, 113.
149. B. Volesky, *Water Res.*, 2007, **41**, 4017.

CHAPTER 5

Hyperaccumulation by Plants

CHRISTOPHER W. N. ANDERSON

Soil and Earth Sciences Group, Institute of Agriculture and Environment,
Massey University, Palmerston North, New Zealand
Email: c.w.n.anderson@massey.ac.nz

5.1 Introduction

Plants that can tolerate and in some cases accumulate very high concentrations of metal have been recognised for hundreds of years. Indigenous knowledge throughout history has correlated the presence of some plant species with mineralisation in the underlying soil[1] but it was only in the 20th century that instrumental techniques evolved to the point where the metal concentration in plants could be accurately determined. Perhaps the first reliable reports of a surprisingly high metal concentration in plants were for the nickel concentration in shrubs growing on ultramafic outcrops in Tuscany, Italy in the 1940s.[2]

The word 'hyperaccumulator' came into scientific usage in 1977, when a research group led by Prof. Robert Brooks from Massey University in New Zealand described the abnormal ability of a selection of plants to accumulate nickel at a concentration greater than $1000\,mg\,kg^{-1}$.[3] In this work, some 2000 specimens of *Hybanthus* and *Homalium* supplied by herbaria around the world were analysed for their nickel concentration and the authors discovered that some species contained a very high concentration of metal.[4] At the conclusion of this study, Brooks and colleagues coined the term hyperaccumulation to describe accumulation of a metal at a concentration 100 times greater than the highest value that would be expected in non-accumulator plants growing in the same environment.

RSC Green Chemistry No. 22
Element Recovery and Sustainability
Edited by Andrew J. Hunt
© The Royal Society of Chemistry 2013
Published by the Royal Society of Chemistry, www.rsc.org

Hyperaccumulator plants can accumulate metal to a truly remarkable level. The highest concentration of metal in any plant component recorded to date is for nickel in the sap of the tree *Sebertia acuminata* (seve blue) which grows on nickel-rich soil in New Caledonia. The sap of this tree has been found to contain about 10% nickel in the fresh material or about 20% on a dry weight basis.[5]

In 1983 a practical use of these plants that accumulate metals to extract metals from contaminated soil was formally described by the US scientist Rufus Chaney.[6] This concept was further developed in 1989 by Alan Baker and Robert Brooks.[7] Since then, the specific application of the physiological trait of hyperaccumulation to extract metals from soil has become an accepted remediation and, more recently, mining technology. This chapter reviews current understanding of the hyperaccumulation of metals and presents an overview for how this technology may fit into 'green chemistry' for the 21st century. In this chapter a specific focus is made on the technological application of plants that hyperaccumulate nickel and gold to the phytoextraction of these metals.

5.2 Hyperaccumulation of Metals

Today in the order of 500 plant species are recognised as hyperaccumulators of a variety of metals and metalloids (arsenic and selenium are defined as metalloids in this context). Around 450 of these species are hyperaccumulators of nickel. The most recent and authoritative review of the number of hyper-accumulators was made by van der Ent *et al.*[8] Reports at conferences and in print describe from time to time the discovery of new hyperaccumulators (especially for nickel) but there is no single database to record their existence.

Set criteria have been defined that allow for the classification of a plant as a hyperaccumulator of a metal. The first criterium concentration to be defined was for nickel ($1000 \, \text{mg} \, \text{kg}^{-1}$) by Brooks *et al.*[3] This criterium was based on a concentration that was 100–1000 times higher than that normally found in plants on non metalliferous soils and 10–100 times than that found in other plants growing on nickel-rich soils. There is significant scientific commentary available on how criteria levels have been set for other metals[7,9,10] but the level has generally been defined as 100, 1000 or 10 000 mg of metal per kg of dry weight biomass. Authoritative commentary on 'the basis for hyper-accumulation threshold criteria' can be found in the recent work of van der Ent *et al.*[8] These authors propose new and lower hyperaccumulation threshold criteria for most metals that better describe the physiological trait of hyper-accumulation (Table 5.1).

Many of the plant species that populate the numbers reported in Table 5.1 will accumulate more than one metal, so the total number of hyperaccumulator plant species is difficult to quantify. There is also some debate about the true classification of some of the species listed in Table 5.1, for example copper and cobalt hyperaccumulation by many of the 34 species listed have not been confirmed beyond field sampling where there is a real risk of airborne

Table 5.1 Number of metal hyperaccumulator species and the redefined hyperaccumulation threshold criteria for each metal as proposed by van der Ent *et al.* in 2013.[8]

Metal	Number of hyperaccumulator species reported	Hyperaccumulation threshold (mg kg^{-1})
Ni	450	1000
Cu	32	300
Co	30	300
Se	20	100
Zn	12	3000
Pb	14	1000
Mn	12	10 000
Cd	2	100
Cr	No reliable record	300
Tl	2	100
As	5	1000

particulate contamination.[11] This is especially true for the widely reported copper hyperaccumulator *Haumaniastrum robertii* found in the Democratic Republic of Congo. Field samples of this plant used to define copper and cobalt hyperaccumulation in this species may have been affected by contamination of the biomass by smoke from a nearby copper smelter. In cases such as this, iron is a good indicator of contamination. Plants do not accumulate high concentrations of iron and thus elevated iron levels would suggest that metals might be adsorbed to the surface of herbage, rather than having been accumulated.

5.3 Reason for Hyperaccumulation

There is general consensus that the reason for hyperaccumulation is a physiological strategy to protect plants against herbivore or pathogen attack.[12,13] Boyd described the evolution of hyperaccumulation from metal accumulation as a defensive enhancement brought about by stepwise increases in metal concentration.[13] Fones *et al.* recently reviewed the evidence both for and against this theory and through their own experimental work verified the accuracy of the hypothesis.[14] They determined that the concentration of zinc, nickel and cadmium in the hyperaccumulator *Thlaspi caerulescens* was sufficient to inhibit the growth of the bacterial pathogen *Pseudomonas syringae* within the plant. They furthermore showed that bacteria which naturally colonised the hyperaccumulator in the field had a higher tolerance to zinc than the same bacteria collected from non-accumulator plants.

Despite the apparent favour of the 'elemental defence' strategy, other reasons to explain the evolution of hypearccumulation have been proposed and these are well described by Rascio and Navari-Izzo.[15] Of the 'other' reasons, the idea of elemental allelopathy is interesting and explains hyperaccumulation and the subsequent leaf fall of highly metal-concentrated biomass to be a mechanism of 'poisoning' the ground adjacent to the hyperaccumulator. This would prevent less metal-tolerant plants from growing and thus competing with the hyper-accumulator for that space in the ecosystem.

What is apparent from relevant literature that attempts to describe the reason for hyperaccumulation is that this is a physiological response of plants to their environment. As such, this is a phenomenon that can and has been exploited for technological purposes.

5.4 Natural vs. Induced Hyperaccumulation

Hyperaccumuation by the plant species in Table 5.1 can be more accurately defined as natural hyperaccumulation. In other words, Table 5.1 describes the numbers of plant species known to have a physiological ability to accumulate the stated heavy metals as a normal function of their growth. However, there are several environmentally and economically important metals missing from Table 5.1 that are also hyperaccumulated by plants. For example, mercury and gold are missing and there is extensive literature published on the hyper-accumulation of both of these metals.

Hyperaccumulation of metals for which no natural hyperaccumulators have been identified can be induced by manipulation of the soil–plant environment. Such manipulation is designed to increase the bioavailability of metal and subsequently to force the plant to take up an increased concentration of a metal. Hyperaccumulation criteria are therefore meaningless in the context of induced hyperaccumulation and although established for metals such as gold (1 mg kg^{-1}),[16] should not be considered to be useful. Induced hyper-accumulation is not a natural physiological process.

Induced hyperaccumulation is generally achieved by adding chemicals to the soil and is described by some authors as assisted hyperaccumulation. The first example to be extensively described was for lead by two US scientists who grew the non-accumulator plant *Zea mays* (maize) on lead-contaminated soil.[17] At the point of plant maturity they treated the soil with ethylenediamine tetraacetic acid (EDTA) which acted as a complexing agent for lead, increasing the concentration of metal in the soil solution, thereby making the metal available for uptake. In this work the maize plants were induced to accumulate lead to a dry weight concentration of over 1%, approximately 1000 times more than control plants that received no EDTA treatment.

In 1999, Anderson *et al.*[16] reported their finding that plants could be induced to hyperaccumulate gold by applying a dilute solution of thiocyanate and thiosulfate to soil. These authors realised that gold from auriferous substrates could be made soluble with dilute thio-solutions and applied this knowledge to plant uptake studies. The mining industry has experimented with or used such chemicals for the past 100 years as lixiviants for gold in heap-leach operations (experimental use of thiocyanate, thiosulfate and thiourea as well as commercial use of cyanide). The translation to induced hyperaccumulation was a progression of this work.

As many of the chemicals used to induce uptake are, in fact, chelating agents, the term chelate-assisted hyperaccumulation is also used in literature. Nowack *et al.* prefer to use the term *chelant*-assisted hyperaccumulation, as the term chelate-assisted infers that chelates are added to soil.[18] A chelate is chemically defined as a metal-chelating agent complex and this complex forms in the soil as a result of the addition of chelating agents (*i.e.* chelants). However, many of the chemicals used to induce hyperaccumulation are not chelates. Chemicals such as citric acid or sodium cyanide which will induce uptake of a range of metals increase the solubility of these metals by simple ligand–metal interactions, rather than by chelation. Therefore, the author of this chapter prefers to use the term 'induced-hyperaccumulation' to describe this chemically assisted process of metal uptake, irrespective of the nature of the chemical–metal interaction in soil. Induced hyperaccumulation has been reliably reported for a range of metals (Table 5.2).

Studies describing the hyperaccumulation of some rare earth elements have appeared in literature since 2000.[23] Uptake in all studies has been from contaminated mining or smelting sites, where there is risk of air pollution and ground water pollution. Hyperaccumulation in these studies cannot, therefore, be definitively classified as natural hyperaccumulation, induced hyper-accumulation or simple contamination of the plant material with particulate metals from the atmosphere. Further research is needed to confirm the uptake pathways of this industrially important group of metals.

A novel study on the hyperaccumulation of rhenium (Re) was reported by Tzvetkova *et al.* in the industry bulletin *Advanced Materials and Processing.*[24] Rhenium is a metal that has received very little attention in hyperaccumulation studies. However, it is an industrially important metal. Tzvetkova *et al.*[24] reported elevated levels of rhenium in plants collected near three copper mines in Bulgaria and hyperaccumulation (defined as greater than $1000 \, mg \, kg^{-1}$) of rhenium by mountain spinach (*Atriplex hortensis*), buckwheat (*Polygonum fagopyrum*), alfalfa (*Medicago sp.*) and white clover (*Trifolium repens*) grown under pot conditions. In all cases however, soluble rhenium was added to the soil at some time after the plants had developed (leaching from the mine, or salting of the pot trials). Hyperaccumulation was therefore more hydroponic or induced than natural and the applicability of these results to current scientific

Table 5.2 Precious and non-precious metals for which induced hyperaccumulation (in the opinion of the author) is viable, useful and reliable.

Metal	Chemicals used to induce uptake	Reference
Pb	EDTA	Huang and Cunningham[17]
Hg	Thiosulfate	Moreno *et al.*[19]; Wang *et al.*[20]
Au	Thiocyanate, thiosulfate, cyanide, thiourea, halide salts	Wilson-Corral *et al.*[21]
Ag	Cyanide	Author's unpublished work
Pd	Cyanide	Walton[22]

knowledge cannot be assured. However, this work is mentioned here as an interesting development of the subject of metal hyperaccumulation and is worthy of further investigation.

5.5 Technological Application of Hyperaccumulation to Metal-rich Soil

The technology-based use of plants to remove heavy metals from soil by hyperaccumulation is termed phytoextraction. In a phytoextraction operation, heavy metals are translocated from roots to leaves, shoots and stems and can be harvested and removed from the contaminated site. Repeated cropping will reduce the soil metal concentration over time.

Phytoextraction has evolved from the pioneering work of the scientists who recognized a practical use for plants that accumulate very high concentrations of some metals and has today become an environmentally and economically attractive metal recovery option for some heavy-metal contaminated soils.[3,6,7] Soil excavation costs have increased dramatically over the past 10 years as international oil prices have increased, and the associated dumping or cleaning costs for soil are also prohibitively expensive. The use of plants to recover metals from soil is seen as a cheaper, more environmentally benign, efficient and a more aesthetically pleasing metal removal operation for some areas of soil metal contamination.

5.5.1 Phytoremediation and Phytomining. What is the Difference?

Where the purpose of phytoextraction is to 'clean' the soil, the specific application of the technology is termed phytoremediation. However, when an economic profit can be recovered by processing the metals from the biomass, application of the technology can be termed 'phytomining'. So a clear distinction between projects that seek either to 'remediate' or 'mine' an environment is difficult to make. Perhaps the greatest value of the technology is in the implementation of a scenario where valuable metals recovered from a plant pay for the clean-up of less valuable metals removed from contaminated land at the same time. Theatres for application of this scenario can be found around the world. The word phytomining is perhaps a slightly unfortunate one, as the name suggests that the technology may be used as an alternative to conventional mining. This is not however a feasible suggestion. Conventional mining exploits ore volumes in the order of millions of tonnes, whereas phytomining can only exploit the topsoil (limited to $3000 \, t \, ha^{-1}$ for a 30 cm rooting depth and a soil density of $1 \, t \, m^{-3}$). The term phyto-reclamation was proposed by the University of British Columbia as an alternative to phytomining.[25] However, this name has not established itself in the literature.

5.5.2 Timeframe

The biggest limit to the viability of phytoextraction for removing metals from soil is the timeframe.[26] Plants that hyperaccumulate heavy metals have both a maximum metal concentration and a biomass yield that will limit the amount of metal that can be removed by each crop. Metal removal from soil is a product of concentration and biomass. As an example, consider the permissible EU limit for nickel in soil, which is set at 75 mg kg^{-1} or ppm. Using *Berkheya coddii*, a high biomass nickel hyperaccumulator that can remove 200 kg of nickel per crop from a hectare of land, soil with a concentration of 250 ppm would be safe after 4 years; 1000 ppm after 18 years; or 10 000 ppm after 138 years.[27] Clearly for lightly contaminated soil, phytoextraction is a viable option. But for land with 1% metal contamination, conventional soil excavation would be a better solution if re-zoning of the land for residential or commercial development is the target land use. If the land is to remain undeveloped then the timescale is less important. Perhaps, under this scenario, long-term management using phytoremediation is a viable option for even mid to high-level pollution; a managed forested ecosystem could lock-up the pollutants in perpetuity. The final land use for a polluted site must therefore be considered during assessment of the merits of phytoremediation for soil clean up or management.

5.6   Phytomining of Nickel and Gold

Throughout history mining activities have created great wealth and have supported the advancement of industry and technology. In fact, today's society, more than any other in history, relies on mineral resources for economic prosperity. The fortunes of many countries are heavily dependent on the mineral demand of others to drive gross domestic profit. For example, the mineral sector boom in Australia throughout the late 2000s was driven by Chinese demand for iron ore. In September 2012, questions arose about the sustainability of Australia's economy in the face of dramatically reduced iron ore prices on the international market highlighting how dependent many nations have become on their mining industry.

Yet are current extraction rates sustainable? Hudson-Edwards *et al.*[28] used the term 'peak metal' (analogous to peak oil) to describe a scenario where humankind has extracted more metal at a point in time than is left in the ground. This scenario is based on the commentary of scientists such as Gordon *et al.*[29] who warned that if all nations consume the same amount of copper, zinc and platinum as developed nations, long-term supplies might not meet demand.

In response to this commentary we need to manage the Earth's metal resources more cleverly. Sustainable management requires better technology for extraction as well as mitigation of the negative environmental consequences incurred by extractive industries in the past, present and future. Hyper-accumulation and phytoextraction have a role to play in achieving this objective. While phytomining will never compete with the scale of conventional mining, the application of phytoextraction to remediate contaminated soil and to recover an economic harvest of metals is technology that can work alongside the conventional mining industry.

The three metals, nickel, thallium and gold were reported as candidates for phytomining in 1999 owing to the relatively high price that each metal commands on international commodity markets.[30] The predicted net revenue yielded from a crop of these metals was expected to cover the costs of their recovery. The phytomining status of these three metals has not changed since the cited publication, although only nickel and gold have received significant research, development and commercial attention for phytomining since about 2000.

5.6.1 Phytomining of Nickel

Nickel has perhaps received more widespread research focus for phytomining than any other metal. This is presumably due to a combination of the existence of numerous plant species that will naturally hyperaccumulate the metal to a concentration greater than 1% dry weight and the relatively high revenue that such a metal concentration would yield on a per hectare basis.

The first experimental evidence of nickel phytomining was described by two USDA scientists who conducted a field trial on ultramafic soil near Chinese Camp in California, using the nickel hyperaccumulator *Streptanthus polygaloides*.[31,32] A nickel yield of approximately 100 kg ha^{-1} was possible in this pioneering study. These authors concluded that it might be possible to use plants to extract nickel from low-grade ores that would not otherwise be economic to mine and that were unsuitable for agriculture owing to their high nickel concentration. Such low-grade nickel ores or ultramafic soil covers large areas of the Earth's surface, for example in Western Australia, parts of South East Asia, Brazil, Canada and Russia.

Robinson *et al.*[27,33] described further tests where the species *Alyssum bertolonii* and *Berkheya coddii* were shown to have potential to yield in excess of 200 kg of nickel per hectare. Li *et al.*[34] then described how yields of 400 kg of nickel per hectare could be achieved with the species *Alyssum murale* and *Alyssum corsicum*, utilising herbicides and appropriate agricultural management techniques. Based on the promise of these initial reports, several patents were taken out concerning the technology (*Method of Phytomining of Nickel, Cobalt and Other Metals from Soil*, Chaney *et al.* US Patent number 5711784, 6786948 issued in 1998, 2004 respectively).

The question is, has the technology developed since this activity in the 2000s? Chaney *et al.*[35] reviewed the concept of nickel phytomining in 2007 but did not report any major advances in the technology. Harris *et al.* [36] reviewed the economic potential of nickel phytomining in 2008 using calculations based on literature data but again did not report any major field advances.

The challenge of proving the commercial potential of nickel was taken on by the US company Viridian Resources LLC throughout the 2000s. Viridian conducted extensive breeding trials with the temperate nickel hyperaccumulator *Alyssum* spp. The aim of this research was to breed selectively a high biomass and high nickel concentration hyperaccumulator that would be suitable for commercial planting on nickel-rich soils around the world. To this end, Viridian secured rights to large areas of land for subsequent planting.

Between 2004 and 2007, Viridian transferred select *Alyssum* spp. cultivars to field locations at Soroako on the Indonesian island of Sulawesi, an ultramafic region with active nickel mining. The objective of this work was to make Soroako the first theatre for commercial nickel phytomining in the world. However, the field results were disappointing. The plants failed to reach either their biomass target or the nickel concentration for which they were selected. No definitive answer has been given to explain this anomaly between predicted and actual success but a key issue may be the inability of the temperate *Alyssum* spp. to grow in the tropical environment of Sulawesi.

Despite the apparent failure of Viridian's Indonesian operation, nickel phytomining remains a subject of research interest in Indonesia as well as the Philippines and Malaysia. This part of South East Asia has thousands of hectares of exposed serpentine (ultramafic) soil and a flora that include many species of nickel hyperaccumulators. Current research projects are seeking to find native species that may prove more suitable for commercial nickel phytomining operations than the temperate species used by Viridian. Such species may be planted on soil not fit for agriculture or on the mine waste left behind by current day mining operations.

The greatest success story for nickel phytoextraction is that reported by the Anglo Platinum Company at their Rustenburg Base Metal Refinery in South Africa.[37] The Rustenburg base metal smelter in the late 1990s had a history of nickel salt spillages and air-fall pollution that had left the land around the smelter contaminated with metal. Fortunately, the nickel hyperaccumulator *Berkheya coddii* is native to this part of South Africa and in an attempt to remediate this historical pollution, the environmental department of the smelter began to investigate the effectiveness of phytoextraction to remove the metal from soil. The first stage of the operation saw local villagers employed to collect and germinate seed. The resulting seedlings were transplanted into the field and irrigated to ensure optimal growth. At the end of the summer when the plants reached maturity the same villagers were employed to collect seed to ensure new plants in the following year. The biomass was then harvested by hand, tied in bundles and fed directly into the modern metal smelter. The biomass had no detrimental effect on the ore smelting and the metal from the harvested plant material became incorporated into the final product for sale. This therefore represents the application of phytoextraction in both phytoremediation and phytomining. The nickel concentration in the soil was slowly reduced during the years the project was run and the metal from the plant was incorporated into a saleable product. During one early operation, the nickel from the biomass was recovered separately, producing small nickel ingots (Figure 5.1). To date this operation remains the best example of the effective application of phytoextraction of any metal to achieve a specific technological purpose.

5.6.2 Phytomining of Gold

The technological use of plants to extract commercial levels of gold from the soil is a relatively recent scientific advancement having been first described by

Figure 5.1   *Berkheya coddii* growing on Ni-contaminated land at the Rustenburg Base Metal Refiners smelter in South Africa in 2003. The inset shows buttons of Ni smelted from harvested biomass.

Anderson *et al.* in 1998.[16] However, the idea of plants accumulating gold was common in the relevant literature throughout the 20th century. Many studies have described the use of trees to indicate the likelihood of subsurface gold mineralisation. This is a subject known as biogeochemical exploration and correlates an elevated concentration of gold in plants with an elevated concentration of gold in soil. Many reviews have been published on this subject and the reader is encouraged to review these for more detail on the theories underpinning gold solubility in the environment and the subsequent uptake of this bioavailable metal by plants.[1,38–40]

No plant species have been identified that naturally accumulate a concentration of gold that would be of interest for phytoextraction studies. Natural gold concentrations in plants are in the order of 10s to 100s of parts per billion,[41,42] but gold can be made soluble in soil and this leads to the possibility of enhancing uptake by the use of induced hyperaccumulation. A recent review was published on gold phytomining that comprehensively describes the history of studies that have established the technical viability of induced gold uptake by plants.[21] The concentration of gold that will be accumulated by plants is a function of both the gold concentration in soil and the effectiveness of the chemical being used to induce uptake. Studies conducted around the world have conclusively shown that cyanide is the most effective chemical to induce uptake and gold concentrations in plants as high as $8000\,\text{mg}\,\text{kg}^{-1}$ (0.8% dry weight) have been recorded for plants growing on soil with $100\,\text{mg}\,\text{kg}^{-1}$ gold after irrigation of mature plants with a dilute concentration of potassium cyanide (author's unpublished data).

5.6.2.1   *Environmental Considerations of Gold Phytomining: Defence of the Use of Chemicals*

Any operation to recover gold from geological media requires the use of chemicals. This is because of the limited solubility of gold under environmental conditions.[39] Therefore, any technology to recover gold, including phytomining, will require the use of chemicals. It is simply not possible to recover gold from rock or soil without them. The environmental challenge in any gold exploitation operation is therefore to use chemicals in such a way that they pose no harm to the environment.

Many concerns have been raised over the persistence of the chemicals applied to soil during induced hyperaccumulation, the potentially damaging effects of these chemicals on soil microbiota, and the possible mobilisation of contaminants to groundwater.[43] As an example, there has been much discussion on the environmental consequences of using the chemical EDTA during phytoextraction.[18] This chemical has been routinely used to induce the hyperaccumulation of lead from contaminated soil and has been proposed for large-scale field application. EDTA is a synthetic chemical that has a relatively long residence time in the environment and therefore the concerns that have been expressed in literature are valid. To achieve the levels of optimal lead concentration reported in literature ($10\,000\,\mathrm{mg\,kg^{-1}}$), Chaney *et al.* estimated that an EDTA application rate of $10\,\mathrm{mmol\,kg^{-1}}$ of soil would be required.[35] It is extremely unlikely that a plant would be able to accumulate all of this applied EDTA and thus such an application rate would present an unacceptable risk of widespread migration of EDTA and metal complexes into the environment. Chaney *et al.* estimated that the cost of EDTA in this scenario would be US$30 000 per ha and based on the economic cost–benefit analysis alone suggested that the use of EDTA for field-scale phytoextraction was a poorly defendable proposition. Unfortunately, the negative consequences of EDTA-induced metal hyperaccumulation have become associated with most, if not all, operations that seek to exploit this technology for the recovery of any metal from soil.

The scenario of gold phytomining is different. Lead has a low unit price and the mass of metal recovered from a crop of harvested plants has low resale value. Gold on the other hand has a very high unit price and even a small amount of gold recovered from plants has appreciable value. This leads to the potential for net positive revenue from an operation. The often quoted target for gold phytomining is a gold yield of 1 kg per ha of land (10 t dry biomass at a gold concentration of $100\,\mathrm{mg\,kg^{-1}}$)[25] and in 2012 this had an average value of approximately US$50 000. The value proposition of a gold phytomining operation allows for appropriate infrastructure that will mitigate the risk of soluble gold or other metal complexes leaching out of the plant root zone and into underlying aquifers or other adjacent water bodies.

The scenario of a 'phyto leach pad' can be imagined, where the gold-rich soil being exploited by phytomining is placed on top of a liner following the design of a conventional heap leach pad. Heap leaching is a common but generally unpopular and much maligned method of conventional gold mining. Both gold phytomining and heap leaching utilise cyanide to dissolve gold from rock. However, a key difference lies in the amount of cyanide used and the timeframe over which this cyanide is used. Heap leaching requires the constant use of chemical over a timeframe of years where the purpose of cyanide use is to leach gold through the rock pile.[44] Phytomining requires the use of cyanide once only, over a timeframe of minutes, where the purpose of the chemical is to promote sufficient gold solubility for uptake. During phytomining, the irrigation of chemical is designed to mitigate the potential for leaching of the chemical and gold out of the root zone. There are decision support systems that can model the water use efficiency of a plant and allow for accurate irrigation volumes.[45] The liner in the scenario of a 'phyto leach pad' could mitigate concerns over the movement of metal complexes into groundwater. There would be a cost associated with this additional layer of operational design but the cost–benefit analysis is such that excavation and transport of soil onto a pad could be economically sustainable.

Chemicals used in gold phytomining like cyanide, thiocyanate and thiosulfate are all naturally occurring chemicals that will degrade in the environment,[39] for example, the degradation of cyanide to thiocyanate and the pathways for further biodegradation of thiocyanate have been well studied.[46] These degradation pathways proceed rapidly in an aerobic biological environment such as would characterise the active root zone of a crop of plants. Ebbs *et al.* [47] in an attempt to defend the use of chemicals during gold phytomining, described the degradation rate of cyanide, thicyanate and thiosulfate in the 24 h after mixing of these chemicals with an oxidic gold ore. They reported at least 50% degradation within 24 h and concluded that these chemicals were not innately persistent and that natural attenuation could greatly reduce their concentration. However, the environment in this study into which these chemicals were applied was sterile and not representative of the biologically active root zone of plants. I propose that the degradation rate of these chemicals in the soil of a field 'phytomine' may be manifold higher. A study of the degradation rate of the chemicals used in gold phytomining in a field soil is one that should be done.

Environmental concerns about the use of chemicals in gold phytomining are real and represent perhaps the greatest drawback to widespread acceptance and use of the technology. Further research is needed to establish more clearly the environmental risks associated with the use of chemical amendments to promote gold uptake by plants and to design strategies to mitigate both the actual and perceived risks.

5.6.2.2   Gold Stored in Plant Biomass: Nanoparticles

The increased availability of electron microscopy instrumentation during the 1990s and into the 2000s allowed direct imaging of the form in which plants store metals such as nickel and gold. Further characterisation of the chemical form of the stored metals was facilitated using X-ray absorption spectroscopy (XAS) which could quantify the speciation of elements in plant tissues and could allow direct analysis of what metals are bound to. Using both of these techniques, Gardea-Torresdey *et al.* in 2002 [48] were the first to describe how alfalfa plants exposed to gold suspended in agar accumulated and stored the metal as nanoparticles of elemental gold. Their finding has been corroborated by a number of research groups working in agar, hydroponic and soil media.

The idea that plants can store metals in a nanoparticulate form has led to the proposal that phytoextraction may be an environmentally friendly route for the synthesis of nanoparticles that are useful in a range of industrial, environmental and medical applications.[48,49] Irvani recently reviewed the potential synthetic advantages of using plants over conventional chemical techniques to synthesise nanoparticles.[50]

In the context of structural chemistry, nanoparticles are defined as objects with at least one dimension in the size range 0.5–50 nm.[51] Metallic nanoparticles behave very differently from their corresponding bulk metals forms. Gold metal, for example, is soft and yellow with a face-centred cubic structure, showing excellent electrical conductivity and has a melting point of 1068 °C. But gold nanoparticles of about 2 nm in diameter (approximately 200 gold atoms) melt at about 350 °C, can show irregular and poorly defined structure, are ruby-red to violet in colour and may act as a semiconductor.[51] The distinctive colour of gold nanoparticles is a function of absorption of light in the green part of the electromagnetic spectrum (about 520 nm). This wavelength corresponds to the frequency at which a plasmon resonance occurs in the gold.[51] Bulk (non-nanoparticulate) gold is yellow owing to reflection of light at a wavelength corresponding to the red-yellow part of the spectrum.[52]

In practical terms, gold-rich dry biomass and gold-rich plant ash will often be a bright purple colour, with the intensity of this colour proportional to the gold concentration of the biomass or ash. One exception is where there is also a high concentration of silver in the plant material; the plant ash that is enriched with both gold and silver is brown. A shift in colour from purple to brown indicates that the silver inhibits the formation of discrete gold nanoparticles, leading to the formation of gold–silver nanoalloys. Electron microscopy observations at Massey University support this hypothesis.

The existence of gold nanoparticles in plants has led some scientists to propose that plants can accumulate gold nanoparticles directly from soil.[48] However, this suggestion appears to contradict the accepted convention that plants accumulate dissolved ions from soil solution.[53] The suggestion that plants will accumulate gold nanoparticles may be based on the perceived lack of reactivity of gold in the environment. This is true for bulk gold but clearly the chemical reactivity of nanoparticulate gold is different. A portion of the gold in plant ash can be dissolved in very dilute, room temperature hydrochloric acid (Table 5.3). During Masterate research at Massey University, as much as 30% of the total gold loading of ash was shown to be soluble in 0.05 M acid with a very small but reproducible percentage being soluble in deionised water.[54] In the reported study, 100% of the gold loading of the ash was soluble in 1.5 M acid. This is very different to the scenario where hot *aqua regia* must be used to dissolve bulk gold.

Nanoparticles of gold may therefore be relatively reactive in the soil environment. It may be that nanoparticles are not in fact accumulated by plants

Table 5.3 Comparison of the effectiveness of variable concentrations of dilute room temperature HCl in dissolving gold from plant ash containing a gold concentration of approximately $55\,mg\,kg^{-1}$ (from the Masterate research of Alice Lamb in 2002).[54]

Concentration of dilute HCl (M)	% Dissolution of gold from plant ash
0	2.6
0.01	8.9
0.05	29.1
0.1	43.9
0.5	92.2
1.0	95.7
1.5	101

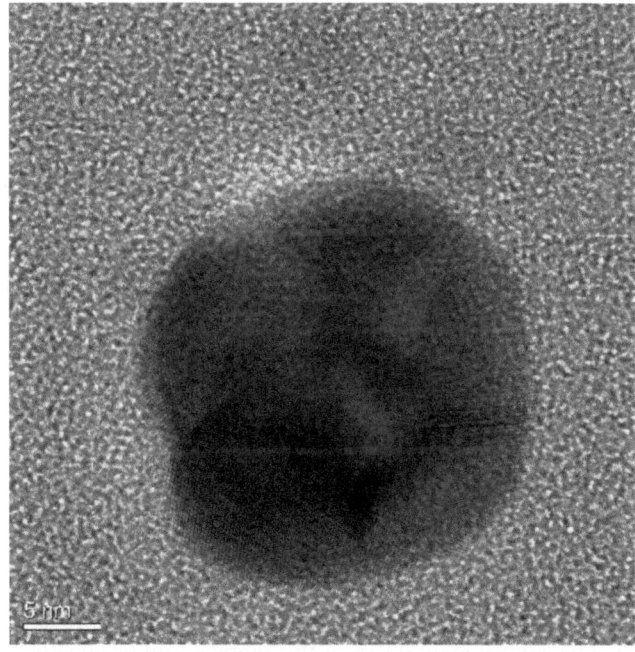

Figure 5.2 High resolution transmission electron microscope photomicrograph of a gold nanoparticle in the leaf material of cyanide-treated *Brassica juncea* biomass. The nanoparticle here is an accretion structure of gold, rather than a discrete particle representing a single reduction event. It seems unlikely that such a complex particle could be accumulated by the plant as the Au(o) would have been readily dissolved by cyanide in the soil.
Image courtesy of Akzo Nobel Chemicals, The Netherlands.

but rather these nanoparticles dissolve in the root zone of plants in the presence of root exudates, are taken up as dissolved ionic species and are then reduced in the aerial biomass. Transmission electron microscopy images of gold nano-particles in plants growing on gold-rich soil and treated with cyanide support this hypothesis (Figure 5.2). Plants that have accumulated gold by induced

Table 5.4 LC-XANES (XAS) fittings for gold in *Brassica juncea* shoot biomass harvested from a gold-only, silver and gold, copper and gold, and copper, silver and gold enriched soil showing the speciation of gold within the biomass after treatment of soil with dilute cyanide solution. The LC-XANES technique can quantify the relative amount of elemental species present within a sample.

Metal in biomass (mg kg^{-1})	*% Au(0)*	*% Au(I)*	*% Au(III)*
Gold (577)	88.5	0.0	14.1
Silver (211)/gold (756)	45.0	0.0	55.0
Copper (435)/gold (1,052)	50.8	0.0	49.1
Copper(146)/silver (86)/gold (195)	51.3	0.0	49.7

hyperaccumulation show evidence of gold nanoparticle accretion structures, indicating that gold may be reduced from Au(III) or Au(I) to Au(0) to form nanoparticles inside the plant. Accretion structures suggest that reduction happens continuously and that the nanoparticles will grow continuously as soluble gold is accumulated and translocated to aerial biomass. The speciation of gold inside cyanide-treated plant material is predominately Au(0) where gold is the only metal present but is a 50 : 50 mix of reduced and oxidised gold where copper and/or silver is also present (Table 5.4). This could provide empirical evidence that reduction of gold to nanoparticles occurs inside the plant; the presence of phytotoxic levels of copper and/or silver in plants may inhibit the *in vivo* physiological reduction of gold.

5.7 Biomass Processing: Recovery of Metal from Harvested Plants – the Big Unknown

Reports of metal uptake by plants have been extensively made but underpinning any technological application of hyperaccumulation is a viable process to manage or recover metals in the biomass. Where the metals have no value, disposal at an appropriate landfill may be an option. Metal from soil is concentrated and partitioned to organic matter by the process of hyperaccumulation. The environmental risk of this metal could be effectively managed at a landfill that was certified to receive hazardous materials. But where the metal content of the plant has economic value (*i.e.* phytomining), processing to recover and sell the metal is the final objective. Such processes have been poorly reported in literature. Chaney *et al.*[35] stated 'the value of nickel in phytomining biomass can exceed US$16 000 per hectare, and methods to recover and market this metal are needed'. The design and implementation of a viable processing system to recover metal from hyperaccumulator biomass is a key step in the phytomining system for which a technology solution is lacking.

Disclosed systems to recover phytomined metal are based upon a primary thermal degradation step (incineration or pyrolysis), followed by solvent extraction, smelting or electrowinning of metal from the plant ash.[21,36,55] Thermal degradation is crude, expensive and can release toxic by-products or air pollution into the atmosphere if emissions are not carefully controlled. This is important where phytomined biomass is enriched in metals such as mercury and arsenic. Thermal degradation necessarily requires large amounts of biomass supplied at a constant rate to maintain the operating economics of the thermal plant. This may not be possible for a plant supplied by small phytomines in widely distributed geographical areas. In such operations, the cost of transport of bulky biomass to a distant thermal plant may preclude the viability of many field operations. Thermal degradation will also destroy the primary structures of metal stored inside biomass. Where such particles have inherent value (*e.g.* nanoparticles), this may represent an opportunity cost on the balance sheet for phytomining.

A range of chemical problems can be encountered in the secondary processing of plant ash. (1) Certain alkaline metals in the ash such as calcium can be enriched by orders of magnitude over the same alkaline metal concentration in conventional ore and high calcium concentrations dictate a reworking of conventional chemical flow sheets.[56] (2) The high pH of plant ash requires the use of large volumes of concentrated acid to reduce the pH to the required acidic operating conditions. (3) Other metals in the biomass that are not recovered can be transformed into an environmentally active state and thus become contaminants, which must be managed or safely disposed of.

The high capital and operating costs and the level of engineering complexity of the proposed biomass processing technology is likely to be a reason contributing to an explanation of why phytomining technology was not generating commercial profits in 2012.

5.7.1 Green Technology Idea for the Future: Supercritical Fluid Extraction

Conventional hydrometallurgical flow sheets that optimise gold recovery from rock are predicated on a low carbon environment, as carbon very effectively scavengers gold from solution. Therefore, conventional hydrometallurgical flow sheets are not immediately suitable to process plant biomass or ash. But plant biomass concentrated in gold represents an unconventional form of high-value material. To capitalise on this unconventional value proposition, unconventional processing techniques may need to be considered.

A supercritical fluid (SCF) is defined as any substance at a temperature and pressure above its critical point. A SCF can diffuse through substances as would a gas, and dissolve substances as would a liquid. Because of these properties, SCFs have received attention for their potential to remove toxic metals from plant and animal cells.[57]

The addition of chelating agents to SCF increases the ability of a supercritical system to target and remove specific metals. Mercury has received much study; Wang and Wai[58] reported 95% efficiency of removal of Hg^{2+} ions from biomass in supercritical CO_2 amended with methanol and a chelating ligand. Indicative results for the extraction for Au^{3+} have also been reported.[59] There has been a commercial interest in developing supercritical CO_2 extraction technology to remove radioactive isotopes of thorium and uranium from contaminated waste at locations such as the Hanford nuclear reservation in the USA.[60]

Could supercritical fluid extraction technology offer an environmentally friendly and economically viable system to recover valuable metals from biomass? Existing literature shows that a SCF system for extracting oxidized metals is a technologically viable proposition, although a system for extracting reduced forms of metal (bulk metal) is less certain. Data and theories from conventional hydrometallurgy could be integrated with appropriate technology development to select chelates and ligands that could target specific metals in both an oxidized or reduced form. An alternative approach could be the extraction of all non-metallic or non-target metallic components of biomass. This would significantly enrich the residual concentration of a target metal and possibly yield a refined product for direct industrial or commercial use.

5.8 Economic Case for Using Nickel and Gold Hyperaccumulator Plants

Limited modelling has been published that investigates the potential economic value of metal-rich biomass and therefore the economic case for phytomining. Nicks and Chambers [32] in their pioneering work assessed the likely economics of phytomining using *Streptanthus polygaloides* in California. Robinson *et al.*[27,33] published some preliminary calculations on the potential profit that may be accrued by the harvesting and processing of field quantities of the Ni hyperaccumuators *Alyssum bertolonii* and *Berkheya coddii*. Chaney *et al.*[35] reviewed the economics of the nickel phytomining conducted by Inco and Viridan Resources in the late 2000s. The most recent economic assessment for Ni phytomining appears to be that published by Harris *et al.*[36] using the hyperaccumulator *Berkheya coddii* in Australia. However, the market price of nickel reported in this model (March 2007) was AU$60/kg. The price has dropped significantly in subsequent years and the cash price was in the order of US$17/kg in October 2012 (London Metal Exchange).

Published economic considerations of gold phytomining appear to be less common. Anderson *et al.*[61] published the first economic assessment of induced hyperaccumulation to recover gold from mine waste in 2003. Wilson-Corral[21] presented an updated version of this economic model for gold phytomining in Mexico. Harris *et al.*[36] presented an assessment of gold phytomining in Australia. In contrast to nickel, the cash price for an ounce of gold steadily increased between 2002 and 2012 and in October 2012 was about US$1750 per ounce, or US$56 per gram. This is a very different price to that which existed when gold phytomining was first proposed in 1999 (approximately US$300 per ounce).

A summary of the most recently published economic data describing the costs, metal value and net revenue that might be expected from nickel and gold phytomining is presented in Table 5.5. The costs in Table 5.5 are partitioned between the agricultural costs of growing the plant, the cost of the chemical necessary to induce hyperaccumulation, the cost of crop management associated with the irrigation of this chemical and the cost of processing the biomass to yield metal for sale. Chemical and management costs are not relevant to nickel phytomining as natural hyperaccumulator species would be used in any operation. These species have the added advantage that they are more tolerant of the edaphically challenging environmental conditions that may exist in metalliferous soils. The use of natural hyperaccumulator species is the reason for the lower agricultural cost for nickel phytomining over gold phytomining. The plants best suited to induced hyperaccumulation are crop plants that have rapid growth rates and high biomass production. However, the agricultural costs associated with growing such a crop in an edaphically challenging environment are relatively higher.

The single greatest cost for phytomining of either metal in Table 5.5 is that associated with processing. For nickel phytomining, the total processing cost of $5445 ha^{-1} is dominated by the cost of ashing ($3480 ha^{-1}) at a unit price of $158 tonne^{-1}. For gold phytomining the reported cost of ashing is $2075 ha^{-1} at a unit price of $250 tonne^{-1}. Burning organic material is an expensive endeavour primarily owing to air pollution mitigation measures that must be met. In the case of both metals, the proposed technique to recover metal from the ash is solvent extraction. The gross profit reported by Wilson-Corral for gold phytomining is recent and, given the costs described in Table 5.5, accurate;[21] the gold price has not shifted significantly since this model was published. However, the described gross profit for nickel phytomining has probably been eroded in the time since the model was published, as the market price for nickel is dramatically lower today than it was in 2007.

The economic models described in Table 5.5 show that the hyper-accumulation of nickel and gold by plants is a physiological trait that can be utilised to generate a gross economic profit. Table 5.5 does not, however, describe the inherent economic value of remediation that might also be

Table 5.5 Summary of the published economic parameters associated with nickel and gold phytomining; after Harris *et al.* (2009)[36] and Wilson-Corral *et al.* (2012)[21].

Economic parameter	US$ ha^{-1}	
	Ni	*Au*
Agriculture	188	1630
Chemical	—	1940
Crop management	—	2200
Processing	5445	3305
sub total	5651	7608
Metal yield	220 kg	463 g
Value(US$ kg^{-1})	60	52
Revenue	13 200	24 076
Gross profit (US$)	7549	20 771

achieved by the implementation of phytoextraction. The economic model for an integrated system where electricity is produced from the incinerated biomass (not considered in Table 5.5), and where jobs are created and remediation is effected, may be more profitable. There is also potential for value-added products to be recovered from the plants, such as precious metal nanoparticles for industrial or medical applications. The application of hyperaccumulation to the phytomining of nickel and gold clearly has the potential to be a profitable venture. However, like any venture that seeks to exploit mineral resources, the economics are dictated by the market price of the final commodity.

5.8.1   Sustainability of Hyperaccumulation with Time

Plants roots, on average, contact less than 1% of the total soil volume available to them.[53] Therefore, there is a physical limit to the amount of metal in soil that plants will be able to remove in any one crop, which in turn is a function of the total metal concentration in the soil as the plants grow. This establishes a law of diminishing returns on the amount of metal removed by subsequent crops as the total soil concentration is continually decreased by accumulation by plants.[36] Robinson *et al.* modelled the diminishing concentration of Ni in a crop of *Berkheya coddii* that could be expected with subsequent cropping of the same piece of land and this data was used to predict the timeframe over which useful metal recovery could be expected.[62]

In gold phytomining, the percentage of gold in the soil removed by any one crop is higher than that for nickel. This is because the bioavailability of gold is increased by the use of chemicals and this soluble gold can be subject to uptake by mass flow. Anderson *et al.*[25] predicted that 15–20% of the gold content of the root zone of plants could be taken up under optimal conditions by any one crop cycle. Harris *et al.* used a value of 20% in their calculations.[36] Gold concentrations in soil are generally orders of magnitude lower than Ni concentrations, so consideration of the sustainability of a gold phytomining operation where both cost and revenue are relatively high is perhaps more essential than for Ni phytomining.

Long-term modelling of the sustainability of hyperaccumulation to generate economic profit in the field is lacking in the scientific literature and this is research that should be conducted. In the case of induced gold hyperaccumulation, actual sustainability data may differ from modelled data as organic matter in the soil that will increase with sequential cropping may adsorb soluble gold and limit the amount that is available for uptake.

Consideration of hyperaccumulation sustainability usefully sets the cut-off grade for which a phytomining operation might be economic. An operation that is not profitable in year one may return a profit over several years (this assumes that capital costs are incurred in year one). However, this cut-off grade is also a function of the market price for metal and the costs associated with the operation. Sustainability should therefore be considered as part of the economic model. Perhaps a useful guideline for a cut-off grade in 2012/2013 for gold phytomining is an initial gold concentration in soil of $1\,mg\,kg^{-1}$.

5.9 Where Might Phytoextraction be Most Usefully Employed?

The case for the hyperaccumulation of nickel and gold and the technological application of this trait for phytoextraction is good, based on the information reviewed in this chapter. The question may therefore be asked, why is this technology not in more widespread use? In my opinion the over-riding answer is hesitation from potential clients who prefer the security of big machines and conventional technology. There are people in the commercial sector who remain incredulous at the idea that plants can be used economically to recover metals from soil. Conventional remediation and mining systems are immediately visible and tangible but plants will take at least one season to grow. The timeframe for phytoextraction and the scale of the economic returns per unit of land are therefore not of the same order of magnitude as for conventional technology.

Lack of education is a further hindrance, as many people are simply not aware that hyperaccumulators can be used in an economically sustainable environmental system. Attitudes should change with time as successful phytoextraction operations become more widely publicised and landowners are forced to take greater responsibility for environmental risk.

Cost is a significant factor. This is not just the cost of phytoextraction but of remediation in general. If there is no money to be made from remediation, then companies primarily interested in their financial bottom line will find reasons to not remediate polluted soil. Revenue can come directly from the sale of metal recovered during phytoextraction. But cost can also be a regulatory instrument. Positive revenue can be generated by government subsidies that reward an environmentally friendly and sustainable company. Alternatively, if the cost of remediation is less than the cost of fines incurred as a result of environmental inaction, then cost–benefit analysis will suddenly favour clean-up. Phytoextraction and regulatory governance are in this fashion inextricably linked. Economic incentives need to be created that actively encourage the use of hyperaccumulation to improve environmental performance.

5.9.1 Phytoextraction as a Community Mining Initiative

The over-riding purpose of using hyperaccumulator plants is to improve the environment. This can be achieved by the technological application of phytoextraction technology. In some circumstances, as has been described in this chapter, net positive revenue can be generated by recovery of valuable metals from biomass. One research centre in Indonesia, the International Research Centre for the Management of Degraded and Mining Lands (IRC-MEDMIND), is working to implement phytoextraction at artisanal and small scale mining locations that are commonly found in developing countries throughout the world. The target valuable metals are nickel and gold. Therefore, phytomining is perhaps an appropriate moniker for this scenario.

The term 'artisanal mining' describes an informal and unregulated system of small-scale mining prevalent in many of the world's poorest countries and communities. Artisanal miners do not make large profits. Instead they strive to make sufficient money to support their immediate families.[63] Artisanal mining is practiced throughout the world and represents a major livelihood for some of the world's poorest people. However, artisanal operations are often illegal and poorly regulated. Miners have no title to the land they are working and thus there is no incentive for sustainable land management. Environmental destruction is the most visible outcome of artisanal mining. Where artisanal mining techniques are employed to mine gold (artisanal and small scale gold mining or ASGM), environmental pollution with mercury and cyanide is commonplace.[64]

Krisnayanti *et al.* assessed the concentration of mercury and gold in ASGM waste on the island of Lombok on Indonesia.[65] They found evidence of significant and alarming mercury pollution but also recorded gold grades above $1 \, mg \, kg^{-1}$ in all tailings analysed. Krisnayanti *et al.*[65] proposed, following the original idea of Anderson *et al.*[25] and Moreno *et al.*,[19] that phytoextraction could be used as a community initiative to manage the emerging risk of mercury contamination of water, plants and food sustainably.

The proposed scenario has three key steps:

1. Metal-rich and contaminated mine waste is first contained within specific farming areas (phytomines) and planted with a fast-growing and high-biomass plant species.
2. When the crop reaches its point of maximum biomass, induced hyper-accumulation is effected by irrigating an appropriate chemical amendment to the soil. This will make a portion of both the mercury and gold in the soil soluble and these metals will accumulate in the roots, shoots and leaves of the crop as it continues to grow. Like gold, mercury can also be made soluble with thiosulate and cyanide and will be subject to induced hyperaccumulation.[19,66]
3. Finally, after one-to-two weeks of metal accumulation, the crop is harvested and processed to recover the metal.

The aim of this system is to remediate mercury-polluted land but there is a crucial advantage, the value of gold in the harvested crop. Commentary from the United Nations Development Programme Sustainable Livelihoods Project shows that artisanal and small scale miners will show little interest in environmental initiatives if there is no quantifiable and immediate payback. An environmental management system based on hyperaccumulation addresses this critical issue. The gold value of the crop may provide a cash incentive to artisanal farmers who clean up their land.

Where implemented, the system would be an agricultural strategy at ASGM locations to manage the environmental burden of contaminants in mine waste. Phytoextraction could generate revenue from what is currently a waste product.

The provision of training for workers with modern agricultural techniques during the operation would lead to a newly educated workforce within ASGM communities that could protect the environment and generate produce for community consumption and external trade from land more suitable for agricultural production. The mechanism by which this change could be created is the extraction of gold and mercury from ASGM waste using plants. The gold value of the crop should provide a cash incentive to artisanal farmers who clean up their land.

Commentators on artisanal and small scale gold mining recognise that the sector can clearly benefit communities while resources are rich, but similarly recognise that an alternative livelihood is needed to sustain the environment during these times and to generate alternative income when resources are poor or depleted. If we apply the trait of hyperaccumulation to this scenario, then a potential livelihood for breaking the poverty cycle is agriculture manifest as the farming of metals. Education and training paid for by gold revenues could empower communities to farm their land efficiently. Farming might then be seen as a more attractive alternative livelihood for migrant workers. Gold could be a catalyst to bring about sustainable agriculture.

The vision of artisanal and small scale miners farming metals is not limited to gold. As early as 1988, Robert Brooks described the scenario of peasant farmers on ultramafic soils in Brazil shifting their crop production from poor yielding soya bean to nickel hyperaccumulators and then selling the biomass to central biomass processing plants.[67] More recently this concept has been driven by researchers at the University of Melbourne for ultramafic areas of the Philippines (Dr. Augustine Doronilla, personal communication) and by researchers at the Centre for Mined Land Rehabilitation at the University of Queensland on ultramafic soil and areas of nickel mine waste on the Island of Sulawesi in Indonesia. In all three of these scenarios, the implementation of phytoextraction would be a community driven livelihood that would create agricultural opportunities from poor yielding or contaminated waste land.

Here is a working scenario:

'Local cooperatives' train artisanal farmers with the agronomic skills necessary to farm metals. Cooperatives then employ and subsidise farmers to carry out the metal recovery operation and purchase the metal-rich biomass after harvest. Environmentally sound processing would recover toxic contaminants present in the biomass for disposal or recycling. Recovery and sale of valuable metal would make the operation economically viable. A newly empowered farming community would eventually utilise the clean soil and their new skills for agricultural production.

- Contaminant removal and productive land use provides environmental sustainability...
- Valuable metal recovery and agricultural development provides economic sustainability...
- Employment and education for the artisanal community provides social sustainability

5.10   Conclusion

Society has an insatiable need for metals for both infrastructure and technology. The source of these metals is the Earth and the concept of mining metals has not changed for thousands of years. The challenge for society in the 21st century is to manage the mining cycle sustainably. This can be achieved by better environmental protection, by the exploitation of non-conventional sources of metal and by non-conventional systems for their recovery.

Phytoextraction is one technology that can assist in meeting the sustainability challenge. Plants can be used to recover metals from waste rock, contaminated land and from low-grade ore and this trait may be usefully applied as a remediation or alternative mining system. The economically important metals nickel and gold can both be reliably concentrated by plants. This concentration occurs through the physiological ability of some species to hyperaccumulate nickel naturally or by manipulation of the soil environment to induce the hyperaccumulation of gold. The potential revenue that can be generated by a crop of nickel or gold-rich plants may pay for the clean-up of toxic metals that are also in the soil or may provide an alternative livelihood for communities struggling to grow food crops on metalliferous soils.

Phytoextraction can be regarded as a non-conventional form of mining that can exploit non-conventional resources. But phytoextraction is also a technology that is associated with both perceived and actual environmental benefits. In combination with green technologies for biomass processing such as supercritical CO_2, and the identification of novel uses for plant-synthesised metallic structures, metal hyperaccumulation by plants is a physiological trait that offers real possibilities to effect sustainable development in the 21st century.

References

1. R. R. Brooks, C. E. Dunn and G. E. M. Hall (eds), *Biological Systems in Mineral Exploration and Processing*, Ellis Horwood, Hemel Hempstead, 1995.
2. C. Minguzzi and O. Vergano, *Atti della Socienta Toscana di Scienze Natruale*, 1948, **55**, 49.
3. R. R. Brooks, J. Lee, R. D. Reeves and T. Jaffre, *J. Geochem. Explor.*, 1977, **7**, 49.
4. R. Brooks, *Plants that Hyperaccumulate Heavy Metals. Their Role in Phytoremediation, Microbiology, Archaeology, Mineral Exploration and Phytomining*, CAB International, Wallingford, UK, 1998.
5. T. Jaffre, R. R. Brooks, J. Lee and R. D. Reeves, *Science*, 1976, **193**, 579.
6. R. L. Chaney, in *Land Treatment of Hazardous Wastes*, ed. J. F. Parr, P. B. Marsh and J. M. Kla, Noyes Data Corp., Park Ridge, NJ, 1983, p. 50.
7. A. J. M. Baker and R. R. Brooks, *Biorecovery*, 1989, **1**, 81.
8. A. Van der Ent, A. J. M. Baker, R. D. Reeves, A. J. Pollard and H. Schat, *Plant Soil*, 2013, **362**, 319.

9. A. J. M. Baker, S. P. McGrath, C. M. D. Sidoli and R. D. Reeves, *Resour., Conserv. Recycl.*, 1994, **11**, 41.
10. R. D. Reeves and A. J. M. Baker, in *Phytoremediation of Toxic Metals: Using Plants to Clean up the Environment*, ed. I. Raskin and B. D. Ensley, Wiley, New York, 2000, p. 193.
11. M.-P. Faucon, M. N. Shutcha and P. Meets, *Plant Soil*, 2007, **301**, 29.
12. R. S. Boyd and S. N. Martens, in *The Vegetation of Ultramafic (Serpentine) Soils*, ed. A. J. M. Baker, J. Proctor and R. D. Reeves, Intercept, Andover, 1992, p. 279.
13. R. S. Boyd, *Plant Soil*, 2007, **293**, 153.
14. H. Fones, C. A. R. Davis, A. Rico, F. Fang, J. A. C. Smith and G. M. Preston, *PLoS Pathog.*, 2010, **6**, e1001093.
15. N. Rascio and F. Navari-Izzo, *Plant Science*, 2011, **180**, 169.
16. C. W. N. Anderson, R. R. Brooks, R. B. Stewart and R. Simcock, *Nature*, 1998, **395**, 553.
17. J. W. Huang and S. D. Cunnigham, *New Phytol.*, 1996, **134**, 75.
18. B. Nowack, R. Schulin and B. Robinson, *Environ. Sci. Technol.*, 2006, **40**, 5225.
19. F. N. Moreno, C. W. N. Anderson, R. B. Stewart, B. H. Robinson, R. Nomura, M. Ghomshei and J. Meech, *Plant Soil*, 2005, **275**, 233.
20. J. Wang, X. Feng, C. W. N. Anderson, Y. Xing and L. Shang, *J. Hazard. Mater.*, 2012, **221–222**, 1.
21. V. Wilson-Corral, C. W. N. Anderson and M. Rodriguez-Lopez, *J. Environ. Manage.*, 2012, **111**, 249.
22. D. C. Walton, *The Phytoextraction of Gold and Palladium from Mine Tailings*, Master of Philosophy Thesis, Massey University, Palmerston North, New Zealand, 2002, 93.
23. Y. Lai, Q. Wang, L. Yang and B. Huang, *Talanta*, 2006, **70**, 109.
24. O. B. C. Tzvetkova, L. Borisova and B. Bryskin, *Adv. Mater. Process.*, 2012, **May**, 34.
25. C. Anderson, F. Moreno and J. Meech, *Miner. Eng.*, 2005, **18**, 385.
26. B. H. Robinson, G. S. Bañuelos, H. M. Conesa, M. W. H. Evangelou and R. Schulin, *Crit. Rev. Plant Sci.*, 2009, **28**, 240.
27. B. H. Robinson, R. R. Brooks, A. W. Howes, J. H. Kirkman and P. E. H. Gregg, *J. Geochem. Explor.*, 1997, **60**, 115.
28. K. A. Hudson-Edwards, H. E. Jamieson and B. G. Lottermoser, *Elements*, 2011, **7**, 375.
29. R. B. Gordon, M. Betram and T. E. Graedel, *Proc. Natl. Acad. Sci. U. S. A.*, 2006, **103**, 1209.
30. C. W. N. Anderson, R. R. Brooks, A. Chiarucci, C. J. LaCoste, M. Leblanc, B. H. Robinson, R. Simcock and R. B. Stewart, *J. Geochem. Explor.*, 1999, **67**, 407.
31. L. J. Nicks and M. F. Chambers, *Discover Mag.*, 1994, **Sept**, 19.
32. L. J. Nicks and M. F. Chambers, *Min. Environ. Manage.*, 1995, **Sept**, 15.
33. B. H. Robinson, A. Chiarucci, R. R. Brooks, D. Petit, J. H. Kirkman, P. E. H. Gregg and V. de Dominicis, *J. Geochem. Explor.*, 1997, **59**, 75.

34. Y.-M. Li, R. L. Chaney, E. Brewer, R. J. Roseberg, J. S. Angle, A. J. M. Baker, R. D. Reeves and J. Nelkin, *Plant Soil*, 2003, **249**, 107.

35. R. L. Chaney, J. S. Angle, C. L. Broadhurst, C. A. Peters, R. V. Tappero and D. L. Sparks, *J. Environ. Qual.*, 2007, **36**, 1429.

36. A. T. Harris, K. Naidoo, J. Nikes, T. Walker and F. Orton, *J. Cleaner Prod.*, 2009, **17**, 194.

37. A. W. Howes, K. A. Slatter, E. A. Sim and A. N. Jones, in *Waste Processing and Recycling in Mineral and Metallurgical Industries III*, ed. S. R. Rao, L. M. Amaratunga, G. G. Richards and P.D.Kondos, The Metallurgical Society of CIM, 1998.

38. R. R. Brooks, *Noble Metals and Biological Systems, their Role in Medicine, Mineral Exploration, and the Environment*, CRC Press, Boca Raton, 1992.

39. C. W. N. Anderson in *Trace and Ultratrace Elements in Plants and Soil*, ed. I. Shtangeeva, WIT Press, Southampton, 2005, p. 287.

40. G. Southam, M. F. Lengke, L. Fairbrother and F. Reith, *Elements*, 2009, **5**, 303.

41. H. V. Warren and R. E. Delavault, *Geol. Soc. Am. Bull.*, 1950, **61**, 123.

42. C. E. Dunn, in *Biogeochemical Prospecting for Metals. Biological Systems in Mineral Exploration and Processing*, ed. R. R. Brooks, C. E. Dunn and G. E. M. Hall, Ellis Horwood, Hemel Hempstead, 1995, p. 371.

43. S. D. Ebbs, S. D. Kolev, R. C. R. Piccinin, I. E. Woodrow and A. J. M. Baker, *Miner. Eng.*, 2010, **23**, 819.

44. R. J. Bowell, J. V. Parshley, G. McClelland, B. Upton and G. Zhan, *Miner. Eng.*, 2009, **22**, 477.

45. B. H. Robinson, J. E. Fernández, P. Madejón, T. Maranón, J. M. Murillo, S. R. Green and B. E. Clothier, *Plant Soil*, 2003, **249**, 117.

46. W. D. Gould, M. King, B. R. Mohapatra, R. A. Cameron, A. Kapoor and D. W. Koren, *Miner. Eng.*, 2012, **34**, 38.

47. S. D. Ebbs, S. D. Kolev, R. C. R. Piccinin, I. E. Woodrow and A. J. M. Baker, *Miner. Eng.*, 2011, **24**, 164.

48. J. L. Gardea-Torresdey, J. G. Parsons, E. Gomez, J. Peralta-Videa, H. E. Troiani, P. Santiago and M. Jose Yacaman, *Nano Lett.*, 2002, **2**, 397.

49. R. Bali and A. T. Harris, *Ind. Eng. Chem. Res.*, 2010, **49**, 12762.

50. S. Irvani, *Green Chem.*, 2011, **13**, 2638.

51. M. B. Cortie, *Gold Bull.*, 2004, **37**, 12.

52. C. Cretu and E. van der Lingen, *Gold Bull.*, 1999, **32**, 115.

53. R. G. McLaren and K. C. Cameron, *Soil Science. Sustainable Production and Environmental Protection*, Oxford University Press, 1996.

54. A. Lamb, *Methods for the Recovery of Gold from Plant Ash*, Thesis submitted in partial fulfilment of the requirements for the degree of Master of Technology in Chemical Technology at Massey University, Palmerston North, New Zealand, 2002, 118.

55. A. H. P. Kirk, *The Recovery of Nickel from Hyperaccumulator Plant Ash*, Thesis submitted in partial fulfilment of the requirements for the degree of Master of Science in Chemistry at Massey University, Palmerston North, New Zealand, 2002, 154.

56. R. Boominathan, N. M. Sah-Chaudhury, V. Sahajwalla and P. M. Doran, *Biotechnol. Bioeng.*, 2004, **86**, 243.
57. J. A. Darr and M. Poliakoff, *Chem. Rev.*, 1999, **99**, 495.
58. S. Wang and C. Wai, *Environ. Sci. Technol.*, 1996, **30**, 3111.
59. S. Wang, S. Elshani and C. M. Wai, *Anal. Chem.*, 1995, **67**, 919.
60. Y. Lin, C. M. Wai, F. M. Jean and R. D. Brauer, *Environ. Sci. Technol.*, 1994, **28**, 1190.
61. C. W. N. Anderson, R. B. Stewart, F. N. Moreno, C. T. J. Wreesmann, J. L. Gardea-Torresdey, B. H. Robinson and J. A. Meech, *Proceedings of the Gold 2003 Conference: New Industrial Applications of Gold*, The World Gold Council and The Canadian Institute of Mining, Metallurgy and Petroleum, Vancouver, Canada, 28th September – 1st October 2003, CD ROM, 2003.
62. B. H. Robinson, R. R. Brooks, P. E. H. Gregg and J. H. Kirkman, *Geoderma*, 1999, **87**, 293.
63. M. M. Veiga and J. J. Hinton, *Nat. Resour. Forum.*, 2002, **26**, 13.
64. M. M. Veiga, D. Nunes, B. Klien, J. A. Shandro, P. C. Velasques and R. N. Sousa, *J. Cleaner. Prod.*, 2009, **17**, 1373.
65. B. D. Krisnayanti, C. W. N. Anderson, W. H. Utomo, X. Feng, E. Handayanto, N. Mudarisna, H. Ikram and Khususiah, *J. Environ. Monit.*, 2012, **14**, 2598.
66. J. Wang, X. Feng, C. W. N. Anderson, G. Qiu, L. Ping and Z. Bao, *J. Hazard. Mater.*, 2011, **186**, 119.
67. R. R. Brooks and B. H. Robinson, in *Plants that Hyperaccumulate Heavy Metals. Their Role in Phytoremediation, Microbiology, Archaeology, Mineral Exploration and Phytomining*, ed. R. R. Brooks, CAB International, Wallingford, UK, 1998, p. 327.

CHAPTER 6

F-block Elements Recovery

LOUISE S. NATRAJAN* AND
MADELEINE H. LANGFORD PADEN

School of Chemistry, The University of Manchester, Oxford Road,
Manchester, M13 9PL
*Email: louise.natrajan@manchester.ac.uk

6.1 Introduction

The f-block elements comprise the 4f series (the lanthanides, cerium to lutetium, Ce-Lu) and the 5f series (the actinides, thorium to lawrencium, Th–Lr) (Figure 6.1).[1-4] The majority of these elements, particularly the lanthanides, that occur naturally are indispensable commodities in today's technologically advanced society with invaluable applications ranging from energy efficient lighting, high colour quality flat-screen televisions, permanent magnets, components in hybrid cars (lanthanides) to nuclear power production (uranium and thorium). The majority of the applications of f-block metals fall under the umbrella of green and low carbon technologies and their use is likely to increase dramatically in the near future. However, all the f-block elements are finite resources and fears over exhaustion of raw materials, security of supply and the supply chain issues are now of immense worldwide concern. Focus has recently turned to recovery and recycling strategies in order to address these issues and to circumvent the pressing matter of supply vulnerability by closing the mining and manufacturing processes.[5]

6.2 The Lanthanide Series

The lanthanide (Ln) series comprises the 14 chemical elements with atomic numbers ranging from 58 (cerium) to 71 (lutetium). Together with the

RSC Green Chemistry No. 22
Element Recovery and Sustainability
Edited by Andrew J. Hunt
© The Royal Society of Chemistry 2013
Published by the Royal Society of Chemistry, www.rsc.org

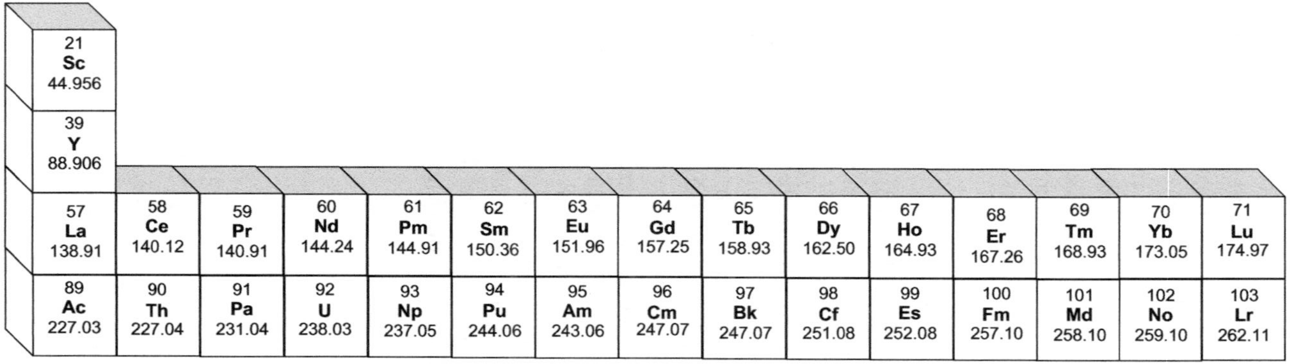

Figure 6.1 The f-block series including Sc, Y, La and Ac.

chemically analogous elements scandium, yttrium and lanthanum, these are often referred to as the 'rare earths'.[1–4] The term 'lanthanoid' generally refers to the lanthanide group and lanthanum and does not tend to include scandium and yttrium since they are formally transition metal elements. The lanthanide elements can be found between barium and hafnium in the periodic table, in the first of the two rows containing the f-block elements (Figure 6.1). All the lanthanide elements are metallic and possess 4f electrons in their valence subshells. The lanthanoid elements all have relatively modest natural abundances, despite often being termed the 'rare earth' elements, except for promethium (Pm), which is radioactive and does not significantly occur in nature. All of the 4f elements, except promethium, were discovered and had successfully been isolated by the early 20th century. Promethium was not discovered until 1947.[6,7]

6.2.1 Discovery and Mining

In 1794, Johann Gadolin, a Finnish chemist, took the first step to discovering the rare earth metals when he extracted a mixed oxide he called 'yttria' from the black mineral ytterbite. Shortly afterwards, another mixed oxide, 'ceria' was extracted from cerite by German chemist Martin Heinrich Klaproth and Swedish chemists Jöns Jakob Berzelius and Wilhelm Hisinger. In 1839, work began to separate these oxides into their elemental components by Swedish chemist Carl Gustaf Mosander (Figure 6.2).[8]

Several of the rare earth elements are naturally found in minerals with abundances comparable to those of other economically important elements such as arsenic, bromine, tin and tungsten (between 1 and 10 µg per gram of mineral) and account for approximately 8–12% of the elemental composition of the Earth's crust.[8] Cerium has an estimated natural abundance of 30 µg per gram in the Earth's crust and is the most abundant of all the lanthanides with the others being found at concentrations between 10 and 100 µg per gram depending on the mineral and location (Figure 6.3).[8] By contrast, promethium

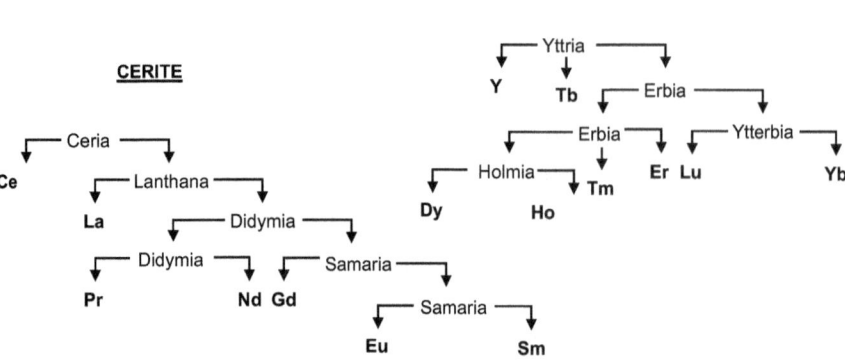

Figure 6.2   Lanthanide separation from the minerals cerite and ytterbite using fractional crystallisation techniques pioneered by Charles James.

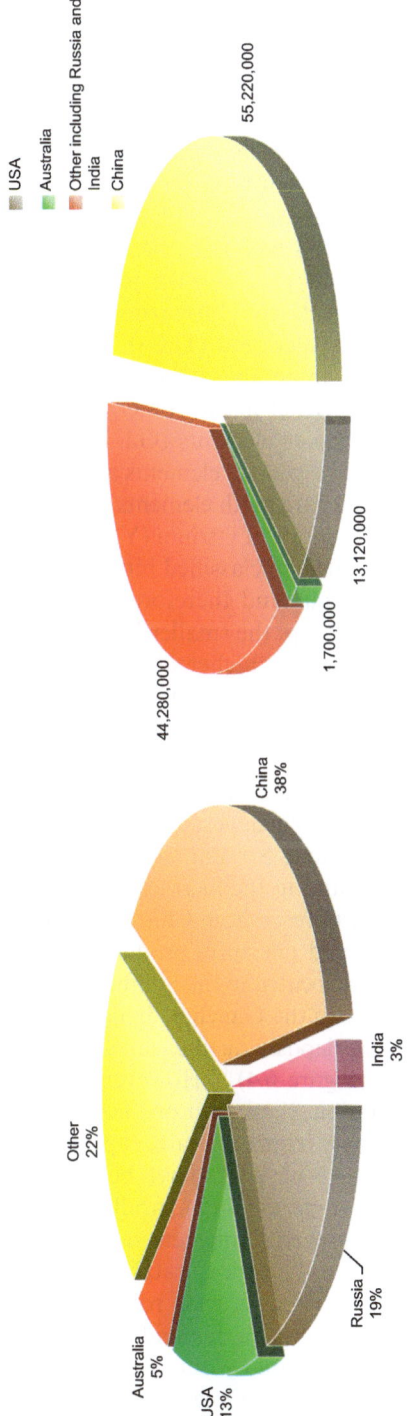

Figure 6.3 Relative percentages of rare earth mineral reserves in the world; together Brazil and Malaysia possess approximately 1% of the total global reserves (left) and amounts rare earth reserves in tonnes in key locations (right).

is only found in trace amounts at best in the Earth's crust as a radioactive fission product of ^{238}U.[9]

Rare earth elements are only found as trace elements in the majority of rock types including sedimentary, igneous and metamorphic rock types. Within these rock types the rare earths are concentrated globally in the rock forming minerals titanite, apatite, zircon, epidote, garnet as well as in clays.[8] Although, there have been over 150 independent types of rare earth containing minerals identified, the naturally occurring lanthanide elements are principally found in their trivalent state in the minerals monazite, bastnäsite and xenotime, which are ores of mixed lanthanide metals and have the general formulae $(Ln,Th)PO_4$, $LnFCO_3$ and $(Y,Ln)PO_4$, respectively. Monazite also contains up to 10% radioactive thorium as the tetravalent ion (ThO_2), so is the less favourable of the two sources of lanthanide minerals to mine since a radiation licence is required. Monazite is generally richer in the lighter elements of the lanthanoid series (La–Eu), whereas the heavier elements (Gd–Lu) are more abundant in xenotime. For convenience, the rare earth elements are classified as two groups, known as the 'Ce group' or light rare earth elements (La–Eu, LREE) and the 'Y group' or heavy rare earth elements and yttrium (Y, and Gd–Lu, HREE). Rare earth containing minerals can also be classified into three groups pertaining to their relative trivalent ionic radius and therefore their relative abundance in crystalline lattices. In the first group, the smaller Y-group ions predominate, in the second, the larger Ce-group has higher abundance and in the third are minerals that possess vacancies that can accommodate both larger and smaller rare earth ions, for example gadolinite-(Y) ($[Y_2FeBe_2Si_2O_{10}]$) and gadolinite-(Ce) ($[Ce_2FeBe_2Si_2O_{10}]$).[10,11]

Despite the fact that the majority of the rare earths were discovered in Scandinavia, the principal monazite deposits are found in Australia, Brazil, India, Malaysia, South Africa and Sri Lanka, whereas the main bastnäsite resources are located in China and in the Sierra Nevada mountain range in the USA. Interestingly, the rare earth reserves in China account for over 70% of the accessible known world total and currently produce 97% of commercial rare earth sources.[8] The majority are located in the Southern Provinces in the form of ionic ores. The Xunwu reserves in the Ganzhou Jiangxi Province are rich in Ce-group deposits, while Longan deposits in the same province are richer in the heavier lanthanide elements (Figure 6.4).[12] However, as a result of decades of excessive mining and exploitation, these environs now suffer from severe environmental damage. Tighter safety regulations are, at present, being implemented.

In recent years, the rare earth elements have become vital components in many areas of modern day life and fulfil an unparalleled role in many advanced 'green technologies' including powerful magnets in wind turbines and in hybrid and electric cars, rechargeable batteries, catalytic converters and energy efficient fluorescent lighting.[13] Their crucial role in these technologies has resulted in an unprecedented increase in demand and price of rare earth oxide resources that in the future could rapidly outweigh supply.

In a number of countries (Canada, Greenland, Malawi, South Africa and Vietnam) there are known reserves but they are not currently mined on an

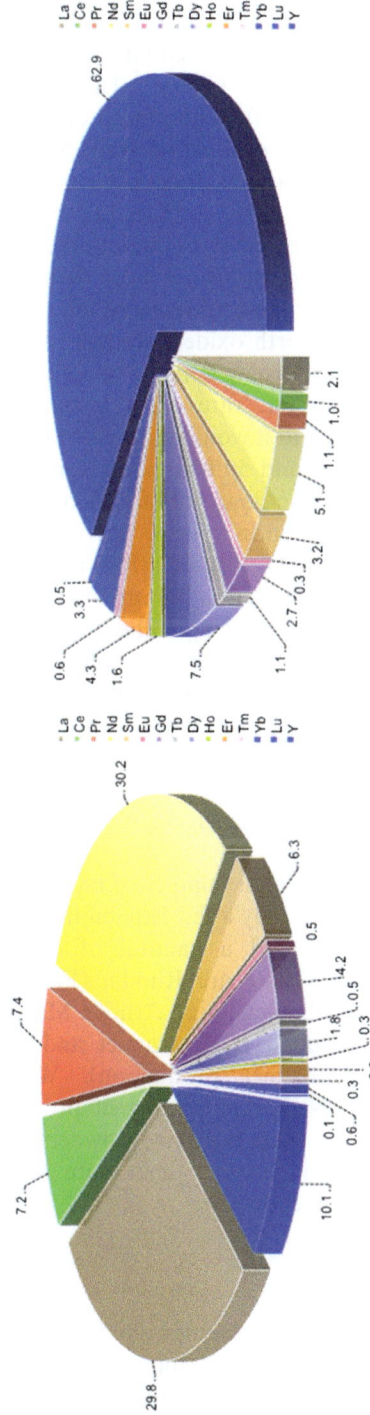

Figure 6.4 Percentages of rare earth adsorption deposits in Jiangxi Province, China, as oxides; composition of the Xunwu reserves (left) and composition of the Longan deposits (right).

appreciable scale. Reserves in ocean floor clay minerals near Japan,[14] in addition to deposits such as the Kvanefjeld deposits in Greenland and the Steenkampskraal mine in South Africa could provide new commercial resources of rare earth elements outside China.[15] The Mountain Pass mine in California, USA (owned by Molycorp)[16] announced operation resumption on a start up basis in 2012 and new mine production at the Mountain Weld deposit in Australia (The Lynas Corporation) has direct potential for the development of light rare earth oxides.[17] In view of the long lead times needed from mining to refined elements, especially if the resources contain the radioactive elements uranium or thorium, constraints on supply are more than likely to be imposed in the short term. The timescale for these mines to be fully operational is 5 to 10 years and in the meantime, rare earth oxide resources may face an uncertain future with sustainability of supply being a major concern.[18,19] Essentially, the current economic uncertainty about rare earth resources stems from the announcement by China in 2010 that it would reduce its rare earth export quota by approximately 60% (from around 50 000 metric tonnes in 2009 to 31 500 metric tonnes in 2012) and increase its export tariffs owing to an increase in domestic demands. This has resulted in volatile raw material prices, with the critical elements Nd, Y, Eu, Tb and Dy particularly at risk.[20,21] As a technology driven society, China not only specialises in mining the rare earth elements but also possesses a highly skilled workforce specialised in the extraction of rare earth elements from ores and in the separation into high purity individual elements necessary for some bespoke applications. Plans to reduce mine output, stockpile reserves by restricting rare earth element exports further, arrest illegal mining activities and tighten regulations regarding the radioactive minerals that have leached from mining exploits into the vicinity may exacerbate the problem of access to a reliable rare earth supply for countries other than China.[22–25]

Outside China there are only a small number of industries involved in rare earth refining and processing, with only a few capabilities in Japan for refining the intermediate alloy products and a handful of companies in Europe producing permanent magnets. These are Rhodia (France; phosphors and automotive catalysts), Magnet Applications, Arnold Magnetic Technologies and Less Common Metals Limited (UK; magnet and alloys production), Vacuumschmeize (Germany; magnet production), Walker Europe and Goudsmit Magnetic Systems (Netherlands; magnet production), Treibacher Industrie AG (Austria; catalysts, glass polishing powders, ceramics, pigments and pharmaceutical products) and finally Silmet Rare Metals (Estonia; rare earth metal separation and production).[5]

6.2.2   Rare Earth End Uses

The rare earth elements are a strategic resource and form essential components in a wide variety of existing and emerging applications.[26] Interestingly, some experts have labelled them 'the fourth most important natural world resource' after water, steel and oil. Their use in high-tech green technologies is chiefly a

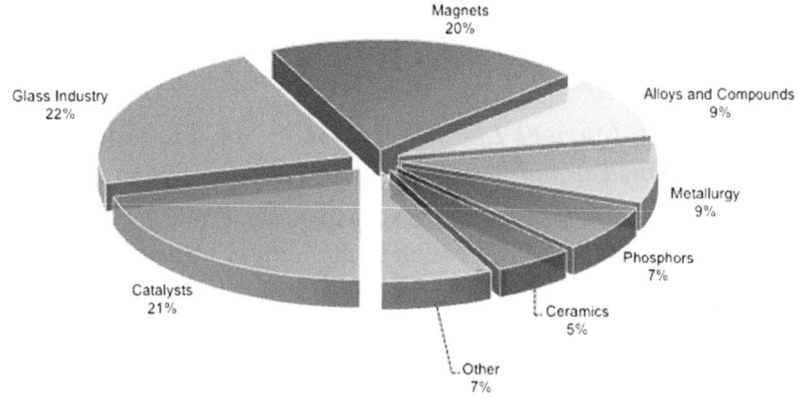

Figure 6.5 Current rare earth consumption per market area.

result of the intrinsic electronic structure of the ions and metals, which confers upon them very distinctive reactivities,[27] magnetic[28] and spectroscopic properties[29–31] that are unrivalled in other systems. In this regard, the rare earth elements are becoming more and more important for these critical applications and devices and their use cannot be substituted without compromising performance. The principal uses of rare earth elements in commercial products are listed below and shown in Figure 6.5 as a percentage share of current market area.

6.2.2.1 Clean Energy Technologies (Pr, Nd, Sm, Tb and Dy)

A substantial share of the rare earth market is devoted to green energy technologies which are designed to be more efficient in terms of energy consumption, are lightweight and address carbon abatement targets by reducing CO_2 emissions compared to more conventional technologies. In particular, these include the use of neodymium–iron–boron ($Nd_2Fe_{14}B$) permanent magnets in wind turbines and electric motors used in electric or hybrid vehicles.[32,33] These strong magnets have enabled the miniaturisation of electronic devices such as hard disks in computing, CD and DVD players and small speakers in headphones. The rare earth materials in these magnets comprise an approximately 30% mixture of neodymium and praseodymium, plus up to 3% dysprosium and terbium additives. The next generation of wind powered mills will employ rare earth permanent magnet generators in direct drive turbines and require a considerable quantity of rare earths by mass (for example 1 metric tonne for a 3 MW turbine); approximately 14% of newly installed wind turbines use Nd-magnets. These Nd-magnets have recently replaced Sm–Co magnets (*e.g.* $SmCo_5$,[34] Sm_2Co_{17}[35]) owing to their enhanced efficiency. Future demand for such magnets is dependent on the forthcoming production of electric and electric hybrid vehicles, electric plug-in vehicles and the number of Nd-magnets required per electric motor (power requirements).

6.2.2.2   Energy Efficient Lighting (La, Ce, Sm, Eu, Gd, Tb, Dy, Er, Tm, Lu and Y)

Perhaps one of the most common and widely used applications of rare earths is their use in energy efficient illumination in the form of phosphors in lighting and displays. Rare earths in compact fluorescent lamps (CFL) have rapidly replaced the incandescent bulb in a large number of households globally. These phosphors benefit from the intrinsic luminescent properties of a number of rare earth ions arising from their 4f electronic configuration, producing both high colour purity and energy efficient light sources. These include three-band phosphor tubes, light emitting diodes (LEDs), organic light emitting diodes (OLEDs), electroluminescent foils, liquid crystal displays (LCDs) and plasma displays; the latter two are rapidly replacing cathode ray tubes in television sets. The rare earth ions europium and terbium are the most widely used for their high quantum efficient red and green luminescence, respectively. Estimates suggest that this market is growing at a rate of 15–20% a year, meaning that additional supply of resources is paramount to support future growth.

6.2.2.3   Advanced Transport Technologies (La, Ce, Pr, Nd, Eu, Tb, Dy and Y)

Besides the use of Nd permanent magnets in electric and hybrid electric vehicles (electricity driven bikes, cars and buses), the rare earths find many uses in state of the art hybrid electric vehicles, for example, the Toyota Prius and Yaris, the Honda Insight and the Ford Focus Electric. The various uses of rare earth containing materials in these new generation cars are illustrated in Figure 6.6.[16]

Of particular note, nickel metal hydride rechargeable batteries (NiMH) used in these vehicles also contain a mixture of lanthanum (*ca.* 25%), cerium (*ca.* 50%), neodymium (*ca.* 15%) and praseodymium (*ca.* 10%) alloyed with iron, also known as 'mischmetal'. Mischmetal is a resource that is most useful in terms of the weight of materials required for many applications (including flints in cigarette lighters and deoxidisers in many alloys) but many others require ultra-high purity rare earth elements as their components, which render recovery and recycling more challenging. Many designer consumer electronics include NiMH rechargeable batteries in their components, such as mobile phones and smart devices including the iPhone and iPad. For example, the iPhone makes use of the elements Y, La, Pr, Nd, Eu, Gd, Tb and Dy in the colour display screen; La, Ce and Pr in glass polishing; La, Pr, Nd, Gd and Dy in the electronic circuitry; Pr, Nd, Gd, Tb and Dy in the speakers and Nd, Tb and Dy in the vibration unit. One interesting commercial product that has arisen from the alloy $Tb_{0.3}Dy_{0.7}Fe_{1.9}$ is Terfenol-D, which is a magnetorestrictive alloy used in the Soundbug speaker device.[36]

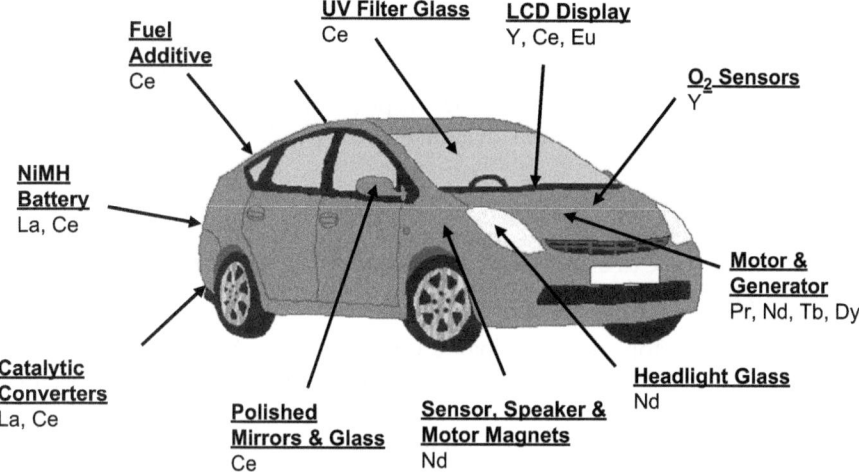

Figure 6.6 Uses of rare earth elements in electric-hybrid cars.

In the automotive industry, lanthanum and cerium are widely used as catalysts in exhaust catalytic converters and oxygen-rich cerium compounds are additionally employed as diesel additives providing more efficient, cleaner combustion. Other uses include fluid cracking catalysts in petroleum refinement and in chemical processing. Together, the demand for rare earths as catalysts constitutes approximately 20% of the total rare earth demand. Outside the automotive field, ceria (CeO_2) is utilised in glass polishing powders in the final stages of optical glass manufacturing. It is also used in prisms, liquid crystal display panels and optical lenses for digital cameras and camcorders. The rare earth components used in optical glasses include lanthanum, gadolinium and yttrium, which bestow the optics with high refractive and low dispersion indices.

6.2.2.4 *Advanced Communication Systems and Defence Technologies (Nd, Gd, Ho, Er and Tm)*

The optical attributes of the rare earth ions erbium and thulium render them extremely useful in amplifier and waveguide technologies in fibre optic communications and telecommunications. These are routinely used in travelling wavetubes and klystrons that use rare earth magnets in their waveguides to generate and amplify microwaves. Additionally, they are used in satellite communications, in pulsed and continuous radar amplifiers and in high capacity fibre optic cables, amongst others. Rare earth containing lasers such as the near infrared Nd-YAG laser (YAG = yttrium aluminium garnet, 1064 nm fundamental) and to a lesser extent the Ho-YAG laser find niche uses in line of sight satellite communications both in the inner and outer Earth orbit, as well as lasers in medicine and photonics, one example being surgical lasers.[37,38]

Regarding defence capabilities, these technologies play a major role in precision guided munitions, detection devices (*e.g.* for underwater mines), communications *via* optical fibres, radar and sonar systems, aircraft control mechanisms, optical equipment, defence display systems and coatings (*e.g.* Gd based coatings/paints have been demonstrated as a defence measure against neutron radiation).

6.2.2.5 Health Care and Medical Technologies (Eu, Gd, Tb, Dy, Tm and Yb)

In this market sector, the principal use of the rare earths is in magnetic resonance imaging (MRI), where molecular based gadolinium contrast agents are routinely used as media enhancing magnetic resonance agents in 30–40% of clinical scans, facilitating the detection and imaging of damaged and diseased tissue in the body (Scheme 6.1).[39–42] More recently, other rare earth containing compounds (Eu^{3+}, Tb^{3+} Dy^{3+}, Tm^{3+}, Yb^{3+}) are being researched for

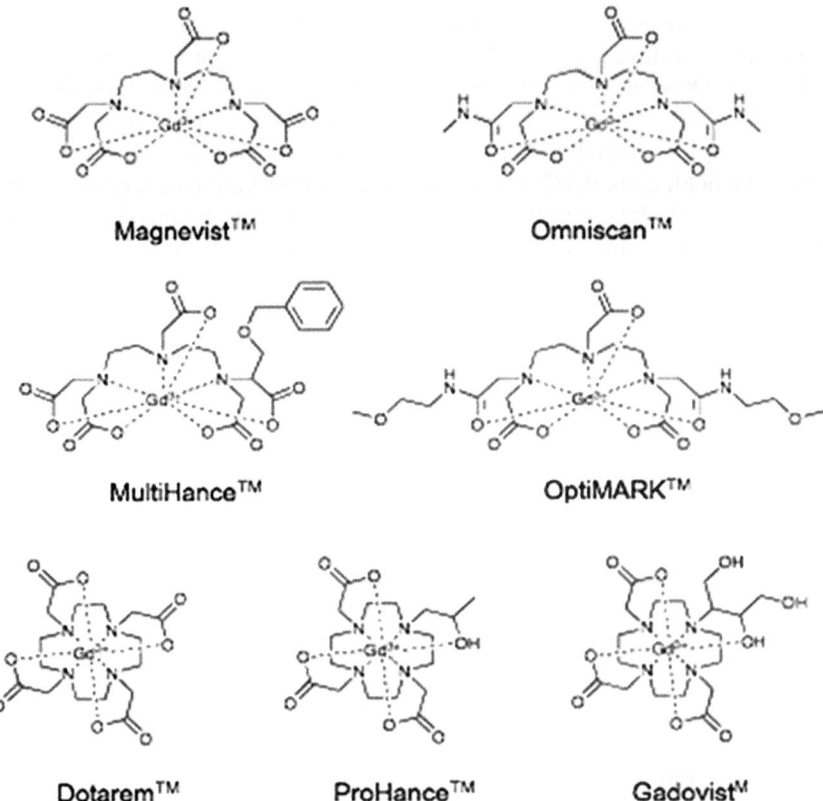

Scheme 6.1   Some commercially available gadolinium-based MRI contrast agents.

alternative imaging procedures as PARACEST (paramagnetic chemical exchange saturation transfer) contrast imaging agents,[43,44] as luminescent dyes for microscopic imaging[45,46] and for optical sensing *in vitro* and *in vivo* biological processes.[47] The radioactive isotopes of several lanthanides such as [86]Y, [134]La and ^{177}Lu are being explored as radiotherapeutic agents in positron emission tomography (PET) and single photon emission tomography (SPECT),[48,49] whereas [153]Gd has been used to good effect to determine the biodistribution of MRI contrast agents *in vivo*.[50–52]

6.2.2.6 Future Technologies

Emerging rare earth future applications include materials for clean water technologies, These consist of phosphorous, heavy metal organic toxin (As, Cr and Se), pathogen, bacteria and virus removal from ponds, basins and mining sites allowing the non-hazardous disposal of toxic elements from effluent streams (*e.g.* SorbX^{TM} manufactured by Molycorp).[16] Other applications may comprise magnetic refrigerant materials,[53,54] single molecule magnets,[55,56] dopants in solar cell photovoltaic technologies,[57,58] upconverting nanoparticles for biological imaging[59–61] and molecular compounds in synthesis and catalysis.[62,63] Regarding biological imaging, rare earth ion based luminescent probes will undoubtedly be in increasing demand since they can be used to develop a comprehensive understanding of cellular and enzyme functionality, structure *in vitro* and detection of genetic disorders caused by mutations on DNA strands amongst others.

6.2.3 Rare Earth Separation, Recovery and Recycling

Current global demand for rare earth oxides is approximately 105 000 tonnes per annum which is anticipated to increase to 160 000 tonnes by the year 2016, a total increase of 52%. This large increase in demand primarily reflects the increasing requirement for rare earth oxides in permanent magnets, NiMH batteries, catalysts and phosphor technologies (forecast growth of 71, 43, 25 and 50%, respectively).[15] Interestingly, if the current market continues to expand with no major recycling plants being operational, the most significant shortages of rare earth oxide supplies will include Eu (36%), Tb (22%), Dy (12%) and Y (25%). Conversely, the rare earth oxides of La, Ce and Sm are expected to be surplus by 41, 25 and 275%, respectively. It is clear that for the rare earth market to be sustainable, extensive recovery and recycling of these important commodities must be achieved. In this regard, there has been a revitalisation of research in this field with corporate enterprises (such as Molycorp,[16] Osram,[64] Solvay,[65] Honda[66]) embarking on recovery and recycling steps. Several recycling flow sheets have been developed and reported in the primary literature but not yet demonstrated on large plant scale.

Historically, the lanthanides have been separated from one another using fractional crystallisation techniques, chiefly the 'James method' which relies on the regular decrease in the ionic radii across the lanthanide series from La^{3+} to Lu^{3+} (Figure 6.1).[1-41] As the lanthanide series is traversed, their salts become slightly less ionic, (which leads to small changes in their chemical properties (*e.g.* $Ce(OH)_3$ is more basic than $Yb(OH)_3$ and $Lu(OH)_3$), the aqueous ions become more acidic and the enthalpy of hydration and complexation (with ligands) increases. Nevertheless, the similar ionic radius and persistence of the +III oxidation state renders chemical separation of the 4f elements rather challenging. In the early 20th century, Charles James developed novel fractional crystallisation techniques which enabled chemists to separate large amounts of ytterbia into ytterbium and lutetium using bromates and magnesium nitrates (Figure 6.2).[67-69] The advent of ion exchange chromatography in the 1940s enabled much larger quantities of the rare earths to be separated and refined using cation exchange resins such as Dowex.[70] The rare earth ions are eluted in accordance with their relative atomic weight and size, where the lighter members of the series are eluted before the heavier elements which possess a smaller hydrated ionic radius. This separation can be enhanced using extractants such as citric acid or citrates by using EDTA as the eluting agent (EDTA = ethylenediaminetetraacetic acid) or lactates, hydroxybutyrate and α-hydroxycarboxylates, facilitating the isolation of kilogram quantities of pure rare earth ions for use in commercial products.

6.2.3.1 Current and Future Separation Strategies and Technologies

To date, post-consumer large scale recycling of the rare earths from their most valuable applications (magnets, batteries, phosphors and catalysts) is virtually non-existent and in order to address mitigation of supply risks, implementation of recovery and recycling processes are of paramount importance. Metal recycling is commonly founded on common mining and metallurgical principles including pyrometallurgical and hydrometallurgical processes. These include physical separation methods (wet and dry), direct leaching, molten salt electrolysis, mechanochemical treatment and separation by solvent extraction methods. Rare earth scrap metal recovery and recycling processes are, in principle, complex and energy intensive and require the effective dismantling of high-tech devices and electronic components. Certain technologies that employ mischmetal as the rare earth working component can be recovered for re-use more easily than those that rely on high purity element oxides such as phosphors and as such, each technology may warrant execution of an individual reprocessing strategy. In contrast to mining processes, many of the rare earths can be refabricated directly into component materials obviating the need to produce rare earth oxides as an intermediate source material. Of all the applications of rare earths discussed, recycling of Nd-containing magnets has received the most attention. Modern processes concerning the recycling of rare earth metals from scrap materials have recently been reviewed in depth and are summarized below.[71]

6.2.3.2   Recycling Neodymium–Iron–Boron Magnets

A general flow diagram of the processes involved in the manufacture and reprocessing of Nd-magnets is outlined in Figure 6.7.[72,73] In the manufacturing procedure, up to 30% of unused starting alloy can be generated as a consequence of the cutting, grinding and polishing steps. Encouragingly, these residues are often currently recycled by the relevant manufacturing companies where up to 95% of solid material is recovered. The process involves roasting steps to oxidise the alloy residues fully, followed by several hydrometallurgical methods (acid treatment and solvent extraction) before conversion to the rare earth oxides and reduction to the metallic forms by either thermal reduction or molten salt electrolysis.

Direct recovery of neodymium from Nd–Fe–B magnet scraps and other Nd containing scraps by continuous extraction with molten magnesium and with silver, was investigated by the group of Takeda, resulting in the formation of Mg–Nd and Ag–Nd alloys.[74,75] The neodymium metal was subsequently obtained in 98% purity by removal of the magnesium by sublimation and, in the case of silver, neodymium was recovered in greater than 90% yield as the oxide (Nd_2O_3) following oxidation in air.

Fractional crystallisation of neodymium and iron has been demonstrated by Sato and co-workers by selective crystallisation of $Nd_2(SO_4)_3.8H_2O$ and $Nd_2(SO_4)_3.5H_2O$ from sulfuric acid–water mixtures in the temperature range 0–80 °C and at 80 °C, respectively.[76] Addition of ethanol sufficiently reduced the solubility of the sulfate salts, permitting recovery of hydrates of neodymium sulfate in 96.8% purity and in 97.1% yield with respect to the original neodymium–iron–boron magnet scrap sample. Sulfuric acid has additionally been employed in whole leaching of Nd-magnet waste by precipitation of neodymium ammonium or sodium sulfate.[77,78] In this process the iron is removed by precipitation as the sulfate mineral jarosite, $K[Fe_3(OH)_6(SO_4)_2]$. It has been suggested that the boron may be precipitated as zinc borate by the

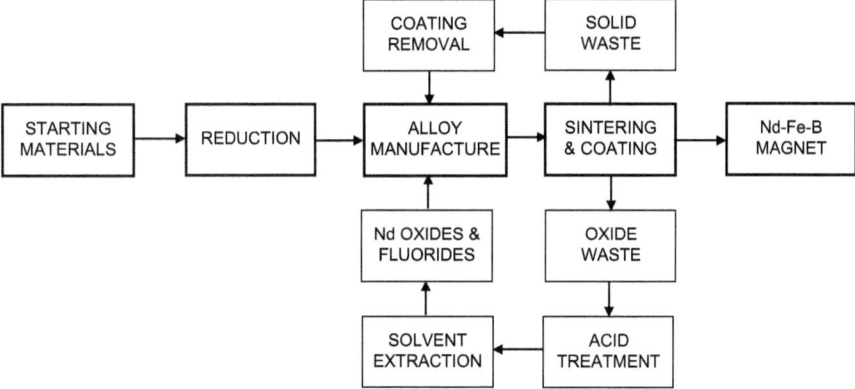

Figure 6.7   Neodymium–iron–boron magnet manufacture and recycling process flow sheet.

addition of zinc at higher pH values.[10,11] Selective leaching of $FeSO_4$ and $Nd_2(SO_4)_3$ has also been reported by a combination of air roasting, sulfuric acid leaching and sulfate precipitation.[79] Neodymium was selectively leached and precipitated owing to the concomitant formation of Fe_2O_3 at higher temperatures, which could subsequently be removed by precipitation as the hydroxide. The final neodymium to iron ratio in the sulfate precipitate was determined to be 18 : 1. In an alternative process, hydrometallurgic treatment of a sample of Nd–Fe–B sintered magnet with hydrochloric acid and oxalic acid enabled selective precipitation of the corresponding neodymium oxalate salt in greater than 99% purity and yield.[80]

Solvent extraction techniques are perhaps better known for the selective chelation and recovery of metal ions (Figure 6.8); the organophosphorous reagents HDEHP (bis(2-ethylhexyl)phosphoric acid) and PC-88A (bis(2-ethylhexyl)phosphinic acid mono-2-ethylhexyl ester) (Scheme 6.2) have been widely studied for the separation of many rare earth metals.[81,82] At pH 1.1, approximately 85% of Dy^{3+} was extracted preferentially over Nd^{3+} from a mixture of their respective chloride solutions using PC-88A, with a high separation factor of 525, demonstrating the utility of this technique.[10,11] The extractant DODGAA (*N,N*-dioctyldiglycol amic acid) (Scheme 6.2) was found

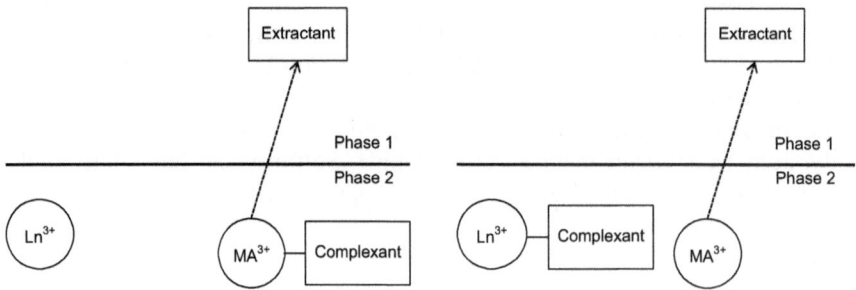

Figure 6.8   Illustration of common solvent extraction processes: use of a complexant to facilitate extraction (left) or use of a complexant to act as a holdback reagent (right).

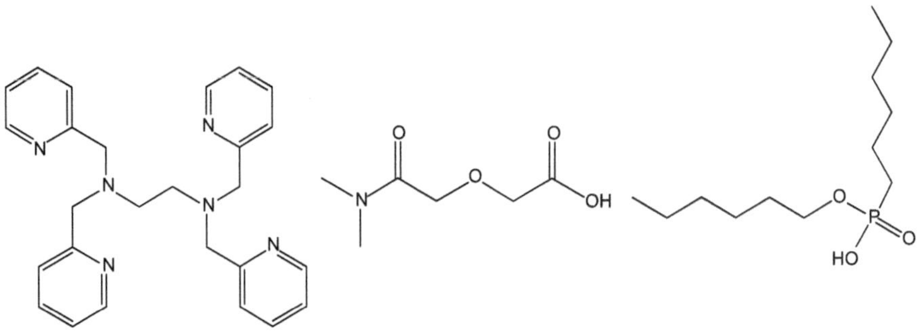

Scheme 6.2   Chemical structures of TPEN (left), DODGAA (centre) and PC-88A (right).

to exhibit a high selectivity between the early rare earths (and over iron and zinc);[83] with quantitative separation of the heavy lanthanides between pH 1 and 3 using 10 mM of DODGAA in 95 : 5 isooctane:1-octanol, a separation factor ($SF_{Lu:La}$) of 2990 was determined. Moreover, the commonly used metal chelator TPEN, (N,N,N',N'- tetrakis(2-pyridylmethyl)ethylenediamine) (Scheme 6.2)[84] selectively separated the lighter lanthanides from an HNO_3 solution into a nitrobenzene solution with a remarkable separation factor between Nd^{3+} and Lu^{3+} of greater than 70 000 using a 5 mM solution of TPEN in the aqueous phase.

Separation of neodymium and samarium by solvent extraction using the organophosphorus acid PC-88A in the systems PC-88A/TOPO (tri-*n*-octylp-hosphine oxide) and PC-88A/TOA (trioctylamine) was achieved, enabling a hydrometallurgical process for mixed rare earth magnet scraps (Nd–Fe–B and Sm–Co) to be proposed.[85] The process is based on oxidising, roasting, acid leaching and solvent extraction steps, with hydrochloric acid which exhibits enhanced leaching performance compared to sulfuric acid.

Further strategies that have been proposed for the selective recovery and recycling of the rare earths in magnets include pyrometallurgical methods such as thermochemical vapour transport using chlorine gas, solid–liquid reactions, molten salt extraction methods and electrochemical methods, which are variants of established procedures. They are described in more detail by Nakai and Kawashima[10] and Miyawaki and Nakai.[11]

6.2.3.3 *Recycling Spent NiMH Batteries*

Rare earth-containing nickel metal hydride batteries are not generally considered hazardous waste, unlike their predecessor cadmium-containing batteries but have been the subject of several studies with the aim of recovering all the metal components.

A molten salt process was proposed to recover La, Ce and Nd in LiCl–KCl eutectic melts using the hydrogen absorbing alloy $LaNi_5$ as the anode. Anodic dissolution of $LaNi_5$ into the molten salt was accompanied by electromigration of the rare earth cations to the anode.[86] The rare earth metals were then recovered by electrodeposition at the cathode using three anode and cathode couples. This process was also suggested to be applicable to the recovery of rare earths in automotive catalysts using room temperature molten salts or ionic liquids.

Hydrometallurgical methods for metal separation from spent NiMH batteries have been the focus of a number of studies, where for example, mixed metal separation from used cylindrical NiMH batteries has been developed.[87] Leaching the metals with 4 M hydrochloric acid at 95 °C over 3 h resulted in near quantitative dissolution of nickel, cobalt and rare earths (98%, 100% and 99%, respectively). Effective separation of the rare earths was subsequently achieved in a single batch extraction employing 25% HDEHP at pH 2.5, leaving behind nickel and cobalt in the acidic raffinate (Figure 6.9). In a separate study, approximately 80% of the rare earths were recovered as the

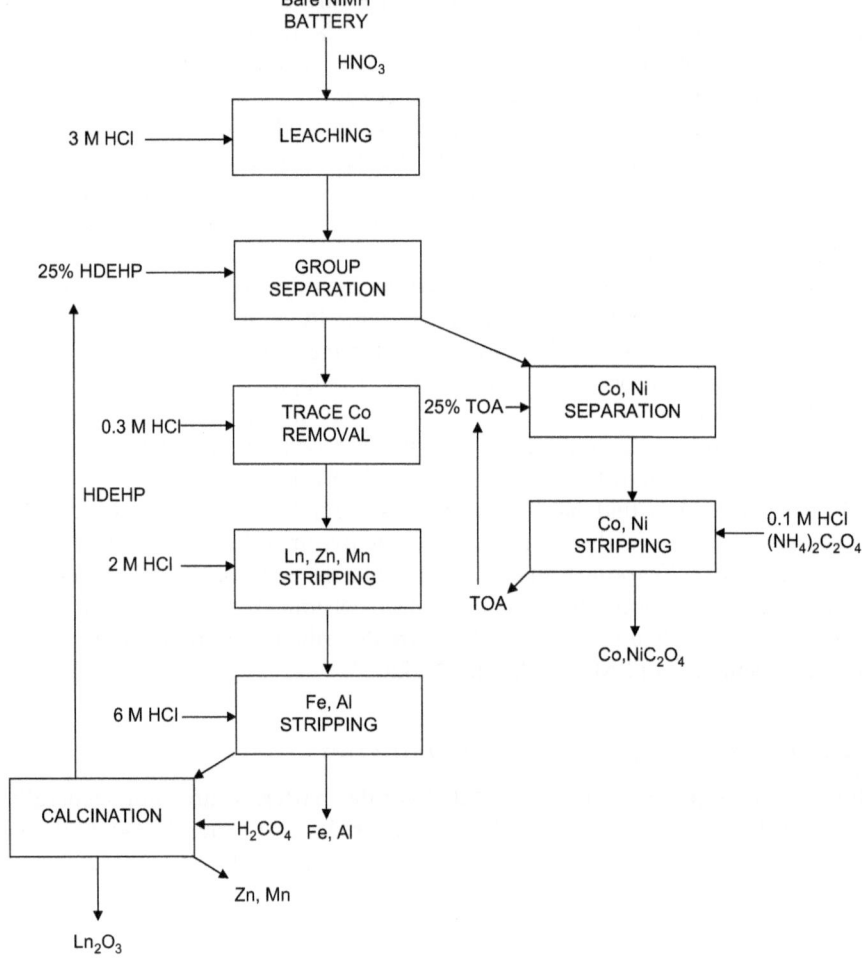

Figure 6.9 Conceptual flow diagram of NiMH spent battery recovery and recycling, redrawn and adapted from reference 71.

sulfate salts from spent NiMH batteries by selective precipitation using 2 M sulfuric acid at ambient temperature.[88] Similarly, Pietrelli *et al.*[89] reported that precipitation of mixed rare earth sulfates from NiMH battery waste could be achieved by addition of NaOH to achieve a pH of <1.5 after the acid leaching step, in order to prevent co-precipitation of ferric hydroxide which occurs at pH values between 2.5 and 3. More recently, a recovery and recycling flow sheet based on an analogous process was proposed by Innocenzi and Veglio.[90] Recovery of rare earths and base metals was achieved using NiMH powders containing 5% by weight lanthanum and cerium with particle sizes of less than 500 μm which were obtained from an industrial scale grinding process. It was found that two sequential leaching steps using sulfuric and citric acid were required to dissolve all of the component metals before selective precipitation of lanthanum and cerium sulfate, following adjustment of the solute pH to 2 with NaOH. In this manner, 99% total recovery of the rare earths was achieved.

6.2.3.4 Recycling Phosphor Waste

Separation of rare earth phosphors in fluorescent lamps first involves removal of the major constituent halophosphates prior to removal of the individual three band phosphors (blue, green and red). Several methods have been described that can achieve a modest separation of rare earths based on relative weights; these include wet separation processes (*e.g.* sink–float separation using heavy media separation with CH_2I_2) and dry separation processes (*e.g.* centrifugal separation).[6,10,11] However, highly precise separation methods are required for these phosphors. In this regard, liquid–liquid extraction techniques have proven more promising. For example, modification of the surface properties of the phosphor particles using surfactants at the liquid–liquid interface (aqueous *N,N*-dimethyl formamide and *n*-heptane) enabled *ca.* 95% recovery of the green phosphor in the first stage followed by *ca.* 92% of the blue phosphor and, in the final stage, *ca.* 91% of the red phosphor with >90% purity in each step.[91,92]

Direct leaching of phosphors from fluorescent lamp waste was investigated using sulfuric acid where, under optimised conditions, 92% yttrium and 98% europium were selectively leached as their oxides. The other rare earth components La, Ce and Tb remained insoluble as the phosphate salts, whereas minor impurity metals (Na, Ca, Mg, Al, Fe, Mn and Sb) were co-dissolved with Y and Eu; these were removed by precipitation of Y and Eu as the hydroxides and then selective precipitation from acidic media as the oxalate salts and subsequently separated in a solvent extraction process with PC-88A and sulfuric acid (Figure 6.10).[93] Other liquid–liquid extraction systems that have been investigated include carboxyl calixarenes, DODGAA in combination with ionic liquids[10,11] and tri-*n*-butyl phosphate (TBP) with supercritical CO_2 and aqueous nitric acid. In the latter, over 99% yttrium and europium were dissolved in the supercritical CO_2 after 120 min at a pressure of 15 MPa and an operating temperature of 60 °C.[94]

In the past few years, in order to address the demand–supply balance of rare earth phosphors, the company Osram has recently developed an industrially viable method for the recovery of rare earths from fluorescent lamps.[64] The steps involved in the process are:

1 mechanical separation of the coarse components
2 separation of the halophosphate phosphor
3 acid extraction of the more acid soluble phosphor components (largely Eu and Y oxides)
4 acid extraction of the less soluble rare earth phosphates
5 decomposition of the remaining rare earth containing components such as aluminates
6 final treatment processes such as calcination.

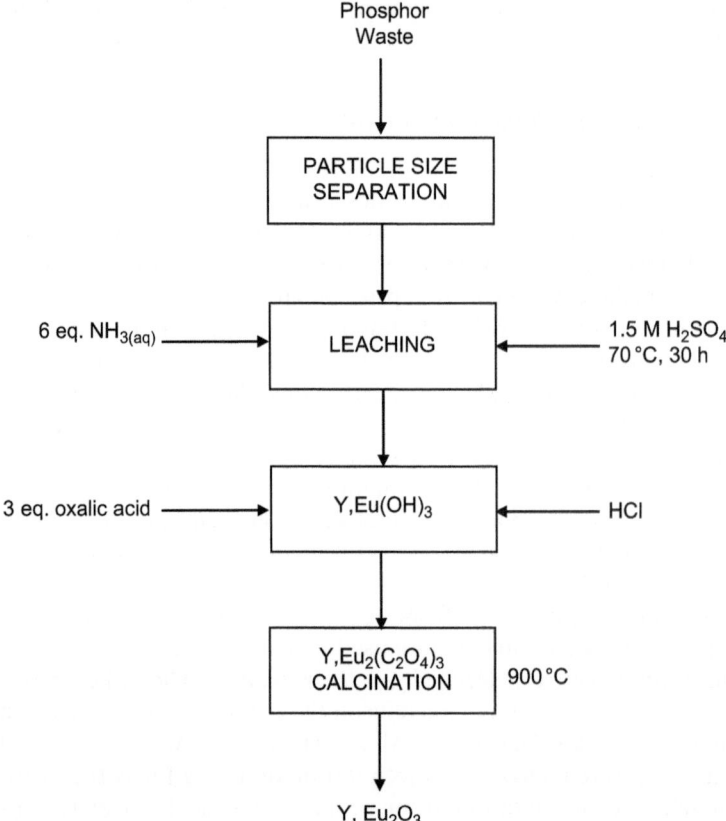

Figure 6.10 Conceptual flow diagram illustrating yttrium and europium recovery from commercial phosphors.

6.2.3.5 Recovery of Polishing Powders and Optical Glass

Despite the fact that ceria based polishing powders form a large proportion of the rare earth market and are generally considered to be one of the simplest rare earth containing materials to recycle, they are currently disposed of by land-filling. Nonetheless, there are several accounts in the literature that effectively address recycling strategies.

Kato *et al.*[95] investigated the possibility of recycling ceria-based glass polishing powder by alkali treatment with 4 M NaOH. It was found that the major silica and alumina components could easily be removed at 50–60 °C with concentrated NaOH by condensation into insoluble zeolite structures, leaving the ceria in the alkaline solution. The same group were also able to show that hydroxysodalite could be obtained from ceria powder waste with the following rare earth oxide composition: 17.8% La_2O_3, 22.1% CeO_2, 2.3% Pr_6O_{11} and 5.1% Nd_2O_3.[96] Shorter separation times for zeolite formation were achieved by heating the composition to 80–100 °C and by using a seed crystal to induce precipitation. A flow diagram for the recovery and recycling process was suggested.

A number of methods for the separation and recovery of rare earth metals from used optical glass have been demonstrated; Figure 6.11 shows a conceptual flow sheet that may be employed based on hydrometallurgical methods. One of these procedures involves initial conversion of the rare earth mixture (a sample of borosilicate glass containing by weight; 43.1% lanthanum, 9.4% yttrium and 4.6% gadolinium) into the corresponding oxides by treatment with NaOH followed by hydrochloric acid leaching of the residues.[97] After sequential treatment of the borosilicate phase with a 55% weight% solution of NaOH at 130 °C for 60 min and 6 M HCl at 95 °C for 30 min, an aqueous mixture of rare earth chlorides was obtained in a recovery yield of 99.4% (La), and 100% (Y and Gd).

In an additional study, the recovery of rare earths from optical glass waste was investigated by alkali treatment, oxalate salt precipitation and alkali leaching followed by solvent extraction. A reasonably high percentage of lanthanum was recovered (71%) in 83% purity by addition of sodium carbonate and potassium carbonate and then ammonium oxalate after the alkali treatment, whereas 99.95% La, 98.65% Y and 95.18% Gd recovery was achieved by alkali leaching followed by sodium hydroxide addition and solvent extraction with HDEHP.[98]

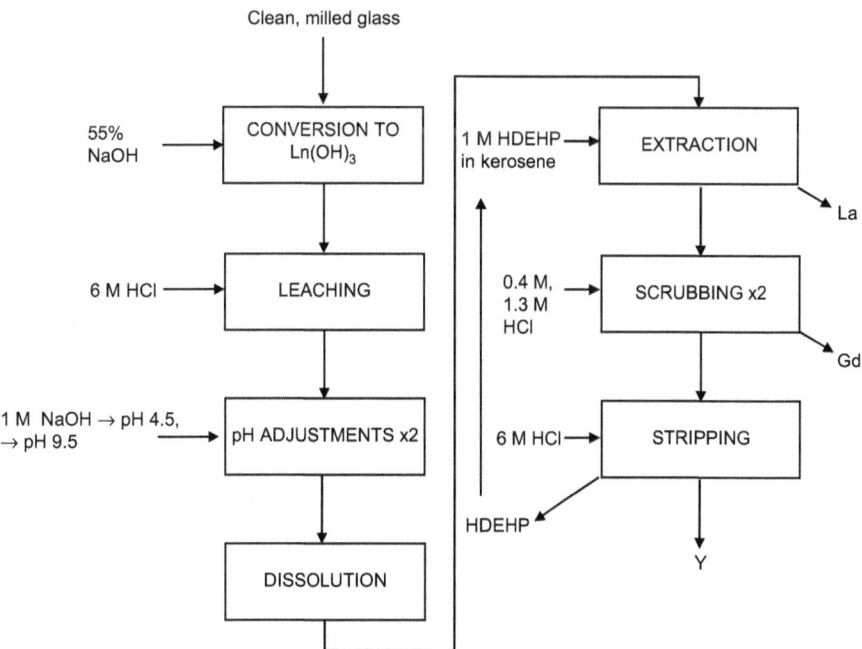

Figure 6.11 Conceptual flow diagram illustrating the recovery of rare earths from used optical glass.

Although it is evident that the landscape of urban mining of rare earth metal components in technological devices is emerging, rapid development of working industrial large scale processes and recycling plants is needed to sustain the demand–supply chain in order to stay abreast of the growing rare earth consumer market. In this regard, room temperature ionic liquids may show great promise for the mutual extraction of individual rare earths[99] along with the development of novel and more selective molecular extraction agents.[100]

6.3   The Actinide Series

All of the elements in the actinide series are radioactive and these unstable nuclei readily undergo radioactive decay into daughter nuclides (another nucleus type) by α-and β'-decay or decay to a lower energy state (γ-decay). As a general trend, the half-lives – time taken for half of the atoms of a radioactive species to undergo decay into another species – of the actinide elements decrease across the series (Table 6.1).[101]

Actinium, thorium, protactinium and uranium (Ac, Th, Pa and U) are the only naturally occurring actinide (An) elements in the 5f series. Uranium is less abundant than thorium on average (2.4 ppm *vs.* 8.1 ppm) in the Earth's crust but is found in concentrated forms in numerous minerals in oxide form, including pitchblende (uraninite) and carnotite. Protactinium (Pa) is one of the most rare elements in the world and is found at trace levels in some uranium ores as an indirect α-decay product of ^{235}U via ^{231}Th.[1–4] The remaining 11 elements in the actinide series (Np–Lr) must be artificially synthesised. The elements neptunium (Np) to fermium (Fm) can be generated by neutron bombardment where a neutron is captured by a heavy element atom and a γ-ray is emitted. This is followed by the emission of a β' particle in a β decay process to form a new element with an increased atomic mass. However, as this is a relatively improbable process, generation of the heavier members of the 5f series is impossible by this method, therefore they are generated by light atom bombardment.

In contrast to the rare earth elements, the actinide elements have a wider range of available oxidation states, particularly for the earlier metals. For the heavier elements, however, the most common oxidation state for the metal ions is +III, having lost both 7s electrons and either a 6d electron (if applicable) or one 5f electron. Ionisation energy values are not available for all of the actinides, although the standard electrode potentials for the reduction of An^{4+} to An^{3+} and An^{3+} to An^{2+} can be used to give an indication of the relative stabilities of the ions.

The ionic radii of the actinide ions in the +III, +IV, +V and +VI oxidation states gradually decrease in size across the series, although the metallic radii do not follow the same trend. The effect of the trend on the chemistry of the elements is not well known as the latter elements cannot be synthesised in large enough yields to study and their spontaneous decay is too rapid. Actinide ions are found as An^{3+} for the later elements in the series and they are often

Table 6.1 Half-lives and mode of decay for the most stable isotopes of each actinide. The half-life times are represented as follows: y = years, d = days, h = hours and m = minutes. EC is electron capture and SF is spontaneous fission.

Element	Mass of most stable isotopes	Half-life	Decay mechanism
90 Th	232	1.4×10^{10} y	A
91 Pa	231	3.25×10^{4} y	α
92 U	234	2.45×10^{5} y	α
	235	7.037×10^{8} y	α
	238	4.47×10^{9} y	α
93 Np	236	1.55×10^{5} y	β^{-}, EC
	237	2.14×10^{6} y	α
94 Pu	239	2.41×10^{4} y	α
	240	6.563×10^{3} y	α
	242	3.76×10^{5} y	α
	244	8.26×10^{7} y	α
95 Am	241	432.7 y	α
	243	7.38×10^{3} y	α
96 Cm	244	18.11 y	α
	245	8.5×10^{3} y	α
	246	4.73×10^{3} y	α
	247	1.56×10^{7} y	α
	248	3.4×10^{5} y	α
	250	1×10^{4} y	α
97 Bk	247	1.38×10^{3} y	SF
	249	320 d	α
98 Cf	249	351 y	β
	250	13.1 y	α
	251	898 y	α
99 Es	252	2.64 y	α
	252	472 d	α
	253	20.47 d	α
	254	276 d	α
	255	39.8 d	α
100 Fm	257	100.5 d	α
101 Md	258	56 d	α
102 No	259	1 h	α, EC
103 Lr	262	3.6 h	α

considered to possess similar physical and chemical properties to the lanthanide series. However, for some of the early actinides, actinyl ions (AnO_2^{+} and AnO_2^{2+}) can be observed in addition to the spherical metal ions and are often more stable, particularly for U in the hexavalent state. Actinyl ions are linear O $=$ An $=$ O moieties, with short An $=$ O 'yl' bond distances, for example 1.78 Å for U $=$ O in UO_2^{2+} and 1.82 Å in UO_2^{+}, suggesting significant multiple bond character.[102,103] Bonding of other ligands is almost always in the equatorial plane, perpendicular to the O $=$ An $=$ O axis.[104] It is primarily this diversity in oxidation state, electronic properties and resultant coordination chemistry that has enabled the development of several strategies mainly based on liquid–liquid extraction for the selective separation and recovery of the actinide elements.

6.3.1 End Uses and Applications

All of the actinides are radioactive as they have no stable isotopes and many of their applications rely on their inherent radioactivity. The actinides are chiefly used for generation of energy by nuclear fission to produce nuclear power, although nuclear weapons can also be produced this way. Some other important uses of actinide radioactivity include the use of plutonium-238 as a power source for the heart pacemaker, which enables the battery to last for up to 10 years (approximately five times longer than conventional battery implants). Plutonium-238 is also used as a fuel for some space shuttles and satellites; both plutonium-238 and curium-244 have been utilised as a power source on the moon, smaller scale uranium-based nuclear reactors have been installed in some ships and submarines and depleted uranium is used in armour piercing munitions and as a ballast in some aircraft.[105]

A more common use of radioactivity is the use of americium-241 in smoke detectors, where nitrogen and oxygen ions are formed by the interaction of α particles. These ions induce a current between two electrodes as a constant voltage is applied. However, smoke absorbs the α particles, preventing the current flowing and activating the alarm.

6.3.1.1 *Nuclear Fuel Cycle*

Uranium is mined in its ore form from the ground, mainly in Middle Eastern countries, Canada, Australia and Africa. The ore is then milled to extract the uranium as 'yellowcake', which is mixed oxides of triuranium octoxide (U_3O_8), uranium dioxide (UO_2) and uranium trioxide (UO_3), by leaching with acid or alkali followed by precipitation. The remaining ore 'tailings' are disposed of as radioactive waste. Radioactive contamination at these mining and processing sites poses an environmental safety issue which has been addressed in a similar manner to spent nuclear fuel.[106] The yellowcake is then further processed, as only 0.7% of uranium is fissile ^{235}U; the dominant isotope is ^{238}U. The uranium oxide is enriched by increasing the ratio of ^{235}U:^{238}U to approximately 3.5–5% ^{235}U. This is done by converting all the mixed oxides into uranium dioxide and subsequently to uranium hexafluoride (UF_6) gas and separating it into two streams; one of which is enriched in ^{235}U and the other depleted (Figure 6.12).

The enriched UF_6 is then converted back to UO_2 which can be pressed and heated to 1400 °C to form fuel pellets. The depleted uranium is treated as waste. The fuel pellets are subsequently encased in metal rods that can then be used in a fuel assembly in a reactor.

After 18–36 months of operation, the build-up of fission products is such that the efficiency of the fuel decreases so the fuel rods are removed and replaced. The used fuel is then stored for months or years in water, which absorbs the

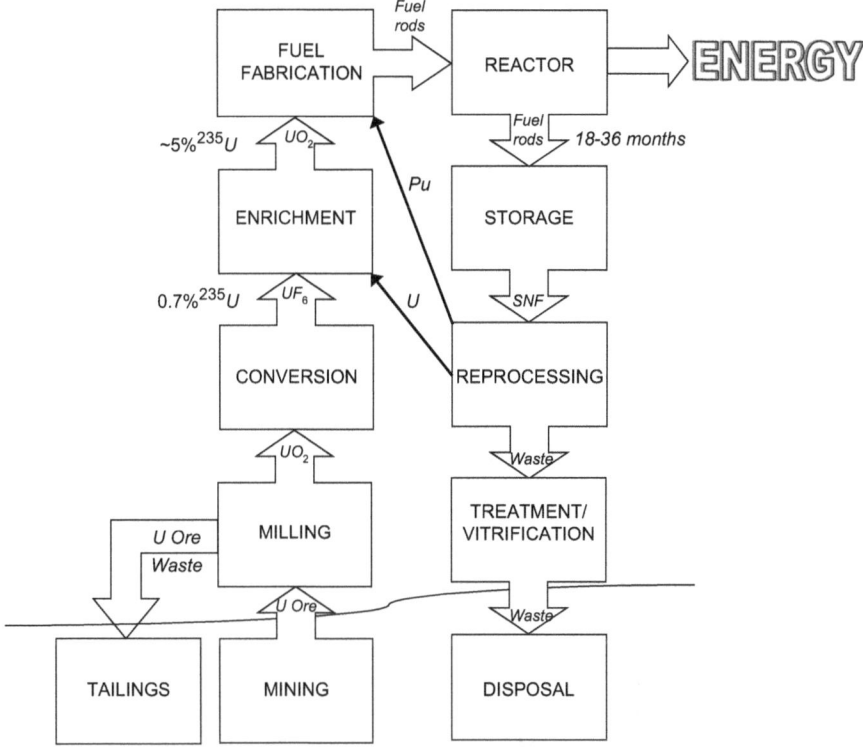

Figure 6.12 The Nuclear Fuel Cycle: an open cycle (dashed arrow) and a closed cycle (solid arrows).

heat, until the radiation levels decrease sufficiently for it to be disposed of or reprocessed. As there are no disposal facilities at present for nuclear fuel waste, it is simply isolated from the environment and left in storage until facilities become available.[107]

The once-through or 'open' fuel cycle, whereby waste is stored for disposal, is presently favoured by a number of countries including Canada, parts of Europe including Germany and the USA, although some research on reprocessing techniques is being carried out in these areas as reprocessing is becoming increasingly important for the future of nuclear power. A 'closed' fuel cycle, where the waste is recycled and reused, is becoming more and more favoured as a result of this and has been done in some parts of the world for many years, including the UK and other parts of Europe, Russia and Japan.[108] Despite the Fukushima incident, many countries in the world are embarking on expanding their nuclear programmes (particularly the UK).

Both decommissioning and new build programmes including efficient separation technologies require the implementation of new protocols to ensure safety and security of supply. The critical problem is managing the increasing volumes of wastes produced by fission and decommissioning which is likely to increase dramatically in the imminent future. Importantly, the UK government very recently announced that it would prefer to re-use the plutonium stock piles as mixed oxide fuels in the future, through partitioning and reprocessing.[109] However, not all plutonium is suitable for recycling and must be stored pending disposal in a suitable geological repository (yet to be identified).

6.3.2 Spent Nuclear Fuel and Reprocessing

A current generation II reactor uses typically 500 kg of uranium in a fuel rod before irradiation, which at the end of its life produces 475–480 kg uranium, 5 kg plutonium and 15–20 kg of corrosion and fission products (Fe, Zn, Mg, Pd, Pt, lanthanides, minor actinides). The reprocessing of spent nuclear fuel (SNF) is therefore essential to prevent the exhaustion of uranium supplies and reduce the volume and radiotoxicity of the waste produced. Based on the known global uranium reserves and the current world usage of 68 000 tonnes per year, estimates suggest that this will lead to exhaustion of supplies in 60 years' time if no recovery is implemented. However, there are at present two commercial reprocessing plants; Sellafield (UK) and La Hague (France), with Rokkasho-Mura (Japan) in development.

Existing reprocessing techniques involve the removal of reusable uranium and plutonium present in the waste which can be recycled and reused together in mixed oxide (MOX) reactors to produce more nuclear power. The amount of waste remaining in storage at present worldwide that is suitable for reprocessing is approximately 200 000 tonnes, with a global reprocessing capacity of around 4000 tonnes per year. Over the last 50 years approximately 90 000 tonnes of spent fuel has been reprocessed.

In addition to the reusable U^{n+} and Pu^{n+} in the SNF, there are also a variety of other fission products (FP) present, such as minor actinides (MA): Np^{n+}, Am^{3+} and Cm^{3+}; Ln^{3+} and transition metals, in addition to corrosion products (CP) as a consequence of radiolysis, erosion and equipment ageing of from steel containers and pipes. The transition metals produced from fission are chiefly cobalt, chromium, iron and manganese. The composition of SNF can be seen in Table 6.2.[110] Recently, research into the removal of the other actinides from the waste has become important in order to transmutate them into shorter-lived radionuclides so that their radioactivity will not persist for as long, making the

Table 6.2 Approximate compositions of SNF in light water reactors (LWR).

Constituent	% of SNF
U	95.6
Stable FP (including Ln)	2.9
Pu	0.9
Cs and Sr (FP)	0.3
I and Tc (FP)	0.1
Other long-lived FP	0.1
MA	0.1

disposal process easier and faster. This, coupled with a similar approach for any remaining plutonium, will ensure the waste is proliferation resistant as it would prohibit recovery of plutonium from storage.

Although MA only make up 0.1% of fission products, they are highly radiotoxic and so it would be beneficial to separate MA from the remaining fission products so that they can be transmuted into shorter lived radio-nuclides by neutron bombardment which is known collectively as the 'partitioning and transmutation scenario'. The necessity of the separation arises from the presence of the lanthanides as Ln^{3+} which are known neutron scavengers,[111] meaning that they possess high neutron capture cross sections, preventing transmutation of other species present.

Neptunium is relatively simple to remove from the mixture of fission products as it has a variety of oxidation states that can be utilised in the recovery process.[112] However, the predominantly trivalent minor actinides Am^{3+} and Cm^{3+} are much more difficult to separate from the remaining lanthanide waste owing to the similarities in the chemistries of the elements. The majority of current research is focussed on developing efficient separation strategies for the selective removal of Am^{3+} and Cm^{3+} from Ln^{3+}.[113,114]

At present, there are no MA–Ln separation techniques employed commercially, although a number of different processes are being developed, particularly in the USA and Europe, with a drive to implement a working process within the next five years. Despite differences in the chemistry between the techniques under development, all of them use solvent extraction as the ultimate separation technique.

6.3.2.1   The PUREX Process

PUREX (Plutonium and Uranium REfinement by EXtraction) is the process used by several nuclear plants which carry out reprocessing to remove U^{n+} and Pu^{n+} from the spent fuel waste in order to reuse it (Figure 6.13). Concentrated nitric acid (2–6 M) is employed to dissolve the waste in an aqueous phase to form hydrated nitrate complexes of the corresponding uranyl(VI) and plutonyl(VI) ions ($UO_2(NO_3)_2.xH_2O$ and $PuO_2(NO_3)_2.xH_2O$). The plutonyl nitrate is then selectively reduced using nitrogen tetroxide (N_2O_4) to the corresponding Pu^{4+} nitrate ($Pu(NO_3)_4.xH_2O$) and the solution is filtered to remove any precipitates.[115] The solution is then contacted with an organic phase (kerosene) containing tributyl phosphate (TBP) as an extracting agent (Scheme 6.3) which forms organic-soluble complexes with the UO_2^{2+} and PuO_2^{2+} nitrates of the form $UO_2(NO_3)_2(TBP)_2$ and $Pu(NO_3)_4(TBP)_2$, facilitating phase transfer.

However, technetium and neptunium are also extracted at this point. This is a disadvantage for the purpose of the PUREX process but is advantageous for subsequent MA–Ln separation processes which could follow. The UO_2^{2+} and NpO_2^+ complexes are separated from the Pu^{4+} and TcO_4^- complexes by reduction of Pu^{4+} to Pu^{3+} with hydrazine (N_2H_4) and extraction back into water.[116,117] The Pu^{3+} and TcO_4^- species are then separated from each other by another extraction cycle and then a 'stripping' solution of nitric acid,

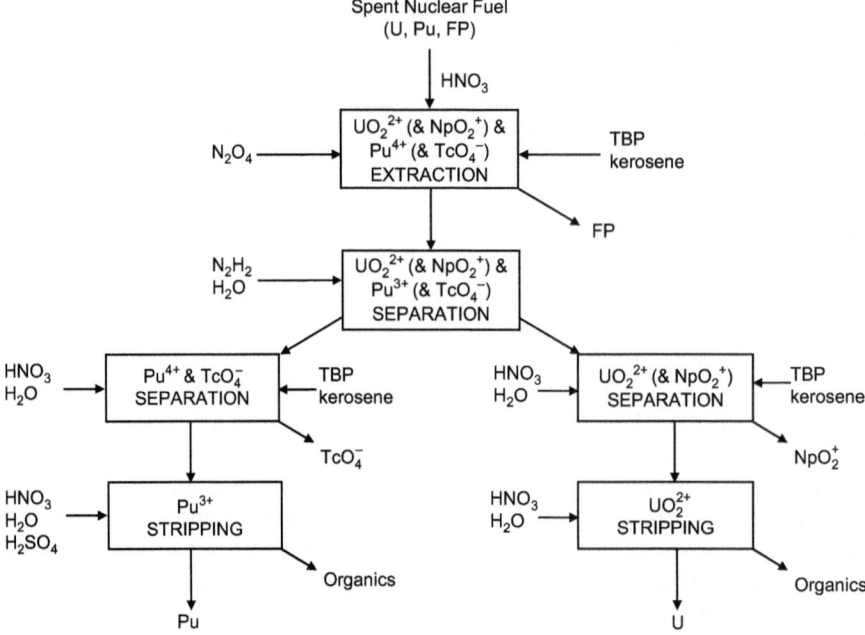

Figure 6.13   PUREX flow diagram.

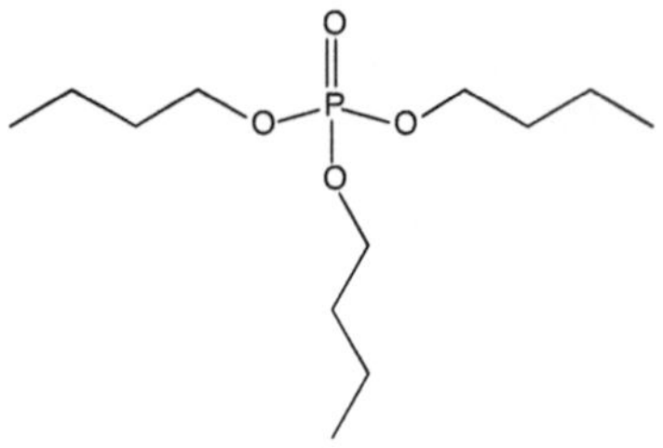

Scheme 6.3   Structure of TBP (tributyl phosphate).

hydroxylamine and sulfuric acid is used to obtain pure plutonium. The UO_2^{2+} and NpO_2^+ are also extracted back into aqueous solution and separated from each other by another extraction cycle. Pure U^{n+} is obtained by using aqueous nitric acid for stripping (back extraction).[118] The process has been proven to work well and it is an advantage that the organic phase can be reused after stripping. Nevertheless, the process has a few drawbacks, mainly that the need for high acid concentrations makes it less environmentally friendly. Also the need for redox control and less stable oxidation states renders the process more complicated and the use of phosphorus reagents make the products more difficult to dispose of, since phosphorus waste cannot be incinerated and so any radioactive waste must be separated from the phosphorus before treatment.[119,120]

6.3.2.2 The TRUEX Process

TRUEX (TRansUranic EXtraction) is an example of advanced reprocessing (removal of MA^{3+} and Ln^{3+}), currently being developed in the USA (Figure 6.14). The principle of the process is to remove Am^{3+} and Cm^{3+} (MA^{3+}) and Ln^{3+} selectively from the other fission products left in the raffinate after the PUREX process. A combination of extractants is used: carbamoylmethylphosphine oxide (CMPO) (Scheme 6.4) and TBP (as in the PUREX process). The benefit of the combined extractant system is that the process is effective over a range of acidities (0.7–5 M HNO_3). The raffinate from the PUREX process is contacted with the extractants in an organic phase of normal paraffinic hydrocarbon (NPH). Oxalic acid is then added to prevent the co-extraction of zirconium and molybdenum with MA^{3+}. An additional wash is also performed using sodium carbonate (Na_2CO_3) to prevent any other fission products from being co-extracted. The extractants selectively remove MA^{3+} and Ln^{3+} into the organic phase, leaving the remaining fission products

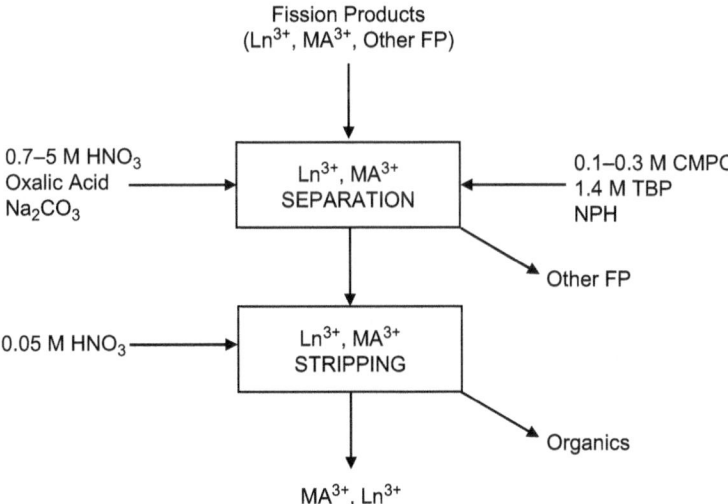

Figure 6.14 TRUEX flow diagram.

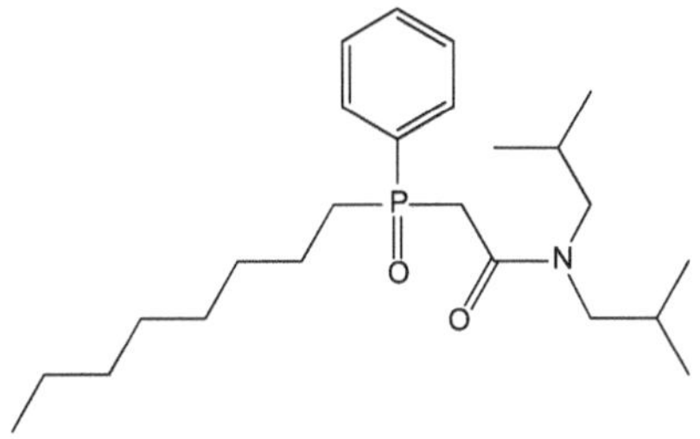

Scheme 6.4 Structure of CMPO (*N,N*-diisobutyl-2-(octyl(phenyl)phosphoryl)acetamide).

in the aqueous phase. The MA^{3+} and Ln^{3+} are finally stripped using nitric acid and can be reprocessed further as required. However, a main drawback of this process is that the lanthanides are still present with the minor actinides so further reprocessing is required.[121]

6.3.2.3   The DIAMEX Process

The DIAMEX (DIAMide EXtraction) process is another example of advanced reprocessing and is currently under development in France by the CEA (Commissariat à l'Energie Atomique et aux Energies Alternatives) (Figure 6.15). It is similar to the TRUEX process as it selectively removes Am^{3+} and Cm^{3+} (MA^{3+}) and Ln^{3+} from the PUREX raffinate. The process is being researched using a variety of different diamides as the extractant, where the most promising of which have been shown to be *N,N'*-dimethyl-*N,N'*-dibutyl-tetradecylmalonamide (DMDBTDMA) and *N,N'*-dimethyl-*N,N'*-dioctyl-hexylethoxymalonamide (DMDOHEMA) (Scheme 6.5).[122,123]

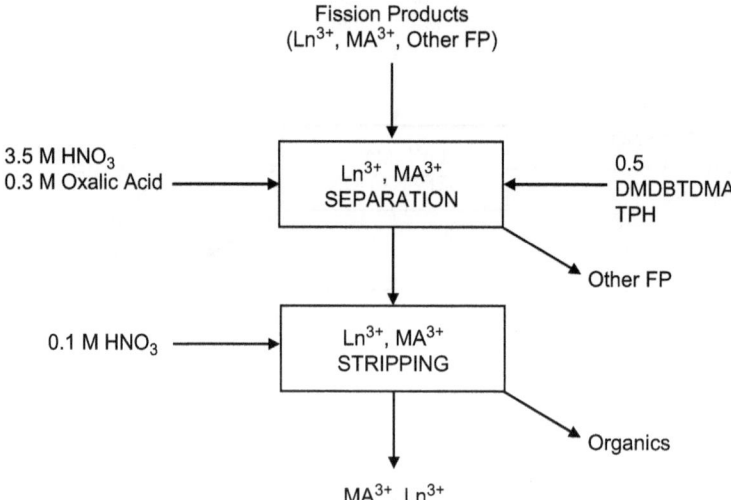

Figure 6.15   Example DIAMEX flow diagram.

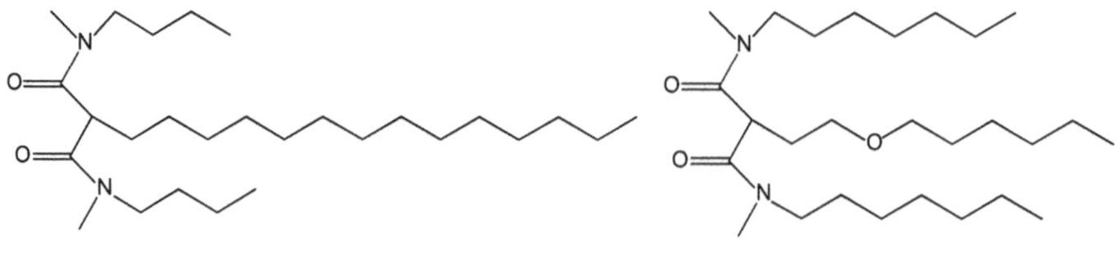

Scheme 6.5   Structure of DMDBTDMA (N_1,N_3-dibutyl-N_1,N_3-dimethyl-2-tetrade-cylmalonamide) (left) and DMDOHEMA (N_1,N_3-dibutyl-2-(2-(hexyloxy)ethyl)-N_1,N_3-dimethylmalonamide) (right).

The nitric acid PUREX raffinate is contacted with the extractant in an organic phase of tetra-propylene-hydrogenated (TPH), a synthetic branched form of dodecane.[124] Oxalic acid is then added to prevent the co-extraction of zirconium and molybdenum with MA^{3+} (as in the TRUEX process) and the extractant selectively removes MA^{3+} and Ln^{3+} into the organic phase, leaving behind the other fission products in the aqueous phase. The MA^{3+} and Ln^{3+} are then stripped using nitric acid and can be reprocessed further as required.

The main benefit of this process compared to the TRUEX process is that the organic waste only contains the elements C, H, N and O. Phosphorus reagents are not used so the waste can be disposed of more easily. However, like the TRUEX process, a main drawback is that the lanthanides are still present with the minor actinides meaning further reprocessing is required.[125]

6.3.2.4 The SANEX Process

The SANEX process (Selective ActiNide EXtraction) is another process being developed by the CEA and is intended to be coupled with a TRUEX or DIAMEX type process as the next step in advanced reprocessing (Figure 6.16). Extractants such as bis-triazinyl-pyridines (BTPs) and their variants have been widely studied, with a more recent extractant tetraoctyldiglycolamide (TODGA) being investigated.[126–128] The extractants have been found to bind preferentially to MA^{3+} allowing only MA^{3+} to be extracted into an organic phase, using TBP, leaving Ln^{3+} in the aqueous phase. Oxalic acid and (2-hydroxyethyl)-ethylenediaminetriacetic acid (HEDTA) are used to prevent the co-extraction of any other fission products (Scheme 6.6).[129] Very high separation factors have been achieved using BTP derivatives. For example, *n*-propyl-BTP displays a separation factor for Am^{3+} over Eu^{3+} of 143. Many

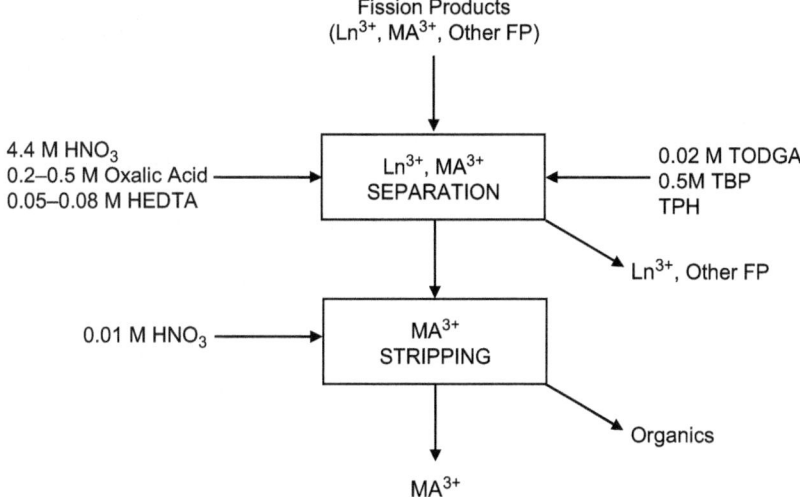

Figure 6.16 SANEX flow diagram for TODGA process.

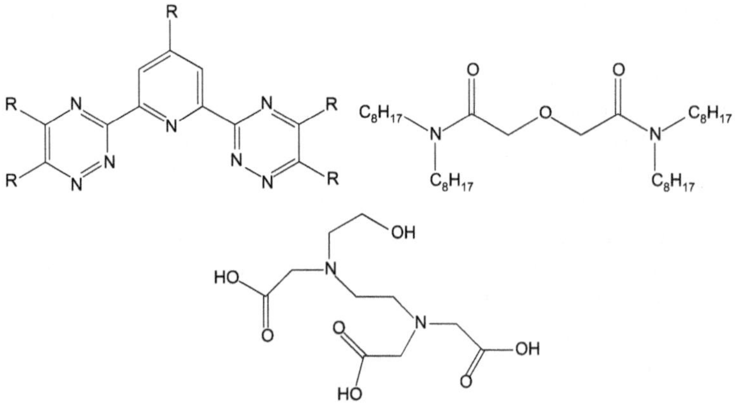

Scheme 6.6   General structure of BTPs (6,6′-di(1,2,4-triazin-3-yl)-2,2′-bipyridine) (top left); structure of TODGA (2,2′-oxybis(N,N-dioctylacetamide)) (top right) and HEDTA (2,2′-((2-((carboxymethyl)(2-hydroxyethyl)amino)ethyl)azanediyl)diacetic acid) (bottom).

significant advances have recently been made in this field, but the origin of the selectivity of the extractants with respect to the nature of the bonding involved is still under debate.

6.3.2.5   The Innovative SANEX Process

The innovative SANEX (or *i*-SANEX) process is also currently under development at the CEA.[130,131] Essentially, it is a modified DIAMEX process with selective back extraction of Am^{3+} and Cm^{3+} from the organic phase. The MA^{3+} and Ln^{3+} ions are initially extracted from the PUREX raffinate using TODGA and then a hydrophilic complexant which is selective for MA^{3+} is employed to back extract the minor actinides from the loaded organic phase into the aqueous phase. In order to retain the lanthanide ions in the organic phase, a nitrate salt is added to the stripping solution. Hydrophilic extractants that have been used to demonstrate this technique are DTPA (diethylenetriaminepentaacetic acid) and the sulfonated BTP derivative 2,6-bis(5,6-di(sulfophenyl)-1,2,4-triazin-3-yl)pyridine; separation factors of up to 1000 are achievable in this process.[132] One other option that has been suggested is to add a second stripping agent such as HDEHP to the organic phase in order to retain the lanthanides in the organic phase at low pH.

One major drawback of this process however, is the limited operative acidity range (*ca.* pH 3), which means that buffering agents need to be added to the aqueous phase in the back extraction step. Another reprocessing concept currently under investigation is the 1-cycle SANEX, with the intention of directly extracting the trivalent actinides selectively from the PUREX raffinate. A system consisting of 0.15 M CyMe4BTBP and 0.005 M TODGA in a mixture of 40% TPH and 60% 1-octanol has been proposed.[133]

6.3.2.6 The GANEX Process

The GANEX (Grouped ActiNide EXtraction) process has been proposed more recently and is a complete separation process, combining the principles of the PUREX and TRUEX processes in order to separate all of the actinides (U^{n+}, Pu^{n+} and MA^{3+}) from the lanthanides and both from the other fission products (Figure 6.17). The extractant bis-triazin-bipyridine (BTBP) and its variants (such as $CyMe_4$-BTBP) (Scheme 6.7) have been tested and found to be effective in selectively coordinating to and extracting MA^{3+} with $CyMe_4$-BTBP exhibiting a separation factor for Am^{3+} over Eu^{3+} of 101. The BTBP derivative is dissolved in cyclohexanone and is used alongside TBP (which extracts U^{n+} and Pu^{n+}). It is stable against radiolysis and hydrolysis, especially the $CyMe_4$ variant. If proven to be successful, this process would simplify reprocessing, but much more work is needed before this process could become operational as co-extraction of fission products is a major problem.[134–136]

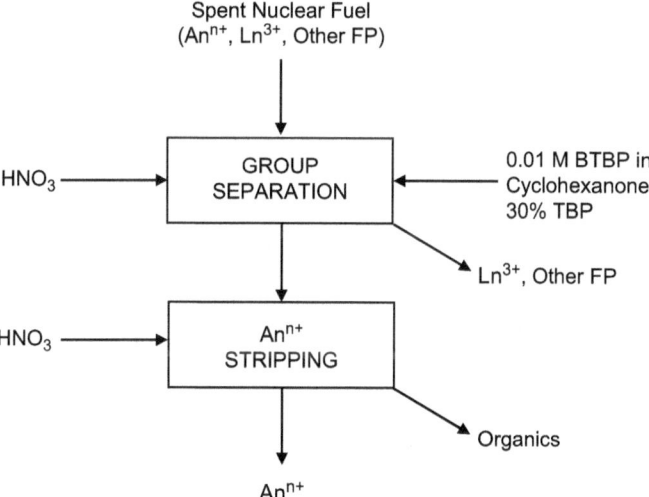

Figure 6.17 GANEX flow diagram.

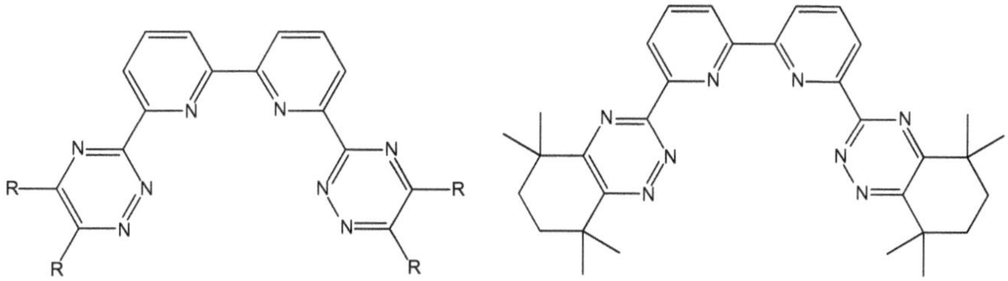

Scheme 6.7 General structure of BTBPs (6,6'-bis(1,2,4-triazin-3-yl)-2,2'-bipyridine) (left) and $CyMe_4$-BTBP (6,6'-bis(5,5,8,8-tetramethyl-5,6,7,8-tetrahydrobenzo[e][1,2,4]triazin-3-yl)-2,2'-bipyridine) (right).

6.3.2.7 The TRPO Process

Another advanced reprocessing extraction process being developed in China is the TRPO (TRialkyl Phosphine Oxide) process which involves the separation of all An^{n+} in stages to remove Np^{n+} and Pu^{n+} together, Am^{3+}/Cm^{3+} and Ln^{3+} together and isolate U^{n+}. There are two processes being researched, both of which use TRPO as the extractant but differ in the other reagents used. One system uses TTHA (triethylene tetramine hexaacetate) as an extractant to bind selectively to different actinides preferentially at different pH values (pH 2.3 and 3.1), allowing selective extraction, buffered by lactic acid (Figure 6.18). The alternative process uses nitric acid to extract MA^{3+} and Ln^{3+}, followed by oxalic acid to extract Pu^{3+} and Np^{n+} (Figure 6.19). Both processes then use sodium carbonate to strip the remaining U^{n+} from solution. The main advantage of the first system is that MA^{3+} and Ln^{3+} can subsequently be separated from each other using CYANEX 301, with the main disadvantage being the need for buffering owing to pH dependence. The main advantage of the second system is that the separation between components is almost quantitative but MA^{3+} and Ln^{3+} cannot be separated from each other using CYANEX 301 in a later step owing to the high acidity of the solution (Scheme 6.8).[137]

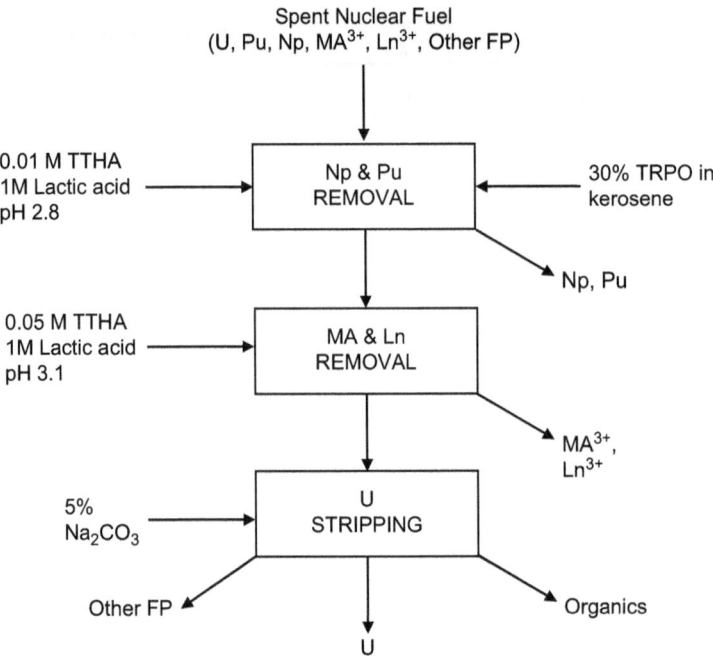

Figure 6.18 TRPO flow diagram using TTHA.

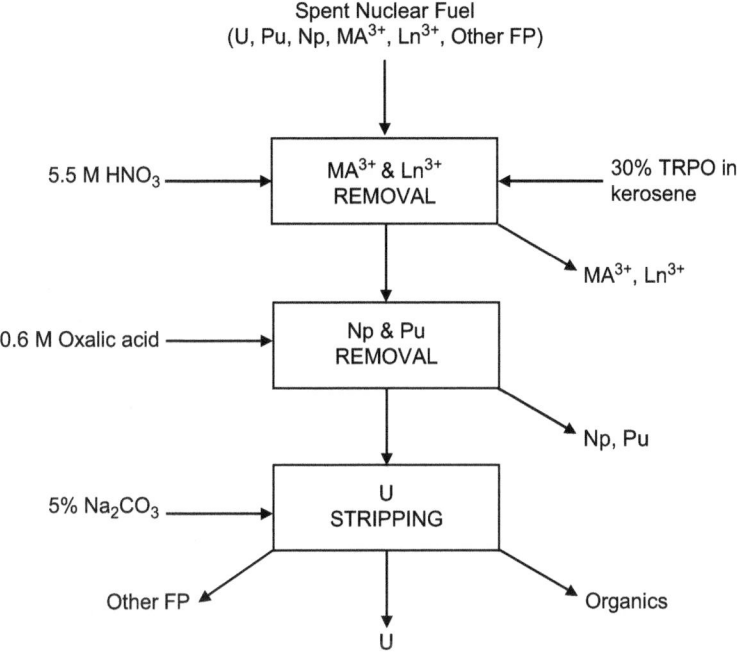

Figure 6.19 TRPO flow diagram using HNO_3 and oxalic acid.

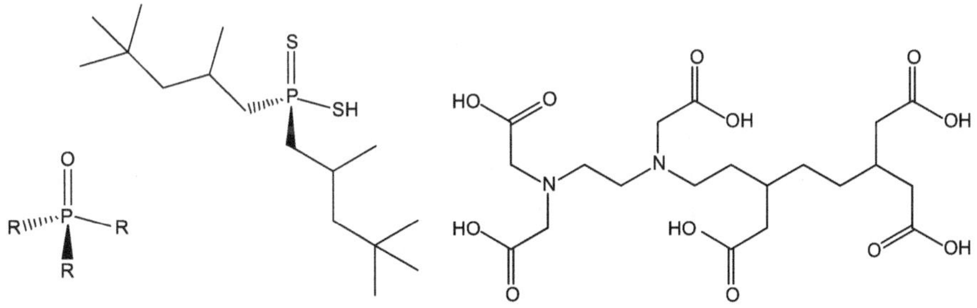

Scheme 6.8 Structure of TRPO (trialkyl phosphine oxide, $R = C_6–C_8$) (left); CYANEX 301 (bis(2,4,4-trimethylpentyl)phosphinodithioic acid) (centre) and TTHA (3-(2-((2-(bis(carboxymethyl)amino)ethyl)(carboxymethyl)amino)ethyl)-6-(carboxymethyl)octanedioic acid) (right).

6.3.2.8 The TALSPEAK Process

TALSPEAK (Trivalent Actinide Lanthanide Separation by Phosphorus reagent Extraction from Aqueous Complexation) is a further, effective method of advanced reprocessing by solvent extraction (Figure 6.20). The process was initially developed at Oak Ridge National Laboratory in Tennessee, USA, during the 1960s and it is still being refined.

The process is designed to allow the separation of MA^{3+} (Am^{3+} and Cm^{3+}) from Ln^{3+} and Y^{3+} to allow MA^{3+} to be reprocessed further by transmutation.

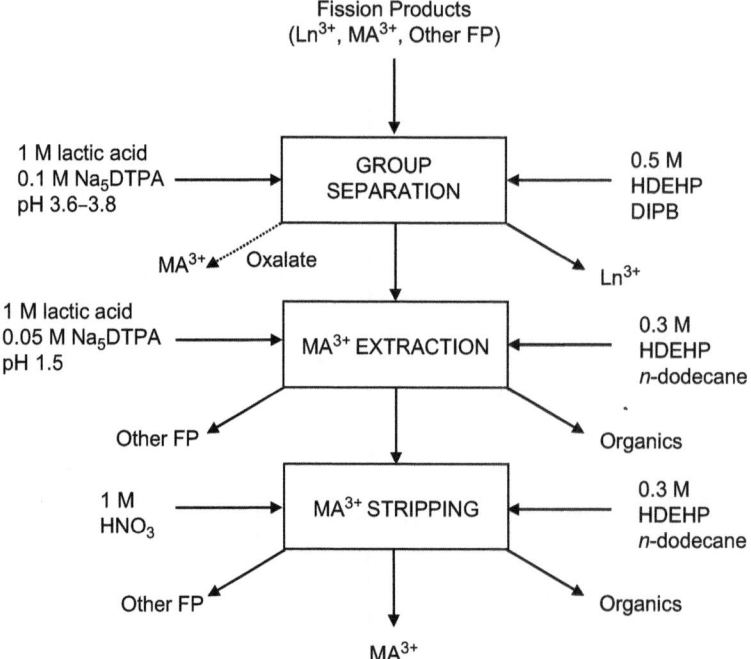

Figure 6.20   TALSPEAK flow diagram.

Essentially, TALSPEAK can be viewed as the reverse operation of SANEX in that the lanthanides are preferentially extracted into the organic phase rather than the actinides. Although it is still under development, the TALSPEAK process has a number of benefits over other similar processes discussed as it is resistant to irradiation and allows the separation to be carried out without the need for high acid and salt concentrations. It also has added benefits in that it has already been performed on a pilot plant scale and uses cheap, readily available reagents (DTPA, lactic acid and HDEHP).[138] Additionally, it can be carried out using affordable stainless steel equipment. The process is very promising despite its potential disadvantage in that it involves removing the major constituent from the minor constituent. However, studies have shown the separation is effective enough for this not to be a problem.

   In the process, the minor actinides form complexes with a poly-aminocarboxylic acid preferentially over the lanthanides. This allows the lanthanide ions to be preferentially extracted by a mono-acidic organophosphate or phosphonate. The most effective complexant (diethylenetriamine pentaacetic acid) in the pH 2.5–3.5 range, giving relatively high SF values (~ 50 for Nd^{3+}, the most difficult to extract Ln^{3+} ion) and other effective extractants are HDEHP (bis-(2-ethylhexyl)phosphoric acid) and HEH[ϕP] (2-ethylhexyl phenyl phosphonic acid) (Scheme 6.9). The extraction can be carried out without the use of an extractant, although the separation is not as

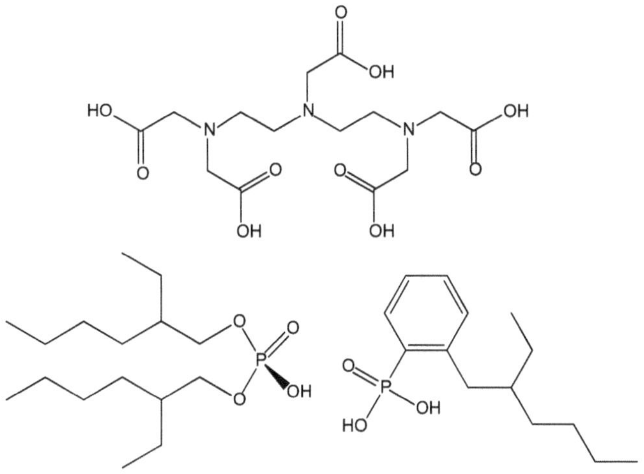

Scheme 6.9 Structure of DTPA (2,2′,2″,2‴-((((carboxymethyl)azanediyl)bis(ethane-2,1-diyl))bis(azanetriyl))tetraacetic acid) (top); HDEHP (bis(2-ethylhexyl) hydrogen phosphate) (bottom left) and HEH[φP] ((2-(2-ethylhexyl)phenyl)phosphonic acid) (bottom right).

discriminative and is significantly enhanced by the addition of an aminopoly-acetic acid, such as DTPA. Other aminopolyacetic acids have been tested, such as TTHA but are not as effective or tend to be less soluble in the raffinate.[139]

6.3.2.9 Separation of Am³⁺ and Cm³⁺ and the LUCA Process

Ideally, mutual separation of the neighbouring radiotoxic MA³⁺ ions Am³⁺ and Cm³⁺ in the waste inventory needs to be achieved to provide a truly optimal reprocessing cycle. However, this poses a much greater challenge than the separation of trivalent actinides over the lanthanides. In principle, both americium and curium could be transmutated together in a fast reactor but the relative high decay and heat generation of curium necessitates the development of a simple, one stage process prior to re-fabrication.

One recent method directed at selective separation of americium from curium takes advantage of the fact that americium can form high oxidation state compounds under oxidising conditions.[140] Oxidation of Am³⁺ using sodium bismuthate in acidic solutions produced hexavalent AmO_2^{2+} quantitatively, which exhibited analogous extraction kinetics to uranyl, neptunyl and plutonyl ions when contacted with TBP. However, at higher temperatures (80 °C) exclusive formation of pentavalent AmO_2^+ was observed.

LUCA (Lanthaniden Und Curium Americium trennung; lanthanide and curium americium separation) is a relatively new process currently being developed in Germany and is designed to follow the SANEX or DIAMEX processes (Figure 6.21). The process involves the selective separation of Am³⁺ from Cm³⁺, Cf³⁺ and Ln³⁺ after co-extraction. A combined

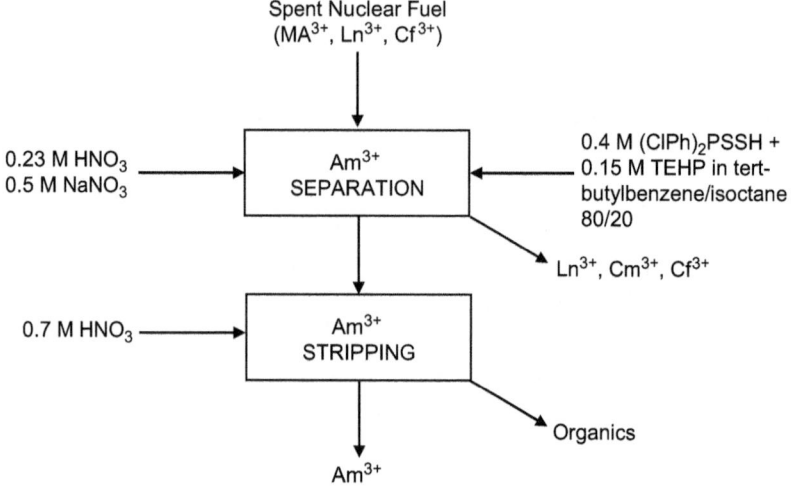

Figure 6.21   LUCA flow diagram.

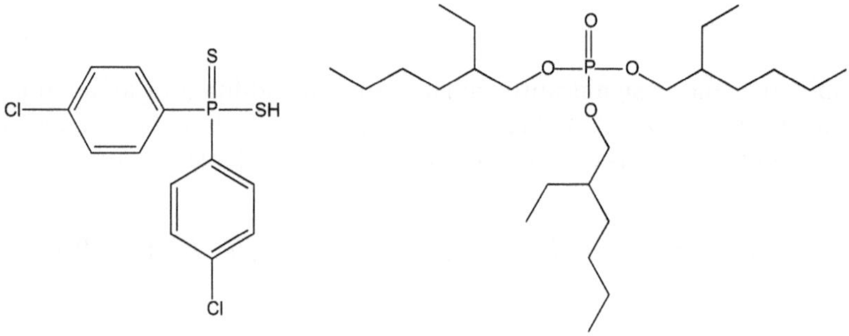

Scheme 6.10   Structure of (ClPh)₂PSSH (bis(chlorophenyl)dithiophosphinic acid) (left) and TEHP (tris(2-ethylhexyl)phosphate) (right).

extractant system of bis(chlorophenyl)dithiophosphinic acid (((ClPh)$_2$PSSH) and tris(2-ethylhexyl)phosphate (TEHP) in isoctane and *tert*-butyl benzene is used (Scheme 6.10). Advantages of the LUCA process include high recovery after stripping and that the phosphinic acid is more stable to hydrolysis and radiolysis than CYANEX 301. However the phosphinic acid was found to be unstable in high HNO$_3$ concentrations.[141] At present, as with the majority of the MA/Ln processes described, the exact origin of the selectivity remains unclear.

6.4   Summary and Future Outlook

The f-block elements are presently a major strategic resource and form essential components in a wide variety of existing and emerging applications. The necessity for reprocessing spent lanthanides from advanced green technologies and the actinides that result from nuclear fission has primarily arisen from increasing awareness and concern over environmental and safety issues, security of supply and proliferation risks. There are a number of different metal recovery and recycling processes currently under development The majority of these are based on liquid–liquid extraction utilising coordination chemistry. However to date, very few of these have been implemented on a large scale, with two exceptions being the recovery of Nd^{3+} from spent Nd–Fe–B magnets and UO_2^{2+} and Pu^{4+} in the PUREX process.

The principles of these separation processes are often very similar, although extraction techniques and reagents vary somewhat. There are a number of factors which must be considered when developing a suitable solvent extraction process for metal ion reprocessing, including the ease of stripping (back extraction); the need for low volatility, non-flammable solvents; the potential of the process to be continuous; minimisation of waste production; the resistance of the process to degradation and radiolysis (in the case of the actinides) and; the practicality and efficiency of the process and economic viability.[142]

Whilst some of the chemistry involved in these reprocessing scenarios is well understood, for example, the redox chemistry in the PUREX process, an in depth understanding of many factors involved in reprocessing cycles under development is lacking, thereby limiting the potential to develop truly efficient processes. For example, the origin of the selectivity of a given extractant in mutual lanthanide extractions is not fully evident and the same is true for MA/lanthanide extractions. Despite this, much advancement in our comprehension of the kinetic, thermodynamic and partitioning parameters involved in a given process is currently being made despite the sheer complexity of the waste content. All of the processes currently under development have advantages and disadvantages but all are ultimately heading towards the same goal, which is the separation of chemically similar elements for the purpose of mitigating raw material supply shortages. In the case of the actinides, this goal also extends to include the reduction in the long-term radiotoxicity of the waste and the storage volume. The urgent requirement to recover and reprocess dwindling supplies of both the lanthanide elements and uranium (and to a lesser extent, thorium) will undeniably lead to the further development and implementation of advanced recycling strategies in the near future. Without advancements in innovative recycling and recovery processes, the sustainability of our current technologically advanced society remains uncertain.

Acknowledgements

The authors would like to thank Professor Koen Binnemans and Professor Laurence Harwood for their kind donation of literature, Dr Leigh Martin and Dr Clint Sharrad for helpful discussions, the EPSRC (UK) for funding a Career Acceleration Fellowship (LN) and the Battelle Energy Alliance (USA) for a studentship (MLP).

References

1. S. Cotton, 1991, *Lanthanides and Actinides*, D. Woolins, R. Crabtree, D. Atwood and G. Meyer, John Wiley & Sons, Chichester UK, 2006, vol. 1, pp. 1–7.
2. P. D. Wilson, *The Nuclear Fuel Cycle from Ore to Waste*, Oxford University Press, USA, 1996.
3. N. Kaltsoyannis and P. Scott, *The f Elements*, R. G. Compton, S. G. Davies, J. Evans, L. F. Gladden, Oxford University Press, USA, 1st edition, 1999.
4. Chemistry in its element podcast – take a tour of the periodic table, a leading scientist or author tells the story behind each element: http://www.rsc.org/chemistryworld/podcast/element.asp, Accessed 24th March 2013.
5. D. Schüler, M. Buchert, R. Liu, S. Dittrich and C. Merz, *Study on Rare Earths and Their Recycling*, Final Report for The Greens/EFA Group in the European Parliament, Darmstadt, 2011.
6. S. Cotton, *Educ. Chem.*, 1999, **36**(4), 96.
7. W. R. Wilmarth, R. G. Haire, J. P. Young, D. W. Ramey and J. R. Peterson, *J. Less Common Met.*, 1988, **141**, 275.
8. A. P. Jones, F. Wall and C. T. Williams, *Rare Earth Minerals: Chemistry, Origin and Ore Deposits*, A. P. Jones, F. Wall and C. T. Williams, Chapman and Hall, London UK, 1966, vol. 1, pp. 6–10.
9. C. H. Evans, *Episodes from the History of the Rare Earth Elements*, Kluwer Academic Publishers, Dordrecht, Netherlands, 1996.
10. T. Y. Nakai and T. Kawashima, *Anal. Sci.*, 1993, **9**, 561.
11. R. Miyawaki and I. Nakai, *Handbook on the Physics and Chemistry of Rare Earths*, K. A. Gschneider and L. Eyring, Elsevier Science Publishers, Netherlands, 1993, vol. 16, ch. 108, p. 249.
12. G. Wang, *Journal of Sichuan Rare Earth*, Current Mining Situation and Potential Development of Rare Earth in China, China Nonferrous Engineering and Research Institute, 2009, vol. 3.
13. T. G. Goonan, *Rare Earth Elements – End use and recyclability*, Scientific Investigations Report, US Geological Survey, 2011, p. 5094.
14. BBC, Japan finds rare earths in Pacific seabed, http://www.bbc.co.uk/news/world-asia-pacific-14009910, 2011, Accessed 24th March 2013.
15. Great Western Minerals Group, http:/ / www. gwmg. ca/ images/ file/ Presentations/ gwmg_aug_28_2012. pdf, 2012, Accessed 24th March 2013.
16. Molycorp, http://www.molycorp.com, 2012, Accessed 24th March 2013.
17. D. J. Kingsnorth, *Meeting the Challenges of Supply this Decade*, http://files. eesi. org/ kingsnorth_031111. pdf, Industrial Minerals Company of Australia (IMCOA), 2011, Accessed 24th March 2013.
18. J. Lifton, *Is The Rare Earth Supply Crisis Due to Peak Production Capability or Capacity*, http:// seekingalpha. com/ instablog/ 65370-jack-lifton/26386-is-the-rare-earth-supply-crisis-due-to-peak-production-capability-or-capacity, 2009, Accessed 24th March 2013.

19. J. Lifton, *The Rare Earth Crisis of 2009 Part I*, 2009, Technology Metals Research, http://www.techmetalsresearch.com.
20. Metal-Pages, www.metal-pages.com, 2012, Accessed 24th March 2013.
21. Asian Metal, www.asianmetal.com, 2005, Accessed 24th March 2013.
22. C. Hurst, *China's Rare Earth Elements Industry: What Can the West Learn?*, Institute for the Analysis of Global Security, Washington, 2010.
23. W. M. Morrison and R. Tang, China's Rare Earth Industry and Export Regime: Economic and Trade Implications for the United States, Congressional Research Service, 2012.
24. J. Lifton, *Implications for Investors of the Dramatically Increasing Chinese Demand for Rare Earths*, http://www.techmetalsresearch.com/2011/06/implications-for-investors-of-the-dramatically-increasing-chinese-virtual-demand-for-rare-earths, 2011, Accessed 24th March 2013.
25. The Chinese Society of Rare Earths, http://www.cs-re.org.cn/en/index.php, 2010, Accessed 24th March 2013.
26. Galissard, http://www.galissard.co.uk/index.php/the-market/what-are-rems, Accessed 24th March 2013.
27. H. C. Aspinall, *Chemistry of the f-block Elements*, D. Phillips, P. O'Brien and S. Roberts, Gordon and Breach Science Publishers, Singapore, 2001, vol. 5.
28. J. Luzon and R. Sessoli, *Dalton Trans.*, 2012, **41**, 13556.
29. S. Faulkner, L. S. Natrajan, W. S. Perry and D. Sykes, *Dalton Trans.*, 2009, **20**, 3890.
30. S. V. Eliseeva and J.-C. Bünzli, *Chem. Soc. Rev.*, 2010, **39**, 189.
31. L. S. Natrajan, *Curr. Inorg. Chem.*, 2011, **1**, 61.
32. M. Sagawa, S. Fujumura, N. Togawa, H. Yamamoto and Y. Matsuura, *J. Appl. Phys.*, 1984, **55**, 2083.
33. J. J. Croat, J. F. Herbst, R. W. Lee and F. E. Pinkerton, *J. Appl. Phys.*, 1984, **55**, 2078.
34. W. M. Hubbard, E. Adams and J. Gilfrich, *J. Appl. Phys.*, 1960, **31**, 368S.
35. Y. Tawara and K. J. Strnat, *IEEE Trans. Magn.*, 1976, **13**, 954.
36. A. E. Clark, *Ferromagnetic Materials: A Handbook on the Properties of Magnetically Ordered Substances*, E. P. Wolfarth, Elsevier Science B. V, Netherlands, 1980, vol. 1, *Magnetostrictive Rare Earth-Fe$_2$ Compounds*, pp. 531–589.
37. T. G. Polanyi, *Clin. Chest Med.*, 1985, **6**, 179.
38. L. R. Janis, R. D. Kravitz and S. S. Wagner, *Clin. Podiatr. Med. Surg.*, 1994, **11**, 483.
39. E. Merbach and E. Toth, *Comprehensive Coordination Chemistry*, M. D. Ward, Elsevier Pergamon, Oxford UK, 2nd edition, 2004, vol. 9, ch. 19, pp. 841–877.
40. P. Caravan, J. J. Ellison, T. J. McMurry and R. B. Lauffer, *Chem. Rev.*, 1999, **99**, 2293.
41. P. Hermann, J. Kotek, V. Kubícek and I. Lukes, *Dalton Trans.*, 2008, 3027.
42. A. J. L. Villaraza, A. Bumb and M. W. Brechbiel, *Chem. Rev.*, 2010, **110**, 2921.

43. S. Viswanathan, Z. Kovacs, K. N Green, J. Ratnakar and A. D. Sherry, *Chem. Rev.*, 2010, **110**, 2960.
44. K. Nwe, C. M. Andolina, C.-H. Huang and J. R. Morrow, *Bioconjugate Chem.*, 2009, **20**, 1375.
45. S. Faulkner and J. L. Matthews, *Comprehensive Coordination Chemistry*, M. D. Ward, Elsevier Pergamon, Oxford UK, 2nd edition, 2004, vol. 9, ch. 21, 913–941.
46. I. A. Hemilla, *Applications of Fluorescence in Immunoassays*, Wiley Interscience, 1991.
47. E. J. New, D. Parker, D. G. Smith and J. W. Walton, *Curr. Opin. Chem. Biol.*, 2010, **14**, 238.
48. M. F. Tweedle, *Acc. Chem. Res.*, 2009, **42**(958), 1319.
49. M. Lubberink, H. Lundqviat and V. Tolmachev, *Phys. Med. Biol.*, 2001, **47**, 615.
50. C. Planchamp, C. M. Pastor, L. Balant, C. D. Becker, F. Terrer and M. Gex-Fabry, *Invest. Radiol.*, 2005, **40**, 705.
51. H.-S. Chong, K. Garmestani, L. H. Bryant Jr., D. E. Milenic, T. Overstreet, N. Birch, T. Le, E. D. Brady and M. W. Breichbiel, *J. Med. Chem.*, 2006, **49**, 2055.
52. K. V. Laere, M. Koole, T. Kauppinen, M. Monsieurs, L. Bouwens and R. Dierek, *J. Nucl. Med.*, 2000, **41**, 2051.
53. M. Evangelisti, A. Candini, A. Ghirri, M. Affronte, E. K. Brechin and E. J. L. McInnes, *Appl. Phys. Lett.*, 2005, **87**, 072504.
54. J. W. Sharples, Y.-Z. Zheng, F. Tuna, E. J. L. McInnes and D. Collison, *Chem. Commun.*, 2011, **47**, 7650.
55. P. H. Lin, T. J. Burchell, L. Ungur, L. F. Chibotaru, W. Wernsdorfer and M. Murugesu, *Angew. Chem. Int. Ed.*, 2009, **48**, 9489.
56. J. Tang, I. J. Hewitt, N. T. Madhu, G. Chastenet, W. Wernsdorfer, C. E. Anson, C. Benelli, R. Sessoli and A. K. Powell, *Angew. Chem. Int. Ed.*, 2006, **45**, 1729.
57. A. Meijerink, R. Wegh, P. Vergeer and T. Vlugt, *Opt. Mater.*, 2006, **28**, 575.
58. J. de Wild, A. Meijerink, J. K. Rath, W. G. J. H. M. van Sark and R. E. I. Schropp, *Energy Environ. Sci.*, 2011, **4**, 4835.
59. W. Feng, L.-D. Sun, Y.-W. Zhang and C.-H. Yan, *Coord. Chem. Rev.*, 2010, **254**, 1038.
60. E. Hemmer, H. Takeshita, T. Yamano, T. Fujiki, Y. Kohl, K. Löw, N. Venkatachalam, H. Hyodo, H. Kishimoto and K. Soga, *J. Mater. Sci: Mater. Med.*, 2012, **23**, 2399.
61. Y. Liu, S. Zhou, D. Tu, Z. Chen, M. Huang, H. Zhu, E. Ma and X. Chen, *J. Am. Chem. Soc.*, 2012, **134**, 15083.
62. N. Yoshikawa, J. M. A. Yamada, J. Das, H. Sasai and M. Shibasaki, *J. Am. Chem. Soc.*, 1999, **121**, 4168.
63. S. C. Coote, R. A. Flowers II, T. Skrydstrup and D. J. Procter, 'Organic Synthesis Using Samarium Diiodide' in *Encyclopedia of Radicals in*

Chemistry, Biology and Materials 2012, C. Chatgilialoglu and A. S. Studer, John Wiley & Sons, 2012, vol. 30, pp. 849–900.

64. R. Otto and A. Wojtalewicz-Kasprzak, *Methods for Recovery of Rare Earths from Fluorescent Lamps, US patent*, 7 976 798 B2, 2011.
65. A. Turley, Solvay Rare Earth Recycling in France, RSC Chemistry World, http://www.rsc.org/chemistryworld/2012/09/rare-earth-solvay-recycling, 2012, Accessed 24th March 2013.
66. CBC News, Honda to Recycle Rare Earth Metals from Hybrid Batteries, http://www.cbc.ca/news/business/story/2012/06/20/honda-rare-earth-metals-recycling.html, 2012, Accessed 24th March 2013.
67. C. James, *J. Am. Chem. Soc.*, 1907, **29**, 495.
68. M. James, *The Life and Work of Charles James*, http://unhmagazine.unh.edu/f10/charles_james.html, UNH Magazine Online, 2010, Accessed 24th March 2013.
69. *Separation of Rare Earth Elements Commemorative Booklet*, American Chemical Society, 1999.
70. G. R. Choppin and R. H. Dinius, *Inorg. Chem.*, 1962, **1**, 140.
71. M. Tanaka, T. Oki, K. Koyama, H. Narita and T. Oishi, *Handbook on the Physics and Chemistry of Rare Earths*, J.-C. G. Bünzli and V. K. Pecharsky, Elsevier Science Publishers, 2013, vol. 43, ch. 255, p. 159.
72. M. Takeda, *Molten Salts*, 2000, **43**, 103.
73. A. Ohta, *Tech. Rep. Sumitomo Spec. Metals*, 2003, **14**, 20.
74. O. Takeda, T. H. Okabe and Y. Umetsu, *J. Alloys Compd.*, 2006, **408–412**, 387.
75. O. Takeda, T. H. Okabe and Y. Umetsu, *J. Alloys Compd.*, 2004, **379**, 305.
76. N. Sato, N. Wei, M. Nanjo and M. Tokuda, *J. MMIJ*, 1997, **112**, 1082.
77. J. W Lyman and G. R Palmer, *US Pat.* 5129945, 1992.
78. J. W. Lyman and G. R. Palmer, *High Temp. Mater. Process.*, 1993, **11**, 175.
79. J.-C. Lee, W.-B. Kim, J. Jeong and I. J. Yoon, *J. Korean Inst. Met. Mater.*, 1998, **36**, 967.
80. T. Itakura, R. Sasai and H. Itoh, *J. Alloys Compd.*, 2006, **408–412**, 1382.
81. N. V. Thakur, D. V. Jayawant, N. S. Iyer and K. S. Koppiker, *Hydrometallurgy*, 1993, **34**, 99–108.
82. M. S. Lee, G.-S. Lee, J.-Y. Lee, S-D. Kim, J.-W. Ahn and J.-S. Kim, *Mater. Trans.*, 2005, **46**, 259.
83. K. Shimojo, H. Naganawa, J. Noro, F. Kubota and M. Goto, *Anal. Sci.*, 2007, **23**, 1427.
84. L. Natrajan, J. Pécaut and M. Mazzanti, *Dalton Trans.*, 2006, 1002.
85. M. Niinae, K. Yamaguchi, N. Ishida, A. Djohari, Y. Nakahiro and T. Wakamatsu, *J. MMIJ*, 1994, **110**, 981.
86. H. Matsuura, H. Numata, R. Fujita and H. Akatsuka, *J. Phys. Chem. Solids*, 2005, **66**, 439.
87. N. Tzanetakis and K. Scott, *J. Chem. Tech. and Biotech.*, 2004, **79**, 919.

88. L. E. O. C. Rodrigues and M. B. Mansur, *J. Power Sources*, 2010, **195**, 3735.

89. L. Pietrelli, B. Bellomo, D. Fontana and M. R. Montereali, *Hydrometallurgy*, 2002, **66**, 135.

90. V. Innocenzi and F. Veglio, *J. Power Sources*, 2012, **211**, 184.

91. A. Otsuki, G. Dodbiba, A. Shibayama, J. Sadaki, G. Mei and T. Fujita, *Jpn. Appl. Phys.*, 2008, **47**, 5093.

92. T. Fujita, G. Dodbiba and A. Otsuki, *Rare Earths*, 2009, **54**, 28.

93. T. Takahashi, A. Takano, T. Saitoh, N. Nagano, S. Hirai and K. Shimakage, *J. MMIJ*, 2001, **117**, 579.

94. R. Shimizu, K. Sawada, Y. Enokida and I. Yamamoto, *J. Supercrit. Fluids*, 2005, **33**, 235.

95. K. Kato, T. Yoshioka and A. Okuwaki, *Ind. Eng. Chem. Res.*, 2000, **39**, 943.

96. K. Kato, T. Yoshioka and A. Okuwaki, *Ind. Eng. Chem. Res.*, 2000, **39**, 4148.

97. Y. Jianga, A. Shibayamab, K. Liuc and T. Fujita, *Hydrometallurgy*, 2005, **76**, 1.

98. M. Mitsuaki, S. Atsushi, M. Keiei, F. Toyohisa and K. Tadashi, *J. Mining and MMIJ*, 2003, **119**, 668.

99. I. Billard, *Handbook on the Physics and Chemistry of Rare Earths*, J.-C. G. Bünzli and V. K. Pecharsky, Elsevier Science Publishers, Netherlands, 2013, **43**, 256, 213.

100. *Critical Raw Materials for the EU*, Report of the RMSG Ad-hoc working group on defining critical raw materials, 2010.

101. J. J. Katz and G. T. Seaborg, The Chemistry of The Actinide Elements, Methuen & Co., The Pitman Press, Great Britain, 1957.

102. M. B. Jones and A. J. Gaunt, *Chem. Rev.*, 2012.

103. L. Natrajan, F. Burdet, J. Pécaut and M. Mazzanti, *J. Am. Chem. Soc.*, 2006, **128**, 7152.

104. C. Fillaux, D. Guillaumont, J.-C. Berthet, R. Copping, D. Shuh, T. Tyliszczak and C. Den Auwer, *Phys. Chem. Chem. Phys.*, 2010, **12**, 14253.

105. M. Betti, *J. Environ. Radioact.*, 2003, **64**, 113.

106. F. Luan and W. D. Burgos, *Envir. Sci. Technol.*, 2012, **46**, 11995.

107. P. E. Hodgson, Nuclear Power, *Energy and the Environment*, Imperial College Press, Great Britain, 1999.

108. P. Dyck and M. J. Crijns, Rising Needs, *IAEA Bul.l*, 1998, **40**, 1.

109. Nuclearmatters.co.uk, *Re-use of Plutonium as MOX Fuel*, http://nuclearmatters.co.uk/2012/02/re-use-of-plutonium-as-mox-fuel, 2012, Accessed 24th March 2013.

110. World Nuclear Association, *Processing of Used Nuclear Fuel*, 2012, http://www.world-nuclear.org/info/inf69.html#a, Accessed 24th March 2013.

111. United States Nuclear Regulatory Commission, *Nuclear Poison*, http://www.nrc.gov/reading-rm/basic-ref/glossary/nuclear-poison-or-neutron-poison.html, 2012, Accessed 24th March 2013.

112. K. L. Nash, *Solvent Extraction and Ion Exchange*, 1993, **11**(4), 729.

113. M. P. Jensen, L. R. Morss, J. V. Beitz and D. D. Ensor, *J. Alloys Compd.*, 2000, **303–304**, 137.
114. P. Panak and A. Geist, *Chem. Rev.*, 2013, **113**, 1199.
115. N. C. O'Boyle, G. P. Nicholson, T. J. Piper, D. M. Yalor, D. R. Williams and G. Williams, *Appl. Radiat. Isot.*, 1997, **48**, 183.
116. C. S. Dileep, P. S. Poonam Jagasia, P. V. Dhami, A. D. Achuthan, U. Moorthy, S. K. Jambunathan, P. K. Munshi, Dey and B. S. Tomar, *BARC Newsletter*, 2007, **285**, 130.
117. H. Schmieder, G. Petrich and A. Hollmann, *J. Inorg. Nucl. Chem.*, 1981, **43**(12), 3373.
118. S. C. Tripathi and A. Ramanujam, *Sep. Sci. Technol.*, 2003, **38**, 2307.
119. G. Thiollet and C. Musikas, *Solvent Extr. Ion Exch.*, 1989, **7**, 813.
120. G. L De Poorter and C. K. Rofer-De Poorter, 720872, 1976, *US Pat.*, 4080273, 1978.
121. E. P. Horwitz, D. C. Kalina, H. Diamond, G. F. Vandegrift and W. W. Schulz, *Solvent Extr. Ion Exch.*, 1985, **3**(1), 75.
122. A. Banc, P. Bauduin and O. Diat, *Chem. Phys. Lett.*, 2010, **494**(4–6), 301.
123. J. Muller, L. Bethon, N. Zorz and J-P. Simonin, *Proceedings of the First ACSEPT International Workshop*, ACSEPT, Lisbon, http://www.acsept. org., 2010.
124. C. Brassier-Lecarme, P. Baron, J. L. Chevalier and C. Madic, *Hydrometallurgy*, 1997, **47**, 57.
125. R. Courson, G. Malmbeck, K. Pagliosa, B. Romer, J.-P. Satmark, P. Glatz, Baron and C. Madic, *Radiochim. Acta*, 2000, **88**, 865.
126. M. Sypula, A. Wilden, C. Schreinemachers and G. Modolo, *Proceedings of the First ACSEPT International Workshop*, ACSEPT, Lisbon, http:// www.acsept.org., 2010.
127. Z. Kolarik, U. Mullich and F. Gassner, *Solvent Extr. Ion Exch.*, 1999, **17**, 23.
128. Z. Kolarik, U. Mullich and F. Gassner, *Solvent Extr. Ion Exch.*, 1999, **17**, 1155.
129. C. Hill, L. Berthon, P. Bros, J-P. Dancausse and D. Guillaneux, *7th Information Exchange Meeting Session II*, Nuclear Energy Agency, Korea, https://www.oecd-nea.org/pt/iempt7/, 2002.
130. S. Bourg, C. Hill, C. Caravaca, C. Rhodes, C. Ekberg, R. Taylor, A. Geist, G. Modolo, L. Cassayre, G. de Angelis, A. Espartero, S. Bouvet and N. Ouvrier, *Nucl. Eng. Des.*, 2011, **241**, 3427.
131. G. Modolo, A. Wilden, A. Geist, D. Magnusson and R. Malmbeck, *Radiochim. Acta*, 2012, **100**, 715.
132. A. Geist, U. Müllich, D. Magnusson, P. Kaden, G. Modolo, A. Wilden and T. Zevaco, *Solvent Extr. Ion Exch.*, 2012, **30**, 433.
133. A. Wilden, C. Schreinemachers, M. Sypula and G. Modolo, *Solvent Extr. Ion Exch.*, 2011, **29**, 190.
134. E. Aneheim, C. Ekberg, A. Fermvik, M. R. St. J. Foreman, T. Retegan and G. Skarnemark, *Solvent Extr. Ion Exch.*, 2010, **28**(4), 437.

135. M. G. B. Drew, M. R. S. J. Foreman, C. Hill, M. J. Hudson and C. Madic, *Inorg. Chem. Commun.*, 2005, **8**, 239.
136. F. W. Lewis, L. M. Harwood, M. J. Hudson, M. G. B. Drew, J.-F. Desreux, G. Vidick, N. Bouslimani, G. Modolo, A. Wilden, M. Sypula, T.-H. Vu and J.-P. Simonin, *J. Am. Chem. Soc.*, 2011, **133**, 13093.
137. M. Wei, X. Liu and J. Chen, *J. Radioanal. Nucl. Chem.*, 2012, **291**, 717.
138. M. Milsson and K. L. Nash, *Solvent Extr. Ion Exch.*, 2009, **27**(3), 354.
139. B. Weaver and T. A. Kappelmann, *Talspeak: A new method of separating americium and curium from the lanthanides by extraction from an aqueous solution of an aminopolyacetic acid complex with a monoacidic organophosphate or phosphonate*, Oak Ridge National Laboratory, 1964, 1–60.
140. B. J. Mincher, L. R. Martin and N. C. Schmitt, *Inorg. Chem.*, 2008, **47**, 6984.
141. G. Modolo, P. Kluxen and A. Geist, *Radiochim. Acta*, 2010, **98**, 193.
142. K. L. Nash, *Actinide Solution Chemistry, Proceedings of the Eighth Actinide Conference Actinides*, Manchester, 2005.

CHAPTER 7

Anthropospheric Losses of Platinum Group Elements

N. T. NASSAR

Center for Industrial Ecology, School of Forestry and Environmental Studies, Yale University, 195 Prospect Street, New Haven, Connecticut 06511, USA
Email: nedal.nassar@yale.edu

7.1 Overview

Ruthenium (Ru), rhodium (Rh), palladium (Pd), osmium (Os), iridium (Ir) and platinum (Pt) are among the rarest elements in Earth's upper continental crust.[1] Despite this geological scarcity, these platinum group elements (PGEs) are utilized today in a wide range of industrial and consumer applications, including automotive catalytic converters (autocatalysts), jewellery, electronics, chemical and petrochemical catalysts, dental alloys, biomedical devices, anticancer drugs, glass manufacturing and laboratory equipment, as well as being instruments of financial investment. Increasing demand for these technologically important elements necessitates a better understanding of how we, as a global society, are utilising these metals. How much is being extracted, processed and used? How much is being recycled and how much is being lost? In the complexity of our technological system, the anthroposphere, these rather simple questions are often not simple to answer, but answering them can provide insight into concerns regarding resource availability, recycling potential and environmental policy.[2]

   This chapter begins to address these concerns by answering the question of how much is being lost. Utilising the principles of material flow analysis

RSC Green Chemistry No. 22
Element Recovery and Sustainability
Edited by Andrew J. Hunt
© The Royal Society of Chemistry 2013
Published by the Royal Society of Chemistry, www.rsc.org

(MFA), global losses at each of the major life cycle stages, from comminution and concentration to end-of-life disposal and recycling, are estimated for five of the six PGEs. Owing to its limited industrial use and related lack of data, Os is excluded from the analysis.

Although MFAs have become relatively common in the literature, a recent review[3] reveals that there are only a few regional (Europe[4,5]) and national (United States,[6] Germany,[7] Japan[8,9] and Russia[10]) studies that focus on Pt, Pd and Rh and only one global-level study that focuses on future stocks and flows of Pt using dynamic scenario analysis.[11] No systematic studies were found for Ru, Ir or Os. This analysis is therefore the first to provide a comprehensive overview of contemporary global losses for all five of the industrially relevant PGEs.

To provide some context, this chapter begins by describing the basic methodology that is used to estimate these losses, before discussing the loss and recovery rates at the various life cycle stages and in turn, providing estimates of global losses for year 2010. It is important to note that the focus of this analysis is on the anthropospheric losses of PGEs (*i.e.* losses of PGEs from the anthroposphere to the biosphere), as opposed to the flows that are controlled by natural phenomena (*i.e.* biospheric) such as volcanic eruptions and meteoritic debris.

7.2   Methodology

Losses invariably occur at every stage of the life cycle of the PGEs. To estimate these losses on a global scale requires data on material flows as well as estimates of recovery rates at each of these stages. In the simplest case when both the flow and recovery rate are known, losses can simply be estimated as follows:

$$Losses_t = \sum_i (I_{t,i} \cdot (1 - R_{t,i})) \tag{7.1}$$

where for process i and for time period t, I is the PGE-content mass input flow and R is the PGE recovery rate.

Input flows for certain processes (*e.g.* purchases by the automotive industry) are often provided in the literature but may need to be adjusted for changes in industry stock to calculate losses appropriately. For some processes (*e.g.* refining production), only output flows (denoted as O), meaning the amount that is recovered from a process, rather than input flows are reported. Losses for such processes are thus calculated as follows:

$$Losses_{t,i} = O_{t,i} \cdot \left(\frac{1}{R_{t,i}} - 1\right) \tag{7.2}$$

In some cases neither the input into nor the output from a process of interest are known. This is often the case for the flows out of use and into the waste management system. Here, instead, one must indirectly estimate the flow into the process of interest by knowing how much has previously entered the process that immediately preceded it and how long it stayed there. Such calculations therefore require historical data on input flows and estimated process residence times (*e.g.* product lifetimes) as illustrated in Equation (7.3), where again, *I* and *O* are inputs and outputs of processes, respectively, δ refers to dissipative losses during the process, subscript *i* refers to the process of interest, subscript *i–1* refers to the process which feeds directly into that process of interest and $L(t,t')$ is a lifetime distribution function that describes the probability that the metal exits the process at time *t* if it entered the process at time *t'* (*i.e.* its vintage year).[12]

$$I_{t,i} = O_{t,i-1} = \int_{t_0}^{t} L_{i-1}(t, t') \cdot (I_{t,i-1} - \delta_{t,i-1}) dt' \tag{7.3}$$

Lifetime distributions for various end-use applications can often be found in the literature and are typically noted as being either normal, log normal or Weibull in nature.[12] Taking the case of a Weibull distribution yields Equation (7.4) where α and β are the shape and scale parameters of the 2-parameter Weibull distribution, respectively:

$$I_{t,i} = \int_{t_0}^{t} \left(\left(\frac{\alpha}{\beta} \left(\frac{t-t'}{\beta} \right)^{\alpha-1} \cdot e^{-\left(\frac{t-t'}{\beta} \right)^{\alpha}} \right) \cdot (I_{t,i-1} - \delta_{t,i-1}) \right) dt' \tag{7.4}$$

Input flows into the use phase, I_{i-1}, often consist of two components: inputs from primary (*i.e.* geological) sources and inputs from secondary (*i.e.* recycled) sources.

The PGEs primary demand (sometimes referred to as net demand) for each of the major sectors is typically known and is reported annually in a number of publications. With the exceptions of autocatalysts and more recently jewellery and electronics, input from secondary sources into the various sectors is typically not reported. Flows for secondary sources into these applications thus have to be indirectly estimated. This is especially necessary in industrial applications in which a significant portion of total demand is fulfilled by secondary sources. In these applications the metals are sent for recovery at the end-of-life and return after processing without exchange of ownership. Input flows into use from secondary sources are thus the output flows from these end-of-life waste management processes, which are assumed to be the output flows from use in the previous time interval multiplied by the recovery rate in the waste management stage, as illustrated in Figure 7.1 and Equation (7.5):

$$I_{t,i-1} = I_{t,i-1}^{p} + I_{t,i-1}^{s} = I_{t,i-1}^{p} + O_{t,i} = I_{t,i-1}^{p} + O_{t-1,i-1} \cdot R_{t,i} \tag{7.5}$$

where superscripts p and s refer to primary and secondary sources, respectively, subscript *i* refers to the process of interest (which in this case is the waste management process), subscript *i–1* refers to the process preceding the process of interest (which in this case is the use phase) and subscript *t–1* refers to the previous time interval.

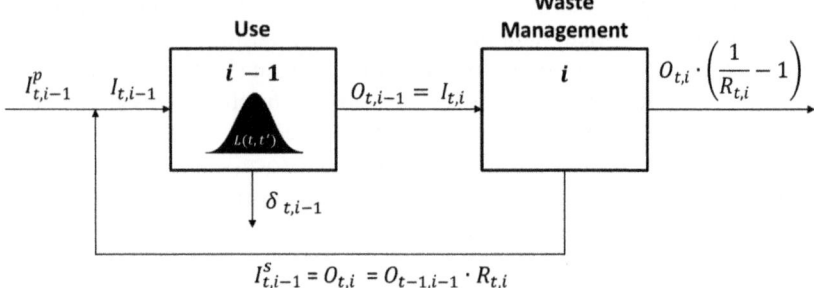

Figure 7.1   Annotated process flow diagram of the use and waste management stages illustrating product life time dynamics and recycling.

Combining the previous set of equations yields the following:

$$Losses_{t,i} = \left(\int_{t_0}^{t} \left(\frac{\alpha}{\beta} \left(\frac{t - t'}{\beta} \right)^{\alpha - 1} \cdot e^{-\left(\frac{t - t'}{\beta} \right)^{\alpha}} \right) \right.$$

$$\left. \cdot \left(I_{t,i-p}^{P} + O_{t-1,i-1} \cdot R_{t,i} - \delta_{t,i-1} \right) dt' \right) \cdot \left(1 - R_{t,i-1} \right)$$

(7.6)

The observant reader will notice that Equation (7.6) contains an output flow from the use phase $(O_{t-1,i-1})$. This variable by itself requires a calculation based on Equation (7.3) for the preceding time period and thereby creates a nested set of equations that are best solved using system dynamics modelling software or other computational means.

Historical supply and demand data necessary for completing such a computation were obtained from a variety of references.[13–18] The remaining necessary variables of lifetime distributions and loss and recovery rates are the topic of the next section.

7.3   Loss and Recovery Rates

In this section rates of losses and recovery at each of the major life cycles stages will be discussed, beginning with the extraction and beneficiation of the PGE-containing ore. Loss rates at manufacturing, use and waste management are then discussed on an application by application basis beginning with auto-catalysts. Losses during fabrication are generally excluded from the analysis owing to lack of data but are assumed to be minimal. Table 7.1 provides a summary of the loss and recovery rates found in the literature.

7.3.1   Mining, Comminution and Concentration, Smelting and Refining

Today major economically viable geological resources and production of PGEs are concentrated in South Africa, Russia and to a lesser degree in Canada, the

USA and Zimbabwe.[19] Several other countries including Australia, Botswana, China, Colombia, Ethiopia, Finland, Indonesia, Japan, the Philippines, Poland and Serbia also produce varying but considerably smaller quantities of PGEs.[17] In the major deposits, PGEs are typically found at concentrations of only a few parts per million, with the specific proportion of each element varying notably by site.[19] Recovering the elements from the Earth's crust at such low concentrations and separating them into individual high-purity metals is no small feat and requires a number of different processes. In general, these processes can be categorized into four main operations: mining, comminution and concentration, smelting and refining. Detailed descriptions of each of these processes can be found in the literature.[20–27]

Recovery rates at these first stages of the life cycle differ from site to site owing to varying geological and technological factors. Several sources in the literature provide some useful estimates. Vermaak suggests a PGE recovery rate of 80–87% at concentration.[27] Recovery rates of 85%, 95–98% and 99% for concentration, smelting and refining, respectively are noted as typical by Jones.[28] Merkle and McKenzie[22] note similar recovery rates of 80% to >90% for comminution and flotation, 95–98% for smelting and converting, >99% for

Table 7.1 Loss rates and product life times by major end-use application.

Metal	End-use category	% of primary demand (2010)[14]	Manufacturing, process or in-use loss (%)	Average useful life (years)	End-of-life recovery rate[a]
Platinum	Autocatalyst	32.8	2[b]	16[80,81,c]	78–83[7,d]
	Chemical	7.2	0.5–79[7,e]	0.33–4[7,e]	95–98[7,e]
	Electrical	3.6	N.A.	10[7,f]	0–38[7,76]
	Glass	6.3	<1[7]	2[7]	>98[7]
	Jewellery	27.7	4[7,g]	30[f]	90–100[76,h]
	Investment	10.8	N.A.		
	Medical & biomedical	3.8	3–100[i]	1–8[82,j]	15–20[76]
	Petroleum	2.8	1–2[7]	1–12[7,83,k]	98[7]
	Other	4.9	N.A.	15[f]	10–20[76]
Palladium	Autocatalyst	54.2	2[b]	16[80,81,c]	78–83[7,d]
	Chemical	4.7	0.5–48[7,e]	0.33–5[e]	95–98[7,e]
	Dental	7.5	3[7]	N.A.	15–20[76]
	Electrical	12.3	N.A.	10[7,f]	5–38[7,76]
	Investment	13.9	N.A.		
	Jewellery	6.3	4[7,g]	30[f]	90–100[76,h]
	Other	1.1	N.A.	15[f]	15–20[76]
Rhodium	Autocatalyst	75.2	2[b]	16[80,81,c]	78–83[7,d]
	Chemical	10.4	10–25[7,e]	0.1–5[7,83,e]	>96[7,e]
	Electrical	0.6	N.A.	10[7,f]	5–10[76]
	Glass	10.5	<1[7]	2[7]	>98[7]
	Other	3.3	N.A.	15[f]	30–50[76]
Ruthenium	Chemical	10.6	N.A.	10–15[84,85]	85[l]
	Electrochemical	13.1		5–8[23]	40–50[76,m]
	Electrical	71.9		10[f]	0–5[76]
	Other	4.4		15[f]	0–5[76]

Table 7.1   (*Continued*)

Metal	End-use category	% of primary demand (2010)[14]	Manufacturing, process or in-use loss (%)	Average useful life (years)	End-of-life recovery rate[a]
Iridium	Chemical	5.3	N.A.	$0.1-5^{83}$	$40-50^{76,m}$
	Electrochemical	23.4		$5-8^{23}$	
	Electrical	59.5		$2^{86,87,n}$	
	Other	11.8		15^{f}	$5-10^{76}$

[a]If a range is reported, the middle value is used in the analysis unless otherwise stated in the following notes.
[b]Value is based on a model-calculated average of in-use losses as a percentage of total demand on an annual basis for the past 10 years. See text for more detail.
[c]Value is a fitted weighted average based on 1990 average domestic service lifespan for vehicles in the United States, Germany, United Kingdom, France, Brazil, Australia, China, India and Japan. Weighting is based on the number of vehicles in operation in those countries in year 2010 using data from Ward's Automotive Group. These countries represent approximately 52% of world light-duty vehicles in operation that year. Autocatalyst life is assumed to equal that of the vehicle. Lifespan database for Vehicles, Equipment and Structures: LiVES by National Institute for Environmental Studies (Japan) was used to find the noted reference. Along with the average value, a Weibull distribution is fitted accordingly. The resultant scale and shape parameters are 17.96 and 2.57, respectively.
[d]Range represents recovery rates once the vehicles are de-registered and includes losses at the shredder, dismantler, collector, decanner and refiner but does not included collection rate losses or unidentified flows.
[e]Process losses, average useful life until irreversible deactivation and end-of-life recycling rates of the PGEs can vary considerably from one chemical application to another. Values in the analysis were based on estimates noted by Hagelüken *et al.*[7] and allocated according to the portion of gross demand per application noted for Germany.
[f]Value is based on author's own estimations unless a reference is given. Value is also assumed to have a Weibull distribution with a shape parameter of 3.5.
[g]Value refers to manufacturing losses.
[h]Manufacturing losses are assumed to be included in this range and so the upper value of 100% is utilized in the analysis.
[i]Net manufacturing losses for dental alloys are relatively small at less than 3%. Anti-cancer drugs are, however, assumed to be completed lost during use.
[j]Anti-cancer drugs are assumed to be consumed in the same year of production. Biomedical devices are assumed to have an average life span of 8 years. A Weibull distribution with a scale parameter of 9 and a shape parameter of 3 is fitted based on data on the longevity of implantable cardioverter-defibrillators. Dental alloys are assumed to be lost with the patient.
[k]An average life of 5 years, noted by Hagelüken *et al.*,[7] is used in the analysis.
[l]Value obtained from personal communication with an industry expert.
[m]Values are noted for industrial applications.
[n]Value is for Ir crucibles which are noted as having a useful life of as low as 2–3 months but also as high as several years in mildly oxidizing atmospheres. A value of 2 years is assumed with a shape parameter of 3.5.

base metal refining and 98–99% for precious metal refining. These figures are for South African operations but overall recovery rates from Russian ores are not dissimilar at 83.4% with ore processing accounting for 71% of the losses.[29] Vermaak [27] notes similar Russian mill recovery rates ranging from 77–92% with an overall average of just under 87% for Pt and 80–88% with an overall average of just over 84% for Pd. In North America, the Stillwater Mining Company notes that mill recovery rates of PGEs have been consistently at about 91% for years 2009–2011 at its Montana operations[30] while in Canada, North American Palladium reported a Pd recovery rate of 80.8% for its Lac des Iles mine in 2010.[31]

As indicated by these estimates, the greatest losses occur during comminution and concentration. As a result of mineralogical differences, ores from the main ore bodies in the Bushveld Complex in South Africa, the Upper Group 2 (UG2), Platreef and Merensky Reef, are often treated separately.[22,25] Concentrator recovery rates are thus different, with ore from Merensky Reef being typically higher than those from UG2 owing to the higher level of chromite.[20] Estimates of concentrator recovery rates from major mining companies,[26,30–34] along with production statistics,[17,19] are used to calculate site-specific losses at comminution and concentration for the year 2010. A recovery rate of 85% was assumed when site-specific information was not available. These recovery rates are assumed to apply to both Pt and Pd. In general, concentrator recovery rates for Rh, Ru and Ir are thought to be lower by an additional 5–10% (Johnson Matthey Plc., personal communication). The middle value of this range is used in the calculations.

As noted earlier, smelting operations maintain recovery rates of around 95–98%. While these recovery rates are relatively high, losses of PGEs at this operational stage are still notable and in some instances result in measurable quantities of PGEs in the environment.[35,36] An overall recovery rate of 96% for smelting is assumed for the purposes of this analysis. Recovery rates at the smelter are believed to not vary significantly between the PGEs (Johnson Matthey Plc., personal communication).

Recovery rates for refining are generally greater than 95% and are often at or above 99% for Pt, Pd and Rh.[20] Element-specific recovery rates vary for a number of reasons including the type of extraction method used. Refineries of several major PGE producers are noted to have typical recovery rates of 99–99.5% for Pt, 98–99.7% for Pd, 93–99.5% for Rh, 98–99.5% for Ru and 90–99.5% for Ir.[20] Refinery recovery rates of 99.5% for Pt and Pd, 98% for Rh, 99% for Ru and 96% for Ir are used in this analysis based on these company-specific recovery rate estimates and the 2010 production data[17,19] of these major producers.

Combining the recovery rates for comminution and concentration, smelting and refining yields overall recovery rates that range from approximately 71–81%. These estimates are similar to those used in analyses elsewhere. As illustrated in their analysis, Saurat and Bringezu[4] use an average recovery of 85% for milling and concentrating, 96% for pyrometallurgy and 100% for refining resulting in an overall recovery rate of 81.6%. Råde and Andersson[37] use the lower-end of loss rate ranges noted in the literature to come up with an overall recovery rate of 79%.

In addition to the losses mentioned here, there are resources in the ground that may never reach the mill owing to geological and extraction losses. Vermaak[27] suggests that "...mining losses due to potholes, replacement pegmatoids, faulting and grade losses..." vary from 20% to 30%. He uses these estimates to calculate the difference between *in situ* and mill-head resources (*i.e.* resources in the ground *versus* those delivered to the mill). As noted by Råde and Andersson,[37] another estimate[38] suggests that for underground mining "10–25% of the in-place resources cannot be extracted". In addition, Merkle and McKenzie[22] suggest that the mine-call-factor which refers to the quantity of PGE delivered to the mill compared to the quantity expected from the mined ore, can be as low as 90% for ore from the Merenesky Reef but is generally higher for the ore from UG2. PGE mining companies perform their own calculations which they use to modify their resource estimates. In evaluting its Crocodile River operations, for example, Eastern Platinum Ltd. (Eastplats) notes modifiying factors of 20% for geological losses which include faults, dykes, potholes and iron rich ultramafic pegmatite (IRUP), 5% for mining losses and 11–15% for pillar losses, while also noting mine-call-factors in the range of 90%.[39] Although these resource adjustments are signficant they are not addressed in this analysis as they are already accounted for in reserve estimates.

7.3.2 Autocatalysts

Since 1974 in the USA and later on elsewhere, PGEs have been used in autocatalysts to reduce the emissions from vehicle exhausts in order to comply with air quality regulations. The amount of Pt, Pd and/or Rh used per vehicle varies by catalyst volume, engine type and size, vehicle type, year of manufacture (owing to changing emissions regulation) and the price of the metals.[7] In the light of continued tightening of emissions regulations worldwide, a combination of these metals in varying proportions is used in an overwhelming majority of new light-duty vehicles, including both those with spark-ignition (petrol or gasoline) and compression-ignition (diesel) engines and increasingly in heavy-duty or non-road vehicles (*e.g.* construction vehicles) as well.[14] Demand for PGEs for autocatalysts is thus an important driver for the entire PGE industry and represents a significant portion of overall primary demand for Pt (33%), Pd (54%) and Rh (75%).[13]

Losses of PGEs occur during the manufacturing, use, disposal and recycling of autocatalysts. No data has been found for losses of PGEs during manufacture of autocatalysts, but these losses are assumed to be relatively small and are thus ignored in this analysis. Small but notable losses of Pt, Pd and Rh do, however, occur during the use of vehicles. These losses to the environment are due to the physical and chemical stresses of the "fast changing oxidative/reductive conditions, high temperature and mechanical abrasion" that occur during normal use,[40] as well as due to damage or destruction of the autocatalyst caused by inadequate or improper maintenance and poor road conditions.[7] Although Ir was previously used in autocatalysts in some Japanese and European petrol (gasoline) vehicles,[14] no estimates of losses have been found in the literature and thus these losses are excluded from this analysis.

Reviews by Ravindra *et al.*,[40] Ek *et al.*[41] and others[42–44] cite numerous studies that find PGEs in soil and roadside dust,[45–49] airborne particles,[47,50,51] vegetation,[52–54] water ecosystems,[55–57] and biological matrices of animals[58–60] and humans[51,61–63] across the globe, including remote areas like the Alps[64] and Greenland,[65] often at levels above those that would otherwise be attributed to natural sources such as meteoritic debris.[42] PGE losses from autocatalysts are likely to be an important source of these emissions to the environment, especially those found near the roadside. Both direct and indirect methods have been used to estimate these losses. Indirect methods rely on developing models based on traffic-related data and roadside PGE sample measurements while direct methods utilize controlled experiments to measure PGE emissions from catalysts for various engines sizes, catalyst configurations, vehicle speeds and catalyst ages.[43]

Using the direct method, early studies on pellet-type autocatalysts measured PGE losses ranging from 0.8–1.9 µg of Pt per km travelled by the vehicle.[66] More recent studies on three-way autocatalysts suggested emissions rates on the order of ng per km. Experiments by Moldovan *et al.*,[67] for example, measured losses of Pt, Pd and Rh from several types of autocatalyst fitted into vehicles with both petrol (gasoline) and diesel engines from new to 80 000 km of travel at 10 000 km intervals. They found that mean emission rates for the four different autocatalysts at the start ranged from 44–255 ng of total PGE emissions per km with an upper and lower bound of approximately 27–700 ng km^{-1}. In most cases less than 10% of those losses were found to be in soluble form which is of importance from a toxicological perspective. After ageing (30 000–80 000 km), particulate PGE emissions decreased by approximately an order of magnitude for the Pt–Pd–Rh and Pd–Rh auto-catalysts fitted onto vehicles with petrol (gasoline) engines, but generally did not decrease for the Pt catalysts fitted onto vehicles with diesel engines. For the most part, soluble emissions of Pt, Pd and Rh also did not decrease after ageing and thus made up a greater percentage of the total losses.[67]

Comparable results for direct measurement experiments were reported elsewhere,[40] but as noted by Farago *et al.*[42] the direct method typically suggests lower emission rates owing to the fact that these experiments are run under controlled conditions and do not take into account factors such as damage from accidents, poor or improper maintenance and extreme driving conditions. This may be especially true for older vehicles often shipped from developed to developing markets as second-hand vehicles, where damage and poor main-tenance lead to greater losses of PGEs during use.[68] Helmers[69] notes that when comparing the results of the direct and indirect methods, the overlapping range is on the order of 0.5–0.8 µg Pt km^{-1}.

Noting that PGE losses decrease with vehicle age, Lloyd *et al.*[70] developed an exponential function to estimate losses of PGEs per km driven based on initial the PGE loading level and vehicle age. The function assumes losses on the order of 1µg km^{-1} during the vehicle's first year of use down to 10 ng km^{-1} during year 30 for vehicles initially loaded with 6 g of PGE.[70] This loss function is utilized in the present analysis to estimate global losses of PGEs from auto-catalysts during use on an annual basis. As the loss function is vehicle age-dependent, losses for each vehicle vintage year are calculated independently and then summed to estimate the total losses in a given year. To determine the initial loading level for each vehicle vintage year, t', the annual amount of Pt, Pd and Rh sold to autocatalyst manufacturers[13] during that vintage year, $I_{t'}$, is first adjusted for any changes in automotive industry stock[71] for that vintage year, $\sigma_{t'}$ and then divided by the number of autocatalyst-equipped vehicles produced during that vintage year, $N_{t'}$. This value is utilized in the loss function noted by Lloyd *et al.*[70] yielding a PGE loss rate per km of vehicle travel for a specific vintage year based on that vehicle's age during the year in question (*i.e.* $t - t'$). This loss rate is then multiplied by the average distance travelled per vehicle, which is also assumed to be vehicle age-dependent. Estimates of US average light-duty vehicle distance travelled, τ, by vehicle age noted by Jackson[72] are multiplied by 77% to approximate a global average.[73] The result is then multiplied by the number of vehicles of that vintage year that are still in operation during the year in question, t, which is calculated using the lifetime distribution function, $L(t,t')$, noted earlier. Summing these results across the relevant vintage years provides an estimate of PGE losses from all operating autocatalysts for the year in question, δ_t:

$$
\delta_t = \sum_{t_0}^{t} \left(\frac{(I_{t'} - \sigma_{t'})/N_{t'}}{6} \cdot 1.2 \times 10^{-6} \cdot e^{-0.16(t-t')} \cdot \tau(t - t') \right.
$$

$$
\left. \cdot 77\% \cdot N_{t'} \cdot (1 - L(t,t')) \right)
$$

$$(7.7)$$

Aside from the assumptions inherent in the PGE-loss function developed by Lloyd *et al.*[70] and those associated with the average vehicle distance travelled as a function of vehicle age, this approach relies on the assumption that all installed PGEs were fitted into autocatalyst-equipped vehicles that were sold and driven that year, which may not be the case as some vehicles remain in dealers' inventories while those purchased may not be driven. There is also an implicit assumption that travel distance is independent of PGE-loading (*i.e.* vehicles with higher PGE-loadings do not on average travel a greater or lesser distance than those with lower PGE-loadings) and that the probability of a vehicle remaining in use is independent of PGE losses. There is also an assumption that no other variables, such as driving conditions or vehicle speed, (which has been noted as being a factor[74]) are significant. Also, the age-dependent loss function developed by Lloyd *et al.*[70] is noted as applicable to both light-duty gasoline and diesel vehicles, although the empirical results noted earlier suggest otherwise.[67] Furthermore, because the use of PGEs in non-road vehicles is very small, both vehicle types are assumed to follow the same rate loss function even though Lloyd *et al.* suggest a higher loss rate function for the non-road vehicles.[70] All of these are less than ideal assumptions but are necessary in order to complete such a calculation on a global scale.

Once vehicles reach the end of their useful lives their autocatalysts can be collected and the valuable PGE content retrieved. As Vermaak noted, it was initially thought that recycling autocatalysts at the end-of-life would be cost prohibitive.[27] This has turned out not to be the case as recycling of auto-catalysts currently provides the largest source of post-consumer secondary supplies of PGEs.[14] As discovered by Hagelüken *et al.*,[7] there are, however, significant quantities (>50%) of PGEs from autocatalysts that never seem to reach the proper waste management system. In their analysis of PGE flows in Germany the most likely explanation of why these PGEs never get recycled is that some are in stolen vehicles while others are in vehicles that were shipped out of the country for resale and which may or may not ultimately be recycled. When a vehicle does reach the proper waste management system there is, however, a good chance that the majority of PGE content will be recovered. Hagelüken *et al.*[7] estimate that approximately 10% of the PGEs that reach the waste management system are lost at the dismantler during the shredding process and around 2–5% are lost at the collector, decanner and refiner stages, resulting in an overall recovery rate of around 78–83% in Germany *circa* 2001.

7.3.3 Jewellery and Investment

Owing to their strength, workability and non-tarnishing qualities,[7,27] as well as their aesthetic appeal and ability to act as portable repositories of wealth, precious metals are often the ideal choice for jewellery fabricators and customers alike. Indeed, jewellery was among the first uses[75] of Pt and has remained one of the largest demand categories.[14] For example, in 2010 jewellery comprised just under a third of primary demand for Pt. Pd use in jewellery is notably less and is largely used as an alloying element although there have been some recent marketing campaigns[30] aimed at establishing Pd itself as a jewellery metal.[7] Ru and Ir are used as alloying elements with Pt in jewellery but overall quantities are presumably relatively small. The majority of demand for Pt jewellery has, since the 1960s, come from Japan. More recently, however, as demographics and socioeconomic conditions have changed, China as well as India, has become a main driver of growth in the industry.[14]

Jewellery is manufactured by both small-scale handicraft enterprises (*i.e.* goldsmiths) and by large-scale industrial manufacturers.[7,27] Notable quantities of production scrap are generated by both, with industrial manufacturers tending to produce significantly larger quantities (40% of material used) compared to goldsmiths (15–20%).[7] Great care is, however, taken by both handicraft enterprises and industrial fabricators to reduce losses and to recycle the production scrap. Irrecoverable losses of Pt and Pd during the fabrication of jewellery are thus estimated to be relatively small at around 2–10%.[7] Therefore, for the purposes of this analysis a loss rate of 4% is assumed for both Pt and Pd.

No information is available regarding losses that occur during the use of jewellery although these are likely to be very small as Pt jewellery is resistant to wear. Until recently, the amount of metal returned after consumer use had not been quantified. However, starting in 2009, Johnson Matthey began publishing estimates dating back to year 2005 of the quantity of Pt and Pd recovered and note that the majority of metals recovered are from Japan and China where the majority of the demand exists and an extensive recycling infrastructure has been recently developed.[14] These estimates also include remanufactured unsold stocks of retailers and wholesalers but do not include the production scrap mentioned above.

Pt and to a lesser degree Pd, are also used as an investment medium in the form of coins, medallions and small and large investment bars with the latter pertaining mostly to Japan. In addition, physically based exchange traded funds (ETFs) in Switzerland, the UK, the USA and elsewhere, which began in 2007, have furthered the use of PGEs in this arena. Net demand for Pt and Pd in investment can fluctuate considerably from year to year but has averaged 9.4% and 5.4% of primary demand for Pt and Pd respectively since 2008.[14] In May 2011 the first-ever physically based Rh ETF was introduced, resulting in a total demand of about 529 kg or less than 3% of primary demand for the year.[14] It is assumed that negligible losses occur during the fabrication and use of the metals in this sector.

7.3.4   Industrial Applications

Owing to their superior catalytic properties, high melting points and corrosion resistance, PGEs play an indispensable role in a wide variety of industrial applications. They are utilized in the form of catalysts in the chemical and petrochemical industries, as alloys in glass manufacturing and laboratory equipment and as coatings in the electrochemical industry, to name but a few. As suggested earlier, there is an important distinction to be made between these industrial applications and those that are more consumer-oriented (*e.g.* auto-catalysts, jewellery and electronics). The use of PGEs in industrial applications occurs in a closed-loop fashion where at the end-of-life the metals are sent to a refinery for recovery in which minimal losses occur. Given such high recovery rates, new demand is typically required only to replenish the amount that was lost during use, recycling and when there are increases in utilisation rates or manufacturing capacity. When losses do occur they are relatively small (typically less than 5%). These losses are described in more detail by Hagelüken *et al.*,[7] and the loss rates noted for Germany, which are assumed to be typical, are used in this analysis.

There are several industrial processes in which notable losses do, however, occur. The use of Pt, Pd and Rh in homogenous catalysis is one such example where overall loss rates vary considerably and are estimated at 79%, 50% and 13%, respectively.[7] In the case of Pt, its use as a catalyst in the cross-linking of silicones results in the loss of the metal in the product from which later recovery is not practical.[7] In contrast, the recovery of Pt used in the production of organosilanes is estimated as being very high at around 90%.[7] Nitric acid production is another example. Here Pt and, to a lesser degree Rh, are used in the form of contact gauzes in the catalytic oxidation of ammonia to nitric oxide. Abrasion and vaporization during the process results in the loss of these PGEs.[7] To reduce these losses, Pd-based "getter" gauzes are placed in the downstream process flow to capture the lost Pt and Rh. The getter gauzes are also recycled and approximately 94% of Pt, 75% of Rh and 70% of Pd are recovered.[7]

Little specific information regarding the loss rates of Ru and Ir in industrial uses is found in the literature. Graedel *et al.*[76] suggest that end-of-life recycling rates of Ru and Ir in industrial applications are of the order of 40–50%. An industry expert suggests that for specific ammonia processes, approximately 85% of Ru catalyst can be recovered by combustion of the carbon substrate (KBR, personal communication). Note that in the present analysis, the use of Ir in the electronics sector is included as an industrial application because its main use in this sector is in high temperature crucibles used for growing sapphire crystal wafers that are ultimately used in light-emitting diode (LED) televisions. The uses of other PGEs in electrical and electronics applications are treated separately below.

7.3.5 Electrical and Electronics

High electrical conductivity, durability and corrosion resistance make PGEs ideal for use in a number of electrical and electronic applications including contact materials used as switches, relays, sensorics such as thermocouples, solders and other components of electronics. Fuel cells are also included in this category but overall demand is still small at about 622 kg Pt in 2010.[14] An important application of Pd in this sector is in multi-layer ceramic capacitors (MLCC) which are used in a variety of consumer electronics including mobile telephones. However, this demand is often vulnerable to miniaturization and substitution with base metals.[7]

No information has been found for losses of PGEs in the manufacture or use of electrical and electronic components. There are several estimates for end-of-life recycling rates that range from virtually no recycling to an upper value of approximately 38%.[7,76] The main reason for such low end-of-life recycling is because a minimum 50% of PGEs in electronics are lost owing to lack of product collection.[7] When the electronic products are collected, losses of PGEs are on the order of 22–25% of the input quantity mostly owing to mechanical processing.[7,77] Johnson Matthey also provide estimates of the quantity of PGEs obtained from recycling of electronics, which are utilized in this analysis.[13]

The subject of waste of electrical and electronic equipment (WEEE) is an important topic that deserves a more in-depth analysis that can be provided in this chapter. Indeed, there is a separate chapter in this book (Chapter 8) dedicated to this topic.[78] There is one WEEE issue relevant to PGEs that should, however, be mentioned here. In the case of Pt and more recently Ru, hard-disk drives have become an important end-use application in which the metals are sputtered onto the magnetic discs to increase their digital storage capacity. The now nearly ubiquitous use of these new techniques has resulted in a surge of demand for these metals and in particular for Ru, as hard disk drive manufacturers have had to build up the quantities necessary to perform the sputtering.[14] With these sputtering techniques, only a small portion, estimated at 4–5%, of the total amount of metal used actually gets deposited on the discs, with the remainder being recovered in close-loop recycling with minimal losses.[7] Production quantities of hard-disk drives,[71] along with estimates of the percentage of these discs that contain Pt and/or Ru, are used to calculate the fraction of the overall demand that is deposited on the discs and enters into use. These values are in turn used to calculate Pt and Ru losses at the end-of-life of this application.

7.3.6 Other Uses

Dental alloys, anticancer drugs, biomedical devices including pacemakers and catheters, non-catalytic automotive applications including spark plugs and oxygen sensors, gas safety sensors, turbine blade coatings, corrosion resistant applications including pipes for the oil and geothermal energy industries and stationary pollution control catalysts are a few of the other applications in which PGEs are used. Most of these applications contribute only a small percentage of overall demand.[14] The use of Pd-based dental alloys, mostly in Japan, accounts for a sizable, albeit decreasing, portion of overall Pd demand.[13] In this application net losses in the production and casting of the alloys and in the milling or grinding of the prosthesis are relatively small as great care is taken to recover the precious metals.[7] A certain portion of crowns and bridges that are removed can be recycled, while the dental prostheses that remain with the patients are lost.[7] As for anticancer drugs it is assumed that the entire quantity is lost to the wastewater system during use.[79] In other applications little information is available regarding losses. End-of-life recycling rates ranging from as low as 0–5% for Ru to 30–50% for Rh are, however, noted as typical in the literature.[76]

7.4 Results and Discussion

Global losses at each of the major stages of the life cycles of Pt, Pd, Rh, Ru and Ir are presented in Figure 7.2. The results indicate that losses are greatest at the first and last stages of the life cycle, namely at comminution and concentration (C) and at the end-of-life waste management (W), with lower losses occurring at smelting (S) and use (U) and minimal losses occurring at refining (R) and manufacturing (M). This U-shaped pattern seems to hold true for all five PGEs but there are notable variations across the life cycle stages and across metals.

On the supply side, losses are greatest at comminution and concentration but this issue is well known and mining companies have been working on improving the recovery rates for some time. For example, Anglo American Platinum Ltd. a major PGE producer in South Africa indicated in its annual report that UG2 recoveries have improved significantly in the past few years, increasing by 6% from 2008 to 2011 and are now over 87% at the mines in Rustenburg and Amandelbult owing to advanced control technologies and optimisation of stirred milling projects.[26] Merkle and McKenzie similarly note that design and operational improvements have increased UG2 recovery by

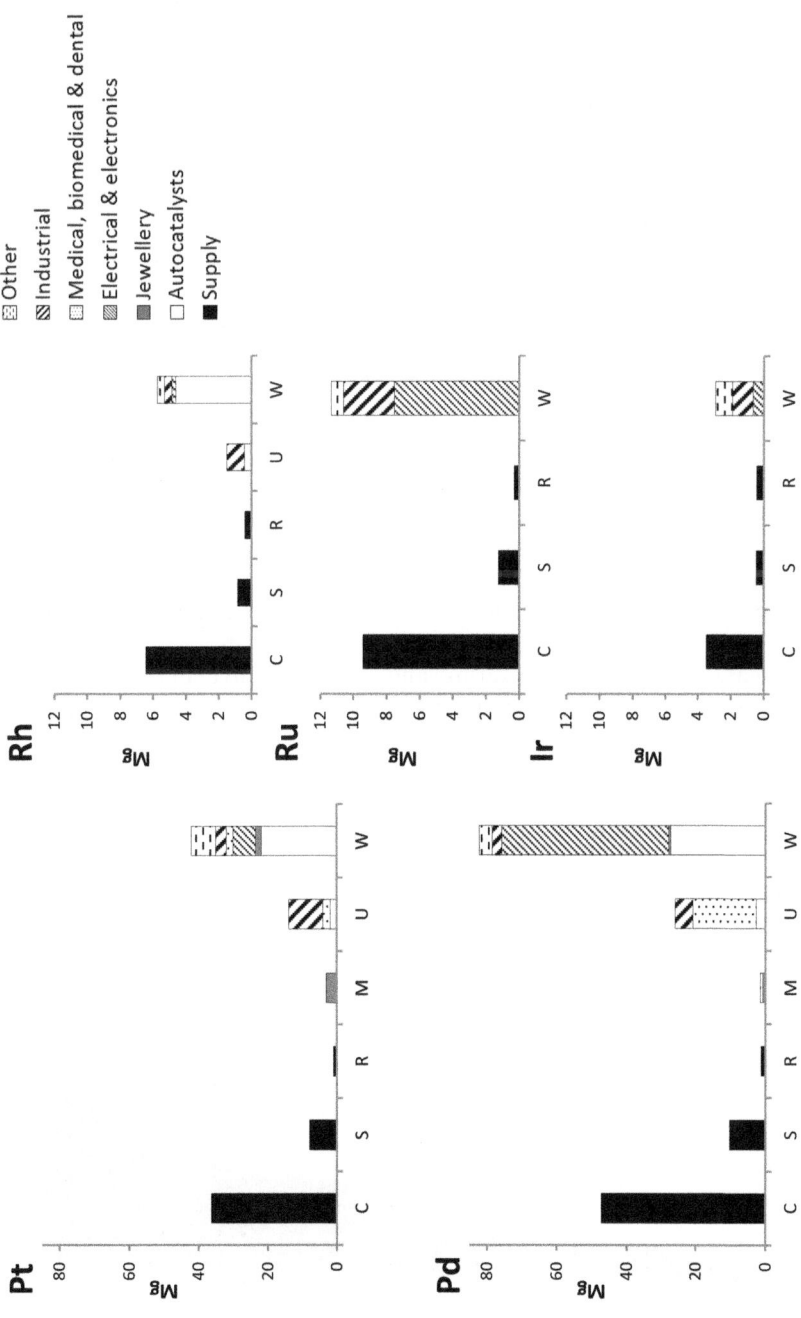

Figure 7.2 Global PGE losses by life cycle stage: comminution and concentration (C), smelting (S), refining (R), manufacturing (M), use (U) and end-of-life waste management (W). Supply-side losses are depicted in black columns, while losses during manufacturing, use, and end-of-life waste management by major end-use application are depicted as stacked columns with varying patterns and shades as noted in the legend. Results are for year 2010 in Mg (metric tonnes).

5%–10%.[22] This is encouraging considering that over the last decade a greater portion of ore being processed was obtained from UG2.[19]

Losses during manufacturing, use and end-of-life waste management are estimated on an application by application basis. Here several important observations can be made. Losses of Pt during the use phase of industrial applications are notably higher than those of Pd. This can primarily be attributed to the losses that occur in the cross-linking of silicones, in which the Pt catalysts remain in the product. The losses for Pd, during the use phase are predominately attributed to dental applications in which the Pd is assumed to be lost with the patients.

Figure 7.3 provides estimates of the quantity of Pt, Pd and Rh lost globally during autocatalyst use since 1974. The results indicate that PGE losses, especially for Pd, have grown substantially in recent years. Overall, these losses are relatively small, averaging at less than 2% of the PGE quantity initially loaded onto the autocatalysts, which is comparable to calculations and suggestions noted elsewhere.[7,65,70]

Overall Pd losses at the end-of-life are notably greater than those of Pt, primarily owing to its greater use in electronics for which little end-of-life recycling occurs. In contrast, minimal losses occur in industrial applications, jewellery and investment, which comprise the vast majority of Pt demand. For Rh, losses during end-of-life are also dominated by its use in autocatalysts and industrial applications. End-of-life losses of Ru are however more similar to those of Pd with electronics accounting for the greatest share.[76]

Loss estimates during manufacturing and use of all five PGEs may be underestimated. This is because manufacturing losses only include losses from

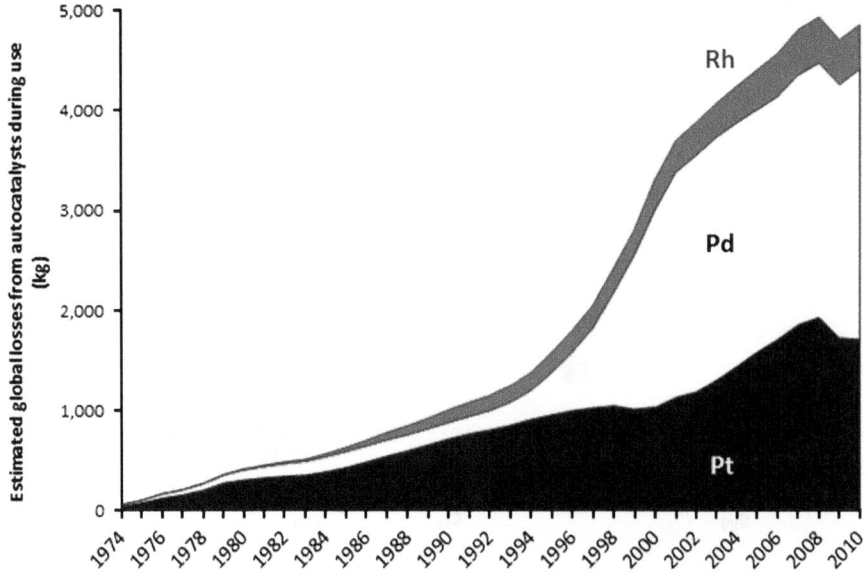

Figure 7.3   Estimated global losses of PGEs from autocatalysts during use, in kg.

jewellery and dental alloys, while losses during the use phase do not include losses that occur in electroplating applications in which very little of the Pt, Pd and Rh that enters the electroplating baths is recovered.[7]

It is important to note that in open-loop non-industrial applications and in particular for autocatalysts, losses at end-of-life include unidentified flows. These flows represent the difference between how much is expected to exit use based on the mathematical model used here and how much enters the waste management system based on what is reported as having been recycled and the assumed recycling rate. These unidentified flows may represent the PGE content of autocatalysts that are discarded without the autocatalyst being removed and recycled. Alternatively, they could represent the PGE content that enters the waste management system but is lost or is being held as stock. These possibilities suggest that the recycling rate in reality is lower than was anticipated in the model. However, these quantities of PGE could still be present in operating or abandoned vehicles, suggesting that the average lifetime used in the model may be too low. The discrepancies could also be due to losses during the manufacture of the autocatalysts, which are not accounted for in the model. Finally, they could be due to inconsistencies in the supply and demand statistics or simplifications in the model. The reality is likely to be a combination of all of these factors but without further information it is difficult to discern which of these possibilities accounts for the majority of these unidentified flows.

Losses on a relative basis, illustrated in Figure 7.4, are comparable across the metals with approximately 30–45% of primary demand being lost during manufacture, use and end-of-life. Viewed another way, the results suggest that approximately 30–45% of demand is used simply to replenish these losses, with the remainder fulfilling net increases in demand. On the supply side, losses as a percentage of primary demand amount to approximately 25–40%, with Rh, Ru and Ir having notably higher losses. Overall, Pt seems to have the lowest

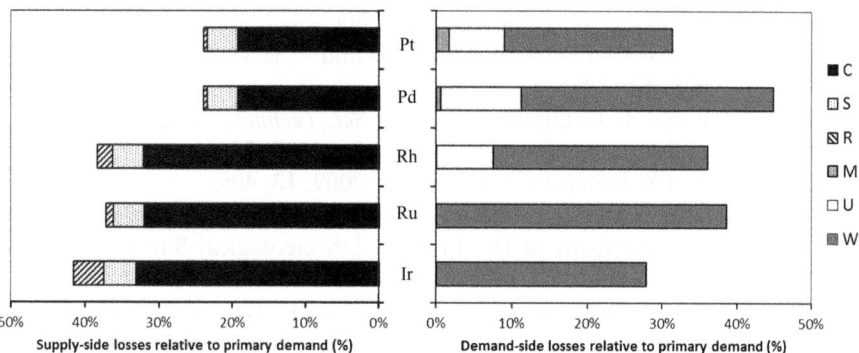

Figure 7.4 Global 2010 PGE supply- and demand-side losses as a percentage of primary demand. Losses from comminution and concentration (C), smelting (S), refining (R), manufacturing (M), use (U), and end-of-life waste management (W) are depicted as stacked bars with varying patterns and shades, as noted in the legend.

relative losses from a combined supply and demand perspective for the reasons noted earlier.

7.5   Conclusions

While there is much inherent uncertainty in analyses like those undertaken here, the overall conclusions that can be drawn are clear. More effort is needed to understand, quantify and ultimately minimize these losses. Owing to the significant amount of utility that is embodied in these metals by the time they reach the end users, it is perhaps most strategic to focus more on the losses that occur at the end-of-life. In general, this analysis shows that in certain applications, such as in the industrial sector, losses of PGE are relatively small and as indicated by others,[7] these processes have been optimized to a point where little further improvement can be expected. There are, however, several end-use applications in which significant improvements are warranted. This is especially the case for electronic applications and autocatalysts where often the most important contribution that can improve the recycling rate is collection. This message is more pertinent now than ever as losses on an absolute level are likely to increase in the coming years as more of the growing stock in-use reaches the end-of-life. It is therefore imperative to continue to work on identifying and reducing these losses so that these precious metals may be readily available for future generations.

Acknowledgement

The author would like to thank Dr. T. E. Graedel for his input.

References

1. R. L. Rudnick and S. Gao, in *Treatise on Geochemistry*, eds. H. D. Holland and K. K. Turekian, Pergamon, Oxford, 2003, vol. 3, pp. 1–64.
2. B. K. Reck, D. B. Muller, K. Rostkowski and T. E. Graedel, *Environ. Sci. Technol.*, 2008, **42**, 3394.
3. W.-Q. Chen and T. E. Graedel, *Environ. Sci. Technol.*, 2012, **46**, 8574.
4. M. Saurat and S. Bringezu, *J. Ind. Ecol.*, 2008, **12**, 754.
5. M. Saurat and S. Bringezu, *J. Ind. Ecol.*, 2009, **13**, 406.
6. H. E. Hilliard, *Platinum Recycling in the United States in 1998* Circular 1196-B, US Department of the Interior, US Geological Survey, Reston, Virginia, 2004.
7. C. Hagelüken, M. Buchert and H. Stahl, *Materials Flow of Platinum Group Metals – Systems Analysis and Measures for Sustainable Optimisation of the Materials Flow of Platinum Group Metals*, GFMS, Ltd, London, 2005.
8. S. Kuriki, I. Daigo, Y. Matsuno and Y. Adachi, *J. Jpn. Inst. Met.*, 2010, **74**, 801–805.
9. K. Halada, *Study on Global Flow of Metals: An Example of Material Recycling*, National Institute for Materials Science, Tsukuba, Japan, 2008.

10. O. A. Babkina and T. E. Graedel, *The Industrial Platinum Cycle for Russia: A case study of materials accounting*, Working Paper Number 8, Yale School of Forestry & Environmental Studies, New Haven, CT, 2005.
11. A. Elshkaki and E. van der Voet, in *Conservation and Recycling of Resources: New Research*, C. V. Loeffe, Nova Science, New York, 2006, pp. 61–92.
12. D. B. Mueller, J. Cao, E. Kongar, M. Altonji, P.-H. Weiner and T. E. Graedel, *Service Lifetimes of Mineral End Uses*, US Geological Survey, 2007.
13. Johnson Matthey, *Market Data Tables*, Johnson Matthey, Royston, UK, 2012.
14. Johnson Matthey, *Platinum*, Johnson Matthey, Royston, UK, 1999–2013.
15. CPM Group, *CPM Group Platinum Group Metals Yearbook 2012*, Euromoney Books, New York, 2012.
16. GFMS/Thomson Reuters GFMS, *Platinum & Palladium Survey 2005–2012*, GFMS/Thomson Reuters GFMS, London, 2005–2012.
17. USBM/USGS, *Minerals Yearbook*, US Bureau of Mines, US Geological Survey, Reston, VA, 1932–2012.
18. USBM, *Mineral Facts and Problems: 1956–1985 Editions*, US Bureau of Mines, Washington, DC, 1956–1985.
19. G. M. Mudd, *Ore Geol. Rev.*, 2012, **46**, 106.
20. F. Crundwell, M. Moats, V. Ramachandran, T. Robinson and W. G. Davenport, *Extractive Metallurgy of Nickel, Cobalt and Platinum-Group Materials*, Elsevier, Boston, 2011.
21. R. Jones, *J. Met.*, 2004, **56**, 59.
22. R. K. W. Merkle and A. D. McKenzie, in *The Geology, Geochemistry, Mineralogy and Mineral Beneficiation of Platinum-Group Elements*, ed. L. J. Cabri, Canadian Institute of Mining, Metallurgy and Petroleum, Montreal, Canada, 2002, vol. 54, pp. 793–809.
23. R. J. Seymour and J. O'Farrelly, in *Kirk–Othmer Encyclopedia of Chemical Technology*, John Wiley & Sons, New York, 2012.
24. H. Renner, G. Schlamp, I. Kleinwächter, E. Drost, H. M. Lüschow, P. Tews, P. Panster, M. Diehl, J. Lang, T. Kreuzer, A. Knödler, K. A. Starz, K. Dermann, J. Rothaut, R. Drieselmann, C. Peter and R. Schiele, in *Ullmann's Encyclopedia of Industrial Chemistry*, Wiley–VCH Verlag GmbH & Co. KGaA, 2000.
25. G. Gunn and A. Benham, Platinum, British Geological Survey, Natural Environment Research Council, Keyworth, Nottingham, United Kingdom, 2009.
26. Anglo American Platinum Limited, *Integrated Annual Report 2011: Building The Foundations for A Future Through Mining*, Anglo American Platinum Limited, Johannesburg, 2012.
27. C. F. Vermaak, *Platinum-Group Metals: A Global Perspective*, Mintek, Randburg, South Africa, 1995.
28. R. T. Jones, in *Nickel and Cobalt 2005: Challenges in Extraction and Production, 44th Annual Conference of Metallurgists*, Calgary, Alberta, Canada, 2005, pp. 147–178.

29. S. M. Kozyrev, M. Z. Komarova, L. N. Emelina, O. I. Oleshkevich, O. A. Yakovleva, D. V. Lyalinov and V. I. Maximov, in *The Geology, Geochemistry, Mineralogy and Mineral Beneficiation of Platinum-Group Elements*, ed. L. J. Cabri, Canadian Institute of Mining, Metallurgy and Petroleum, Montreal, Canada, 2002, vol. 54, pp. 757–791.

30. Stillwater Mining Company, *2011 Annual Report*, Stillwater Mining Company, Billings, MT, 2012.

31. North American Palladium, *2011 Annual Report*, North American Palladium Ltd., Toronto, 2011.

32. Aquarius, *Annual Report 2012*, Aquarius Platinum Ltd., 2012.

33. Lonmin, *Annual Report and Accounts*, Lonmin Plc., 2011.

34. Implats, *Implats Annual Report 2009*, Impala Platinum Holdings Ltd., 2009.

35. D. Gregurek, C. Reimann and E. F. Stumpfl, *Environ. Pollut.*, 1998, **102**, 221.

36. S. Rauch and O. Fatoki, *Water, Air, Soil Pollut.*, 2012, **224**, 1.

37. I. Råde and B. A. Andersson, *Platinum Group Metal Resource Constraints for Fuel-Cell Electric Vehicles*, Department of Physical Resource Theory, Chalmers University of Technology and Götegborg University, Göteborg, Sweden, 2001.

38. D. M. Sutphin and N. J. Page, *International Strategic Minerals Inventory Summary Report — Platinum-group metals* US Geological Survey Circular 930-E, US Department of the Interior, US Geological Survey, Washington, D.C., 1986.

39. B. Montpellier, *Technical Report Update on the Crocodile River Mine, Eastern Platinum Limited, North West Province, South Africa, December 2010*, Eastern Platinum Ltd., 2010.

40. K. Ravindra, L. Bencs and R. Van Grieken, *Sci. Total Environ.*, 2004, **318**, 1.

41. K. H. Ek, G. M. Morrison and S. Rauch, *Sci. Total Environ.*, 2004, **334–335**, 21.

42. M. Farago, E. Hutchinson, P. Simpson and I. Thornton, *Appl. Earth Sci.*, 2005, **114**, 182.

43. A. Dubiella-Jackowska, ż. Polkowska and J. Namieśnik, in *Reviews of Environmental Contamination and Toxicology*, ed. D. M. Whitacre, Springer Science + Business Media LLC, 2009, vol. 199, pp. 111–135.

44. F. Zereini and F. Alt, *Anthropogenic Platinum-Group Element Emissions: Their impact on man and environment*, Springer, New York, 2000.

45. V. F. Hodge and M. O. Stallard, *Environ. Sci. Technol.*, 1986, **20**, 1058.

46. M. Müller and K. G. Heumann, *Fresenius J. Anal. Chem.*, 2000, **368**, 109.

47. B. Gómez, M. Gómez, J. L. Sanchez, R. Fernández and M. A. Palacios, *Sci. Total Environ.*, 2001, **269**, 131.

48. E. Helmers, M. Schwarzer and M. Schuster, *Environ. Sci. Pollut. Res.*, 1998, **5**, 44.

49. E. J. Hutchinson, M. E. Farago and P. R. Simpson, in *Anthropogenic Platinum-Group Element Emissions.Their impact on man and environment*, eds. F. Zereini and F. Alt, Springer-Verlag, Berlin, 2000, pp. 57–64.

50. F. Zereini, C. Wiseman, F. Alt, J. Messerschmidt, J. Müller and H. Urban, *Environ. Sci. Technol.*, 2001, **35**, 1996.
51. O. Nygren and C. Lundgren, *Int. Arch. Occup. Environ. Health*, 1997, **70**, 209.
52. T. Hees, B. Wenclawiak, S. Lustig, P. Schramel, M. Schwarzer, M. Schuster, D. Verstraete, R. Dams and E. Helmers, *Environ. Sci. Pollut. Res.*, 1998, **5**, 105.
53. F. Alt, H. R. Eschnauer, B. Mergler, J. Messerschmidt and G. Tölg, *Fresenius J. Anal. Chem.*, 1997, **357**, 1013.
54. K. Hoppstock, F. Alt, K. Cammann and G. Weber, *Z. Anal. Chem.*, 1989, **335**, 813.
55. C. B. Tuit, G. E. Ravizza and M. H. Bothner, *Environ. Sci. Technol.*, 2000, **34**, 927.
56. E. de Vos, S. J. Edwards, I. McDonald, D. S. Wray and P. J. Carey, *Appl. Geochem.*, 2002, **17**, 1115–1121.
57. C. Chen, P. N. Sedwick and M. Sharma, *Proc. Natl. Acad. Sci. U. S. A.*, 2009, **106**, 7724.
58. S. Artelt, O. Creutzenberg, H. Kock, K. Levsen, D. Nachtigall, U. Heinrich, T. Rühle and R. Schlögl, *Sci. Total Environ.*, 1999, **228**, 219.
59. K. H. Jensen, S. Rauch, G. M. Morrison and P. Lindberg, *Arch. Environ. Contam. Toxicol.*, 2002, **42**, 338.
60. S. Zimmermann, F. Alt, J. Messerschmidt, A. von Bohlen, H. Taraschewski and B. Sures, *Environ. Toxicol. Chem.*, 2002, **21**, 2713.
61. G. Philippeit and J. Angerer, *J. Chromatogr., Biomed. Appl*, 2001, **760**, 237.
62. O. Nygren, G. T. Vaughan, T. M. Florence, G. M. P. Morrison, I. M. Warner and L. S. Dale, *Anal. Chem.*, 1990, **62**, 1637.
63. G. T. Vaughan and T. M. Florence, *Sci. Total Environ.*, 1992, **111**, 47.
64. K. Van de Velde, C. Barbante, G. Cozzi, I. Moret, T. Bellomi, C. Ferrari and C. Boutron, *Atmos. Environ.*, 2000, **34**, 3117.
65. C. Barbante, A. Veysseyre, C. Ferrari, K. Van De Velde, C. Morel, G. Capodaglio, P. Cescon, G. Scarponi and C. Boutron, *Environ. Sci. Technol.*, 2001, **35**, 835.
66. R. F. Hill and W. J. Mayer, *IEEE Trans. Nucl. Sci.*, 1977, **24**, 2549.
67. M. Moldovan, M. A. Palacios, M. M. Gomez, G. Morrison, S. Rauch, C. McLeod, R. Ma, S. Caroli, A. Alimonti, F. Petrucci, B. Bocca, P. Schramel, M. Zischka, C. Pettersson, U. Wass, M. Luna, J. C. Saenz and J. Santamaria, *Sci. Total Environ.*, 2002, **296**, 199.
68. M. E. Kylander, S. Rauch, G. M. Morrison and K. Andam, *J. Environ. Monit.*, 2003, **5**, 91.
69. E. Helmers, in *Anthropogenic Platinum-Group Element Emissions: Their impact on man and environment*, eds. F. Zereini and F. Alt, Springer, New York, 2000, pp. 133–144.
70. S. M. Lloyd, L. B. Lave and H. S. Matthews, *Environ. Sci. Technol.*, 2005, **39**, 1384.
71. GFMS, *Platinum & Palladium Survey 2010*, GFMS Limited, London, 2010.

72. T. R. Jackson, *Fleet Characterization Data for MOBILE6:Development and Use of Age Distributions, Average Annual Mileage Accumulation Rates and Projected Vehicle Counts for Use in MOBILE6* EPA420-R-01-047, M6.FLT.007, US Environmental Protection Agency, Washington, DC, 2001.

73. L. Fulton and G. Eads, *IEA/SMP Model Documentation and Reference Case Projection*, IEA/WBCSD (International Energy Agency and the World Business Council for Sustainable Development), 2004.

74. S. Artelt, H. Kock, H. P. König, K. Levsen and G. Rosner, *Atmos. Environ.*, 1999, **33**, 3559.

75. D. McDonald and L. B. Hunt, *A History of Platinum and its Allied Metals*, Johnson Matthey: Distributed by Europa Publications, London, 1982.

76. T. E. Graedel, J. Allwood, J.-P. Birat, M. Buchert, C. Hagelüken, B. K. Reck, S. F. Sibley and G. Sonnemann, *J. Ind. Ecol.*, 2011, **15**, 355.

77. P. Chancerel, C. E. M. Meskers, C. Hagelüken and V. S. Rotter, *J. Ind. Ecol.*, 2009, **13**, 791.

78. A. Matharu, in *Elemental Recovery and Sustainability*, ed. A. Hunt RSC, Cambridge, 2013.

79. K. Kümmerer, E. Helmers, P. Hubner, G. Mascart, M. Milandri, F. Reinthaler and M. Zwakenberg, *Sci. Total Environ.*, 1999, **225**, 155.

80. M. Sano, *A Convenient Estimation Methodology for Survival Rate of Automobiles*, ITEC Working Paper Series 08-06, 2008.

81. Ward's Automotive Group, *Vehicles in Operation by Country*, Penton Media, Michigan, 2011.

82. B. A. Schaer, M. T. Koller, C. Sticherling, D. Altmann, L. Joerg and S. Osswald, *Heart Rhythm*, 2009, **6**, 1737.

83. C. Hagelüken, *Chim. Oggi*, 2006, **24**, 14.

84. Casale, Haldor Topsøe and KBR, in *Nitrogen + Syngas*, BCInsight Ltd., London, 2011, p. 14.

85. H. E. Hilliard, *Platinum-Group Metals*, US Geological Survey, 2001.

86. H. S. Fang, Y. Y. Pan, L. L. Zheng, Q. J. Zhang, S. Wang and Z. L. Jin, *J. Cryst. Growth*, 2012, **363**, 25.

87. C. Couderc, P. Williams and D. Coupland, *CG Iridium – Metal for the 21st Century*, Johnson Matthey, Royston, UK, 2010.

CHAPTER 8

WEEE Waste Recovery

AVTAR S. MATHARU

Green Chemistry Centre of Excellence, Department of Chemistry,
The University of York, Heslington, York, YO10 5DD, UK
Email: avtar.matharu@york.ac.uk

8.1 Introduction: EEE and WEEE

Modern day society is increasingly reliant on electrical and electronic equipment (EEE) as a means of communication both at home and in the workplace. Production of modern EEE is resource intensive, using up to 60% of the elements in the Periodic Table including those deemed *critical*: cobalt; gallium; germanium; indium; platinum group metals (platinum, palladium, iridium, rhodium, ruthenium and osmium); rare earths (including yttrium and scandium) and tantalum.[1–5] For example, over 300 tonnes of gold is used in electronics each year. Manufacture of mobile phones and personal computers uses 3% of the global gold and silver output (interestingly silver is not deemed critical), 13% of palladium and 15% of cobalt. Gold concentrations range from 300–350 g t^{-1} for mobile phone handsets to 200–250 g t^{-1} for computer circuit boards.[6,7] In 2005, approximately 10.3 million tonnes of new EEE entered the EU27 and sales of electronic products in emerging economic countries and/or BRIC (Brazil, Russia, India and China) countries are expected to rise significantly in the next 10 years as they become more cash-rich.[8] However, as technology changes so does societal need, whether for pleasure or business, the volume of waste electronic and electrical equipment (WEEE) is increasing at an alarming rate. Thus, modern day EEE is both resource intensive and resource wasteful. Robinson estimates approximately one billion computers will be out of use or changed in the next few years.[9]

RSC Green Chemistry No. 22
Element Recovery and Sustainability
Edited by Andrew J. Hunt
© The Royal Society of Chemistry 2013
Published by the Royal Society of Chemistry, www.rsc.org

Globally, WEEE is estimated at 40 million metric tonnes per year and predicted to rise to approximately 60 million metric tonnes by 2013.[10–12] Computer-derived WEEE is expected to increase by almost 300% from 2007 to 2020 in South Africa and China and by a staggering 500% in India.[12] WEEE is the fastest growing segment of municipal solid waste (MSW) accounting for 3–5% of incoming materials. The United States of America (USA) and China are the biggest producers of WEEE, generating approximately 3 million and 2.3 million tonnes, respectively. In the EU-27, Norway and Switzerland, WEEE is estimated to increase by 11% from 2008–2014 owing to rapid technological advancement accompanied by reduced prices.[12] The volume of WEEE in circulation is based on the average life cycle or obsolescence rate of electrical and electronic equipment, which is cultural and geographic-dependent.[13–16] The average life cycle or obsolescence rate is defined as the sum of three parts:[12]

1 active life–the number of years an electrical/electronic item can be effectively used;
2 passive life–the period in which it can be refurbished or reused, and;
3 storage–constitutes both storage time before disposal and storage time at repair shops prior to dismantling.

The average life cycle or obsolescence rate of electrical and electronic equipment in developed nations is generally equivalent to "active life", while in developing and emerging countries it is a sum of active life, passive life and storage. The market economics for resale and reuse of secondhand electronic goods is more prevalent in developing countries because electronic waste stays longer in circulation prior to either landfill or incineration.

8.2   WEEE Urban Mine

WEEE serves as a lucrative above ground "urban mine" for future supply of critical elements. The concentration and purity of certain metals is significantly richer above ground than that available in the primary ore. Recycling rates and technologies for precious and platinum group metals are well established with rates well above 50% and some approaching over 90%. Although recycling rates of platinum group metals exceed 50%, importantly, recycling rates of critical metals found increasingly in modern EEE such as tantalum, gallium, indium and rare earths are less than 1%.[2,5] According to Pike Consultants, approximately 14 million metric tonnes of raw materials could be available for new product manufacturing provided recycling rates reach 50% (of the 60 million metric tonnes of WEEE estimated in 2013) with a 45% recovery rate of valuable materials.[17]

Good recovery and recycling rates are dependent on sorted, screened or segregated WEEE, otherwise processing of mixed WEEE drastically reduces recovery rates. For example, recovery rate of gold drops to 26% and to 12%

for silver.[18] There are three general methods of recycling, each of which differs in complexity based on appliance type, environmental risk and resource value:

1 crushing and mechanical separation of complete appliances – predominantly used for appliances with low resource value and of limited environmental risk;
2 direct feed of almost complete appliances to metallurgical processes – this may be a very good solution for small appliances such as mobile phones that contain significant amounts of precious metals. Any recycler operating this regimen will have to ensure very good flue gas abatement technology;
3 partial dismantling with downstream processes (methodology best adopted for IT appliances and servers) – appears to be the best approach for maximum resource efficiency of critical metals where high value individual components are isolated as well as those posing environmental risk such as mercury lamps or batteries.

Unfortunately, the lure of high metal prices and concerns over future security of supply have resulted in millions of tonnes of WEEE being exported illegally to developing countries where it is being sourced for its precious metals by backyard recyclers with little or no regard for personal and environmental safety. The benefits of recycling, as opposed to primary production of critical metals, in terms of CO_2 footprint and energy savings are well-documented. Primary production is very energy intensive and has a large CO_2 footprint because certain critical metals either have a very low natural abundance and/or are found as by-products of other metal production processes. For example, primary production of silver or indium expends approximately $140\,kg\,CO_2$ per kg of metal whereas gold production is 100 times more intensive; $14000–17000\,kg\,CO_2$ per kg of gold.[1]

8.3   Case Study: Liquid Crystal Displays

8.3.1   Success of LCDs

Liquid crystal displays (LCDs) play an important part in new EEE owing to their widespread occurrence in mobile phones, laptop computers, satellite navigation systems and increasingly in large area flat-screen televisions but their success is now problematic. LCDs have superseded the traditional cathode ray tube (CRT) because of their versatility: flat panel, space saving, lightweight, portable and do not emit harmful radiation. The success of LCDs, at the expense of CRT, is probably best exemplified by the rapid decline in shipments of CRT computer monitors to developed countries where in some cases it is almost zero. LCD technology is so successful that it accounts for approximately 87% of the total market share of the global display industry estimated at US$93 billion (2010), which includes mature and emerging technologies such as organic light-emitting diodes (OLEDs).

Future growth of LCD sales is application (televisions, desktop monitors, notebooks and personal computers, and public displays) dependent. Shipment of large area thin film transistor–liquid crystal displays (TFT–LCD) from 2011–17 is predicted to be 9.4% compounded annual growth rate (CAGR), whilst notebooks and personal computers are expected to accelerate at a rate of 15%. By 2017, analysts predict that the industry will ship approximately 1.25 billion large-area TFT–LCDs, of which notebooks and personal computers will account for at least 50%. Shipment of large area TFT–LCD televisions will near a staggering 250 million units. Globally, China is a hot-bed for large area TFT–LCD TV received a shipment of 52 million units in 2012. As LCDs continue to penetrate the flat panel display market with increasing year-on-year sales of LCD TVs forecasts, the LCD industry must adopt strategies to minimise waste disposal and maximise recovery and reuse. Future volumes of waste will be significant and thecurrent practice of incineration and/or landfill of LCD panels may need to be reconsidered, as it not only leads to loss of liquid crystal and glass but also to lodd of indium metal.

The basic construction of an electro-optical LCD device shown in Figure 8.1 comprises a thin film of liquid crystal (nematic mesophase) sandwiched between two glass substrates whose inner surfaces have been coated with a transparent electrical conductor, indium tin oxide (ITO), allowing a voltage to pass across the medium. The nematic mesophase is prealigned depending on the display mode aided by preferentially rubbed alignment layers (polyimide, approximately 100 nm thickness), which are also coated on the inner surfaces of the glass. The two glass substrates are kept apart by a few micrometres using spacer beads or spacer-posts. Polarisers are attached to the outer surface of the glass substrates and their arrangement dictates the appearance of the display which is either dark or bright in the off state. The role of the nematic mesophase is two-fold: (i) to modulate incoming polarised light and (ii) to respond to an electrical impulse by aligning with the applied field.

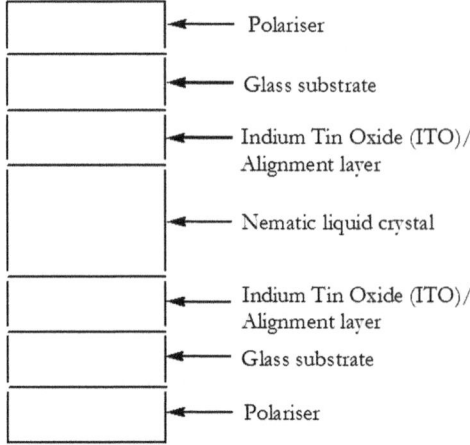

Figure 8.1   Basic construction of an electro-optic LCD device (side view).[19]

8.3.2 Demanufacturing and Resource Recovery

An LCD device has many sub-components as shown in Figure 8.2. The LCD module can be further disaggregated in two main components: backlight unit and the LCD panel. The latter comprises a colour filter, sealant, liquid crystal mixture, glass and TFT array.

Glass used in TFT–LCD panels has a high melting point (approximately 1150 °C) comprising: 50% SiO_2, 15% B_2O, 10% Al_2O_3 and 25% BaO. Since glass is a major component of an LCD panel (approximately 90% by weight) and with large volumes of glass produced annually for the TFT–LCD industry, it is feasible to consider its reuse or diversion from landfill. Provided that the mercury backlight is removed from an LCD panel, the remaining constituents, including the liquid crystal mixture, are classified as non-hazardous. Merck reported the use of LCD glass instead of silica to line the inner wall of incinerators thus protecting it from aggressive substances.[20]

Removal of liquid crystals by catalytic decomposition is well documented in the literature. For example, Sharp Corporation reported the decomposition of liquid crystals using photocatalysts (*e.g.* TiO_2). The decomposition uses low energy and does not require waste water treatment.[21] Many solutions for recovering LC rather than using incineration are now being disclosed in patents. Dainippon Ink & Chemicals (DIC)[22,23] have patented a method for removal and purification of liquid crystal components from waste liquid crystal such that the isolated components can then be reused as additives in mixtures used for active matrix displays. Isolation and purification is achieved by

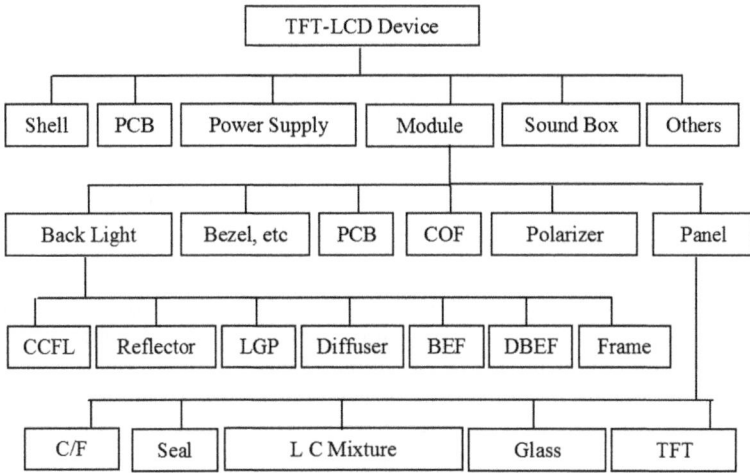

Figure 8.2 A TFT-LCD Device disassembly (PCB - Printed Circuit Board, COF - Chip on Film, LGP - Light Guild Plane, BEF - Brightness Enhancement Film, DBEF - Dual Brightness Enhancement Film, TFT - Thin Film Transistor, CCFL - Cold Cathode Fluorescent Light and C/F - Colour Filter/Film).[19]

column chromatography and followed by distillation. The isolated components are most likely to be ester and cyano-group containing liquid crystals.

Sumoge and Hiruma studied the thermal decomposition, gasification and melting of LCD containing WEEE allowing for a process of isolating indium, gold, rare-metals and metal-free glass.[24] Unbroken LCDs were furnace heated in the absence of oxygen. The exhaust gases were heated at elevated temperatures in order to decompose toxic substances and minimise dioxin formation. The tar is reused as fuel. The decomposed glass is treated with chemicals to release metals (indium, gold and rare metals). Both glass and metals can be then be reused although not necessarily for the electronics industry.

LCDs are a valuable source of above ground indium. Indium is a silvery-white metal with an estimated abundance of 240 parts per billion by weight in the Earth's crust. Indium is found mainly as a by-product or "hitch-hiker element" in sphalerite, the mineral ore from which zinc is extracted. Even though sphalerite contains indium in the range 1–100 parts per million it is still economically more viable to extract indium from zinc rather than any other metal ore. In 2011, the United States Geological Survey (USGS) estimated global production of primary indium to be 640 metric tonnes.[25,26]

Over 55% of the total indium extracted is used by the LCD industry in the manufacture of transparent indium tin oxide (ITO) electrical conductors found on the inner surfaces of glass panels. The ITO consists of 90% In_2O_3 and 10% SnO_2 corresponding to 78% (by weight) of indium and is sputtered on the glass substrate to a thickness of a few hundred nanometres. The sputtering process is relatively inefficient as the sputtering target is replaced frequently even though 50–70% still remains active and a considerable amount of ITO deposits on the walls of the chamber rather than the glass substrate. However, approximately 70% of indium is generally recovered from sputtered indium tin oxide (ITO) waste. Newer LCDs with light-emitting diode (LED) backlights also use indium, albeit to a much lesser extent, in the form indium gallium nitride (GAN) as the semiconductor chip. LCD manufacture requires ITO to be deposited on both panels that make the LCD display whereas, OLED displays only require one substrate to be ITO-coated.[1,2,26]

Several methods have been reported for recovery of indium from ITO[26]: pyrometallurgical or hydrometallurgical processes, solvent-assisted extraction[27] and precipitation.[28] In Sweden, a three year, multi-partner, project "Sustainable Recycling of Flat Panel Displays" is investigating several aspects of LCD recycling with aims to develop (i) sustainable recycling processes (in particular indium from ITO); (ii) applications for recovered components (in particular optical components); (iii) a model for the collective assessment of all sustainability aspects of recycling scenarios and (iv) guidelines for ecodesign, environmental certification and legal requirements concerning LCD products. In their process indium is recovered from ITO by first dissolving ITO in acid (H_2SO_4 or HCl) followed by extraction of the resultant metal salts (either sulfates or chlorides) with di-2-ethylhexyl-phosphoric acid (DEHPA) in kerosene. DEHPA preferentially extracts indium salts from the mixture leaving behind tin salts in the aqueous phase as shown in Figure 8.3.

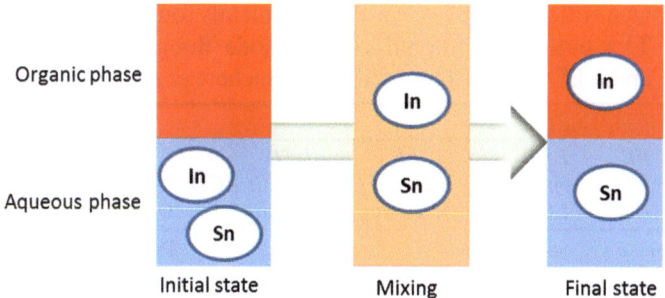

Figure 8.3 Extraction protocol for indium recovery from LCD ITO (adapted from Felix *et al.*[29]).

The best conditions identified in their study gave indium concentrations in the range 0.20–0.25 g kg^{-1} glass at a recovery yield of nearly 90%. In addition, the project recovers optical components, constituting 25 wt% of an LCD monitor, that are reused in energy-efficient flat LED luminaries.[29]

Dodbiba *et al.*[26] have investigated two different pretreatment methods: electrical disintegration and mechanical pulverisation, for isolation of indium from waste LCD modules in the context of a life cycle assessment (LCA). Electrical disintegration was able to disjoint the two glass panels of the module and therefore enable a high leaching capacity for indium. LCD modules processed by conventional grinding required a larger amount of acid solvent to achieve a relatively high leaching rate, since indium was not fully liberated from the glass pieces. Their research concluded that electrical disintegration is the most effective and environmentally friendly preprocessing method for leaching indium.

8.3.3 Distribution and Quantification of Critical Metals in LCDs

Printed circuit boards (PCBs) are a more lucrative source of precious/critical metals. In analyses conducted by Buchert *et al.*,[2] summarised in Table 8.1, the quantities (mg) of silver, gold and palladium on a PCB from an LCD monitor (approx. weight 0.4 kg) are 520, 196 and 40, respectively. Similar amounts are found in printed circuit boards (PCBs) from LCD TVs (approx. weight 2.3 kg): silver, 575 mg; gold, 138 mg, and palladium, 44 mg.

The rare earths contained in the luminescent materials are not presently recycled.[2] Currently, there are no suitable separation and refining processes for recovering the indium from display units and the rare earths from the illumination source. Rare earths are also used in permanent magnets (neodymium iron boron, NIB) found in the following components of a modern day laptop computer (see Table 8.2): voice coil accelerator in the hard disk drive, spindle motor for the hard disk drive, spindle motor for the optical drive and loudspeakers.

Table 8.1 Quantities of precious and critical metals found LCD monitors and TVs–both conventional cold cathode fluorescent lamps (CCFL) and LED backlit (adapted from Buchert *et al.*[2]).

| | *Content (mg)* | | | | |
| | *CCFL backlit LCD TV* | *LED backlit LCD TV* | *CCFL backlit LCD Monitor* | *LED backlit LCD Monitor* | |
Metal					*Occurrence*
Ag	580	580	520	520	PCB and contacts (100%)
In	260	260	79	82	Internal coating, ITO, on display (100%)
Au	140	140	200	200	PCB and contacts (100%)
Y	110	4.8	16	3.2	Background illumination (100%)
Pd	44	44	40	40	PCB and contacts (100%)
Eu	8.10	0.09	1.20	0.06	Background illumination (100%)
La	6.80	0.00	1.00	0.00	CCFL background illumination (100%)
Ce	4.50	0.30	0.68	0.20	Background illumination (100%)
Tb	2.30	0.00	0.34	0.00	CCFL background illumination (100%)
Ga	0.00	4.90	0.00	3.3	LED illumination (100%)
Gd	0.63	2.30	0.096	1.50	Background illumination (100%)
Pr	<0.13	0.00	0.00	<0.05	CCFL background illumination (100%)

Table 8.2 Distribution of rare earths from different magnets found in laptop computers (adapted from Buchert *et al.*[2]).

| | *Quantity of rare earth (%)* | | |
Source of magnet	*Nd*	*Pr*	*Dy*
Voice coil accelerator	29[a]		2.3
Spindle motors	29	0	0
Loudspeaker	31[a]		0

[a]A combined total is given as some manufacturers replace neodymium with much cheaper praseodymium (adapted from Buchert *et al.*[2])

Rare earths are being increasingly used as luminescent materials in direct displays such as plasma and OLED displays. Approximately 2% of rare earths are also found in mercury-based cold cathode fluorescent lamps (CCFL) used in conventional backlit LCDs.[2] A greater percentage of rare earths are used in LED backlighting found in the latest LCD LED televisions comprising oxides and mixed metal oxides of yttrium (Y_2O_3), cerium ($CeMgAl_{11}O_{19}$) and lanthanum (La_2O_3). In addition, europium, terbium and gadolinium are used, albeit in lesser amounts. The estimated content of rare earths found in LCDs with conventional mercury-based CCFL backlighting is shown in Table 8.3.

Table 8.3 Quantification and distribution of rare earths found in CCFLs of LCD notebooks, monitors and televisions (adapted from Buchert *et al.*[2]).

Device type	Mean number of tubes per device	Estimated mean weight (mg) of rare earth					
		Y	Eu	La	Ce	Tb	Gd
Notebook	1	1.80	0.13	0.11	0.076	0.038	0.011
Monitor	6	16.0	1.20	1.00	0.680	0.340	0.095
Television	15	110.0	8.10	6.80	4.500	2.300	0.630
Approximate weight of material (%)		83.1	6.0	5.1	3.5	1.8	0.5

The magnetic disk-drive is also source of precious/critical metals.[2] Depending on the type of hard disk drive, aluminium- or glass-based, gold, silver, platinum (glass-based only), palladium, rhodium (glass-based only) and ruthenium can be sourced. Aluminium-based hard disk drives contain significant amounts of silver (850 mg kg^{-1}) and to a lesser extent gold (21 mg kg^{-1}) and palladium (14 mg kg^{-1}) whilst ruthenium is found at less than 7 mg kg^{-1}. Glass-based hard disk drives predominantly contain platinum (38 mg kg^{-1}) and less than 6 mg kg^{-1} of the other earlier listed metals.

8.4 Environmental Legislation: WEEE, RoHS, REACH, EuP, ERP

Global, national and regional environmental legislative measures for WEEE are increasing. The most apt for end-of-life LCDs is the European Union (EU) WEEE Directive (2002/96/EC) adopted by the EU on January 2003 which recommended that Member States should transpose this directive into national law by 2004.[30] The WEEE Directive aims to divert the amount of electronic waste entering landfill or incineration by encouraging recovery and reuse. Recent revisions of the Directive which came in to force on 13 August 2012, which have to be implemented by all Member States by 14 February 2014, take into account economic variability across the Member States. Under the new revision the current national WEEE collection rates of 4 kg per capita will remain for four years, representing about 2 million tonnes per year, from around 10 million tonnes of WEEE generated annually in the EU. For the following three years collection rates will be assessed at 45% of the weight of equipment entering the market. After that (2019), member states can chose a collection target of either 65% of the weight of equipment entering the market or 85% of the weight of waste equipment. By 2020 it is estimated that the volume of WEEE in the EU will increase to 12 million tonnes and if, collected at the 85% level, approximately 10 million tonnes or roughly 20 kg per capita will be saved.[31]

In the WEEE Directive, electronic equipment is divided into 10 broad categories, of which Category 3C (LCD monitors), 4C (flat panel TVs) and 5B (lighting equipment) are pertinent to LCDs. Importantly for LCDs the WEEE

Directive (2002/96/EC) stipulates that "LCD-containing WEEE with a surface area greater than 100 cm² and those containing mercury backlights must be isolated from the waste stream". Detailed studies have shown that the liquid crystal mixture itself is non-hazardous and the volume per display is miniscule so that incineration is feasible. Small displays such calculators, watches and mobile phones may be treated as non-hazardous waste but the WEEE Directive has major significance for the large growing area of the TFT–LCD market. Many WEEE recyclers are stockpiling thousands of TFT–LCD panels waiting for a viable recycling alternative to landfill or incineration.

Parallel relevant measures impact on initial manufacturing include:

1 the RoHS (Restriction of Hazardous Substances, 2002/95/EC, implemented in UK July 2006) Directive[32] restricting use of lead, mercury, cadmium, hexavalent chromium and brominated flame retardants (polybrominated biphenyl (PBB) and polybrominated diphenyl ether (PBDE) in electronic goods.
2 REACH[33] (Registration, Evaluation, Authorization and Restriction of Chemicals, implemented UK April 2007) ensures a high level of protection of human health and the environment as well as the free circulation of substances on the internal market while enhancing competitiveness and innovation.
3 the EuP[34] (Energy using Products Directive) Directive (EuP Directive 2005/32/EC) is a framework for eco-design to ensure gains in energy efficiency and the free movement of electronic goods in the EU, which has now been superseded by the Energy Related Products (ERP) Directive 2009/125/EC.[35]

Future combined legislation will probably force the electronics industry, for example the LCD industry, to adopt the principles of sustainability across its entire supply chain including at end-of-life. Individual Producer Responsibility (IPR), already in place as a policy within the WEEE Directive (Article 8.2), will become even more prominent. Sustainability or the "triple bottom line" and the 3Rs principle (reduce, recycle and reuse) will be achieved through waste minimisation and resource efficiency, reduced greenhouse emissions, substitution of hazardous chemicals with less hazardous ones and design for active disassembly will be as important as the technology or application itself.

8.5 Summary and Future Outlook

As the electronics industry becomes more resource intensive, there is a huge need for new technologies that are able either to substitute for current materials or to recycle from existing feedstocks which are located in the so-called urban mine. Above ground recycling and extraction of critical metals is more beneficial to the environment than extraction from below ground of primary ore. The success of liquid crystal displays is a classic example of a modern day electronics success story but they are reliant on indium, gold, silver and increasingly rare earths. The future will see an electronics industry based on sustainability.

As for the LCD industry, the LCD WEEE problem is here to stay and will increase in the future. It is highly unlikely that an automated unit dedicated to the complete disassembly of TFT–LCDs, independent of source and size, will ever be conceived unless industry adopts standardisation. The best approach will be to combine manual with automated processes. In the future, the recycling of TFT–LCDs will be considerably less hazardous for the operator as current practices, for example, the move towards LED backlights in place of mercury-filled CCFLs and banning of toxic materials through REACH and RoHS, take effect.

The TFT–LCD manufacturing industry from conception to end-of-life of the display will be increasingly forced into sustainable practices and policies. With global resources of precious minerals and metals becoming limited, more effort will be diverted to resource efficiency.

References

1. K Hieronymi, R. Kahhat and E. Williams (eds), *E-Waste Management from Waste to Resource*, Routledge, Oxon, 2013, 1–272.
2. M. Buchert, A. Manhart, D. Bleher and D. Pingel (eds), *Recycling Critical Raw Materials from Waste Electronic Equipment*, North Rhine-Westphalia State Agency for Nature, Environment and Consumer Protection, Oko-Institut, Darmstadt, 2012, 1–80.
3. M. Buchert, D. Schüler and D. Bleher (eds), *Critical Metals for Future Sustainable Technologies and Their Recycling Potential*, United Nations Environment Programme & United Nations University, 2009.
4. D. Schüler, M. Buchert, R. Liu, S. Dittrich and C. Merz (eds), *Study on Rare Earths and Their Recycling*, Oko-Institut, Darmstadt, 2011, 1–140.
5. B. K. Reck and T. E. Graedel, *Science*, 2012, **337**, 690.
6. C. Hagelüken and C. W. Corti CW, *Gold Bull.*, 2010, **43**, 209.
7. P. Chancerel, C. E. M. Meskers, C. Hagelüken and V. S Rotter VS, *J. Ind. Ecol.*, 2009, **13**, 791.
8. *E-Waste Volume 1 Inventory Assessment Manual*, United Natios Environmental Programme Division of Technology, Industry and Economics International Environmental Technology Centre, Osaka/Shiga, 2007.
9. B. H. Robinson, *Sci. Total Environ.*, 2009, **408**, 183.
10. *The European Environment – state and outlook 2010 Materials Source and Waste*, European Environment Agency, Copenhagen, Denmark, 2010.
11. O. Tsydenova and M. Bengtsson, *Waste Manage*, 2011, **31**, 45.
12. M. Schluep, C. Hageluken, R. Kuehr, F. Magalini, C. Maurer, C. Meskers, E. Mueller and F. Wang, United Nations Environment Programme and United Nations University, *Sustainable Innovation and Technology Transfer Industrial Sector Studies: Recycling from E-waste to resources*, 2009, 1–90.

13. M. Schluep, D. Rochat, A. W. Munyua, S. E. Laissaoui, S. Wone, C. Kane and K. Hieronymi, *Electronics Goes Green 2008 +*, 2008, 1–6.
14. P. Parthasarathy, K. A. Bulbule and K. S. Murthy, *Res. J. Chem. Environ.*, 2008, **12**, 93.
15. B. Steubing, H. Böni, M. Schluep, U. Silva and C Ludwig, *Waste Manage.*, 2010, **30**, 473.
16. N. Ciocoiu, S. Burcea and V. Tartiu, *Theor, Empirical Res. Urban Manage.*, 2010, **6**, 5.
17. B. Boggio and C. Wheelock, *Executive Summary: Electronics Recycling and E-Waste Issues Recycling and Responsible Disposal of Consumer Electronics, Computer Equipment, Mobile Phones, and Other E-Waste*, Pike Research LLC, Boulder, USA, 2009.
18. P. Chancerel, C. Meskers, C. Hagelüken and S. Rotter, *J. Ind. Ecol.*, 2009, **13**, 791.
19. A. S. Matharu and Y. Wu, *Electronic Waste Management, Issues in Environmental Science and Technology*, ed. R. E. Hester and R. M. Harrison, RSC Publishing, Cambridge UK, 2008.
20. R. Martin, B. Simon-Hettich and W. Becker, *IDW 04 Proceedings of the 11th International Display Workshop*, Niigata, Japan, 2004, 583–586. http://www.lcdtvassociation.org/images/Proceeding_New_EU_Legislation_WEEE_Compliant_Recovery_Processes_for_LCDs-Merck_September_2008n.pdf. Last accessed 5th November 2012.
21. http://www.sharpdirect.co.uk/environment/recycling-technologies/page/recyclingtechnologies, Last accessed 5th November 2012.
22. H. Hiroshi, T. Kiyobumi and T. Haruyoshi, Dainippon, JP2006091266.
23. H. Hiroshi, T. Kiyobumi and T. Haruyoshi, Dainippon, JP2006089519.
24. I. Sumoge and F. Hiruma, Densho Engineering Co Ltd, Saitama, Japan, *Denki Kagakkai Gijutsu, Kyoiku Kenkyu Ronbunshi*, 2005, **12**, 33.
25. *US Geological Survey, Mineral Commodity Summaries*, 2012. http://minerals.usgs.gov/minerals/pubs/commodity/indium/mcs-2012-indiu.pdf, Last accessed 5th November 2012.
26. G. Dodbiba, H. Nagai, L. Pang Wang, K. Okaya and T. Fujita, *Waste Manage.*, 2012, **32**, 1937.
27. S. Virolainen, D. Ibana and E. Paatero, *Hydrometallurgy*, 2011, **107**, 56.
28. J. Jiang, D. Liang and Q. Zhong, *Hydrometallurgy*, 2011, **106**, 165.
29. J. Felix, H. Tunell, B. Letcher, S. Mangold, J. Yang, T. Retegan, A. Grammatikas, T. Rydberg and H. Ljungkvist, personal communication, 2012.
30. Directive 2002/96/EC of the European Parliament and of the Council of 27 January 2003 on Waste Electrical and Electronic Equipment (WEEE) – Joint Declaration of the European Parliament, the Council and the Commission relating to Article 9; http://ec.europa.eu/environment/waste/weee/index_en.htm, Last accessed 6th November 2012.
31. http://eur-lex.europa.eu/JOHtml.do?uri = OJ:L:2012:197:SOM:EN:HTML, Last accessed 5th November 2012.
32. Directive 2002/95/EC of the European Parliament and of the Council of 27 January 2003 on the Restriction of the Use of Certain Hazardous

Substances in Electrical and Electronic Equipment; http://ec.europa.eu/environment/waste/weee/index_en.htm, Last accessed 6th November 2012.

33. Regulation (EC) No 1907/2006 of the European Parliament and of the Council of 18 December 2006 concerning the Registration, Evaluation, Authorisation and Restriction of Chemicals (REACH), establishing a European Chemicals Agency, amending Directive 1999/45/EC and repealing Council Regulation (EEC) No 793/93 and Commission Regulation (EC) No1488/94 as well as Council Directive 76/769/EEC and Commission Directives 91/155/EEC, 93/67/EEC, 93/105/EC and 2000/21/EC; http://ec.europa.eu/environment/chemicals/reach/reach_intro.htm, Last accessed 6th November 2012.

34. Directive 2005/32/EC of the European Parliament and of the Council (2005). Official Journal of the European Union: L191/29-L191/58. http://www.energy.eu/directives/l_19120050722en00290058.pdf, Last accessed 6th November 2012.

35. http://eur-lex.europa.eu/LexUriServ/LexUriServ.do?uri = OJ:L:2009:285:0010:0035:en:PDF, Last accessed 6th November 2012.

CHAPTER 9

Mining Municipal Waste: Prospective for Elemental Recovery

J. DODSON[a] AND H. L. PARKER*[b]

[a] Institute of Chemistry, Federal University of Rio de Janeiro, Cidade Universitária, Rio de Janeiro, 21941-909, Brasil; [b] Green Chemistry Centre of Excellence, Department of Chemistry, University of York, York, YO10 5DD, United Kindom
*Email: h.parker@york.ac.uk

9.1 Introduction

9.1.1 The Material Cycle and Material Flows

A discussion about elemental sustainability would be incomplete without a focus on end-of-life (EOL) waste, one of the largest potential areas where it is possible to 'close the loop'. 'Waste' is produced at every stage of the material cycle in resource extraction, manufacturing, distribution, consumption and disposal via emissions to the atmosphere, water and land (Figure 9.1).

Currently around 47–59 billion tonnes of construction minerals, ores and industrial minerals, fossil fuels and biomass are extracted globally per annum. This material consumption correlates with both gross domestic product (GDP) and population increases (Figure 9.2) and is inequitably distributed, with 20% of the world's population accountable for 86% of consumption, whilst the poorest 20% are responsible for only 1.3%.[1] Conventional projections indicate that from 2000–2050 global material throughput is likely to triple.[2] However,

RSC Green Chemistry No. 22
Element Recovery and Sustainability
Edited by Andrew J. Hunt
© The Royal Society of Chemistry 2013
Published by the Royal Society of Chemistry, www.rsc.org

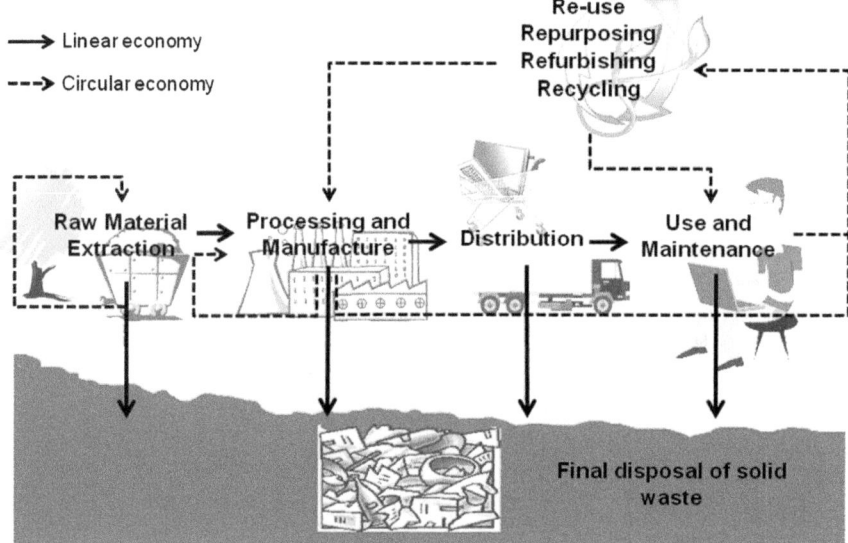

Figure 9.1 Diagram of the through-flow of materials and elements during extraction, processing, distribution, use and EOL showing the traditional linear material economy and the potential for a circular material economy.

unless this level is dramatically reduced by an absolute decoupling of resource use, economic activity and population size, the environmental impact of any economic growth and increasing material consumption will be severe. Whilst some decoupling between economic growth and material consumption has occurred, total material consumption and per capita levels of consumption are still on the rise (Figure 9.2).

Currently, it is estimated that between 50–75% of all materials flowing through industrialised economies are emitted to the environment within a year (including CO_2), whilst the remainder becomes part of the physical stock of society as buildings, roads and household equipment, until eventually these too become waste.[3] As previously highlighted in Chapter 1, to reach an ideal vision of complete elemental sustainability we must change from 'one use' or linear material systems to one where the retention of materials within the economy via reuse and recycling, a 'circular economy', reaches a point such that the need for extraction of resources no longer exceeds natural global deposition processes (Figure 9.1).

These are finite resources, with unique properties including thermal and electrical conductivity, ductility and performance at high temperature, resulting in their use in machinery, cars, computers, mobile phones and making them crucial for low-carbon technologies such as wind turbines and solar cells.[4] This ideal obviously becomes an imperative when natural concentrated deposits of these elements become increasingly scarce and when the quantities of waste produced during the extraction of these resources are considered.

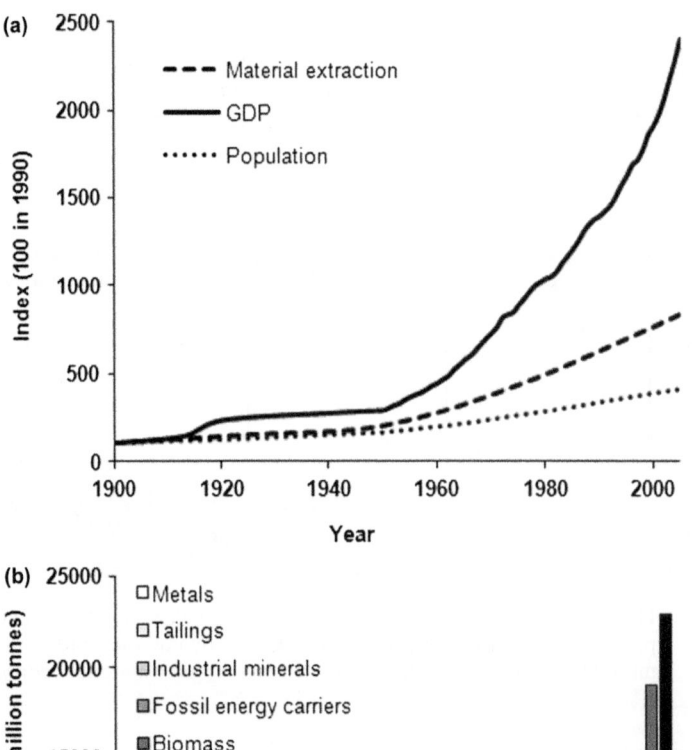

Figure 9.2   (a) Increase in relative material consumption, GDP and population growth and (b) variation in extraction of different materials globally, between 1900 and 2005, using the year 1900 as a baseline.[5,6]

Estimates by Halada *et al.*[7] showing the projected consumption of metals up to 2050, based on GDP growth and population growth in the BRIC (Brazil, Russia, India and China) and the G6 (Japan, USA, UK, France, Germany and Italy) nations, indicate that the overall consumption of metals will be five times greater in 2050 than at present, driven primarily by demand for iron. In addition, the demand for some metals including gold, silver, copper, indium, nickel, tin, zinc, lead and antimony is expected to be several times greater than the amount of their respective reserve bases (total amount of element technically extractable but not due to economic, political or other reasons), whilst demand for cobalt, iron, lithium, manganese, platinum and tungsten is expected to exceed the current reserves (economically extractable using current technology).[7] Further consideration must be made of the waste and environmental damage produced during mining of these elements, with, for example, 125 tonnes of material excavated for every tonne of copper produced.[8] However, hope lies in the fact that metals are also inherently recyclable, meaning that if systems and processes can be developed to recover them, they can be reused again and again.

9.1.2 End-of-life Waste

In comparison to the dispersed emissions of CO_2 from the combustion of fossil fuels or the leaching of potassium to water courses following application of potash as a fertiliser, elements in solid waste have for many years, particularly in industrialised nations, been collected, separated and concentrated, making municipal solid waste (MSW) or EOL a prime target for the recovery and recycling of elements. This is the waste that is collected from municipalities including households, small businesses and local government and includes paper, plastic, organic waste, glass, textile and metals. Exact figures for the amount of MSW produced globally are difficult to obtain owing to the use of different metrics and a lack of data collection in many countries or the presence of informal waste collection and recycling.[8] Data from the United Nations Environment Programme for some of the countries (45 of 162) signed up to the Basel Convention shows that 338 million tonnes of MSW were produced annually in 2001,[8] whilst 540 million tonnes were estimated to be created within the Organisation of Economically Developed Countries (OECD).[9] Nevertheless, although some areas of the world have reduced the levels produced, data indicates that globally MSW and EOL levels are on the rise as income levels increase (Figure 9.3).

The majority of MSW generated has, in the recent past in industrialised nations, ended in landfills as contained and (generally) controlled management systems. However, as a result of environmental concerns related to greenhouse gas emissions and leaching of toxic compounds, land use pressures and the economic costs of waste disposal, alternative waste disposal methods are being adopted. Levels of recycling and composting are on the rise, however, incineration or waste combustion has expanded quickly in many countries which can afford the infrastructure costs, both for reasons of waste reduction and energy recovery, despite concerns about emissions, whilst in many countries land-filling continues (Figure 9.4). Metals make a minor contribution in terms of the individual household MSW composition, with a typical Swiss annual rubbish bag containing 2.6 kg non-ferrous metal, 3.3 kg iron and 1.2 kg of electronic and electrical waste a year, in total 3.5% of the waste.[8] However, these are some of the most valuable and important elements that are available to 'close the loop'.

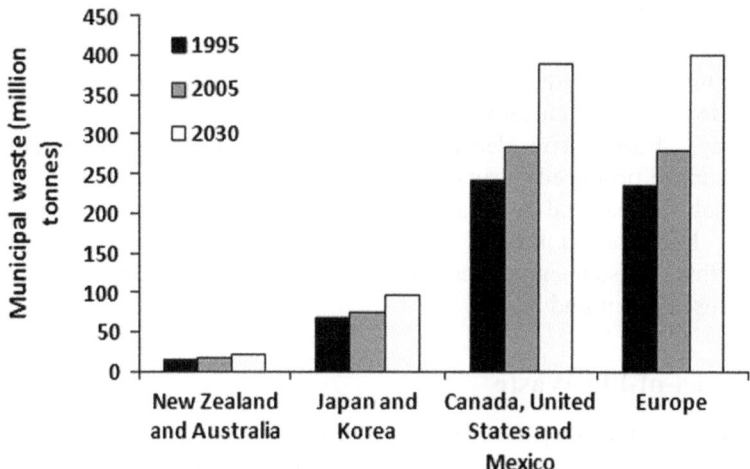

Figure 9.3   Past, current and future projected trends in municipal solid waste levels in four areas of the world.[9]

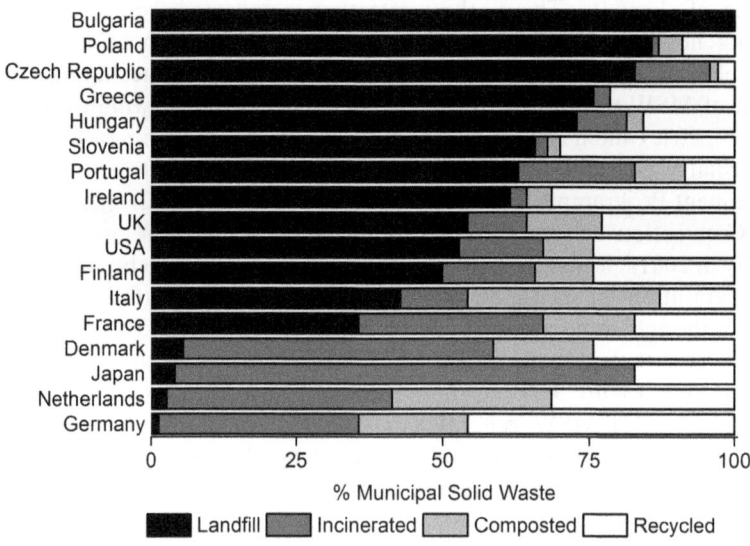

Figure 9.4   Contribution of different treatment methods to municipal solid waste disposal in different countries around the world.[10,11]

In terms of elemental sustainability, EOL waste is therefore a vital source for the capture of non-carbon elements. Current research areas include:

1 source-separation and capture of elements through recycling programmes, 'urban mining'
2 excavation of existing landfill sites, 'landfill mining'
3 use of ashes from incineration plants.

9.2 Elemental Stocks in Society

When people buy products such as televisions, cars or mobile phones, or when houses are built, elements flow from ores below ground to above ground stocks. These elements may remain in use for anywhere from a few months to years before they are eventually discarded, becoming 'EOL' products. To enable a thorough understanding of the potential for recovering elements from EOL products it would be useful to know the stocks of elements in society, 'anthropogenic stocks', how long they remain in society, where they are flowing to when they become waste (into landfills, incinerators or recycling systems) and what stocks are already present in landfills from the waste we have thrown away over the past 30–40 years. Material flow analyses (which follow the flow of elements during their extraction, manufacture, use and waste management, *i.e.* Figure 9.1) show that generally the material stocks within different countries are growing (*i.e.* the input of materials is greater than the output), causing a potential waste challenge for the future.[12] Unfortunately, information on individual stocks of elements in society is much more scarce than estimated geological resources which are publicly reported, for instance by the US Geological Survey. This is largely because the estimation of quantities of stock in society is challenging and is limited by the lack of historical data, the reliability of in-flow data from government or industry sources and the lack of knowledge about the specific elemental content of different products.[13]

9.2.1 Metal Stocks

A review on metal stocks in society by the Global Metals Flows Group of the UNEP's International Resource Panel found that reliable estimates of stocks and their lifetime could only be made for aluminium, copper, iron, lead and zinc.[13] These elements have some of the highest annual production levels. However, other elements with high volumes including manganese, silicon and chromium have not been analysed in detail, along with elements produced in smaller quantities but with significant technological value such as the rare earth and platinum group elements.[14]

Based on the analysis of the in-use stock, the per capita in-stock levels of the major engineering metals are shown in Figure 9.5 for both the more economically developed and less economically developed countries. This highlights the high metal resource basis upon which industrialised economies are built, with an in-stock metal usage of around 5–10 times higher on a per capita basis, compared to developing countries, at around 10–15 metric tonnes per person, mainly made up of iron, aluminium, copper and zinc.[13] This suggests that if developing country populations were to have the same products and infrastructure as the developed world, using existing technologies, total global in-use metal stocks would have to increase by between 3–9 times their current level.[14]

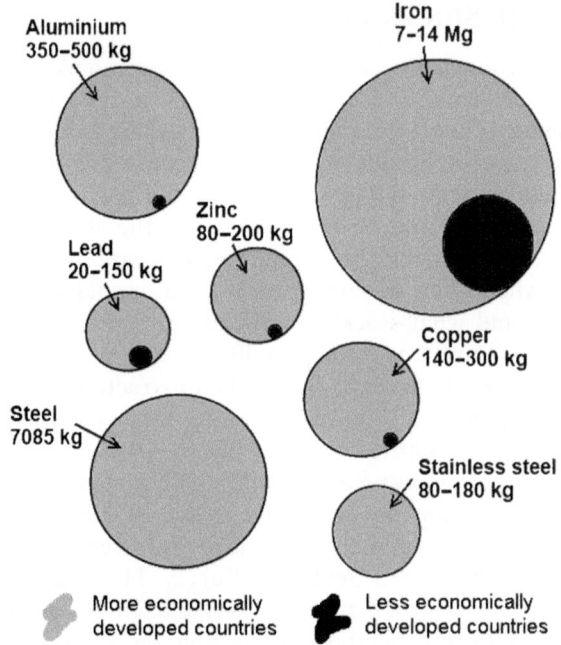

Aluminium
350–500 kg

Iron
7–14 Mg

Zinc
80–200 kg

Lead
20–150 kg

Copper
140–300 kg

Steel
7085 kg

Stainless steel
80–180 kg

More economically
developed countries

Less economically
developed countries

Figure 9.5   Per capita in-stock metal quantities of the major engineering metals
highlighting the differences between the more economically developed
countries (grey) and the less economically developed countries (black).
Area indicates per capita use, with the largest extent of the range used. No
data were available on the level of steel used in less economically
developed countries. The extent of the area shows the maximum estimate
for all elements.[13]

9.2.1.1   Rare Earth Elements

Recently there has been a huge amount of concern about the criticality of
supply of rare earth elements, which are currently indispensable in various
applications especially in some low carbon technologies (Table 9.1). Data show
that production levels of most rare earth elements have at least doubled over
the past ten years, with the main elements used being La, Ce, Nd and Y.[15]
Recent work to estimate stocks of rare earth elements in-use have calculated the
various products used in different countries, their rare-earth metal content and
the lifetime of the products, focusing on specific sectors including catalytic
converters, computers, batteries, audio systems and wind turbines.[15] The
estimated stocks in society of each element are around 4–10 times that of the
yearly flow into use, totalling 440 Gg (10^9).[15] After metallurgical applications,
computers, followed by car catalytic converters, are estimated to have the
highest in-stock content of rare earth elements at 55–60 Gg each.[15] Overall, this
indicates that once these products reach their EOL a large quantity of elements
will be available for reuse and recycling. However, these may not be available

Table 9.1 Estimates of global flows into use and in-use stocks of rare earth elements and their applications.[15]

Element	Flow into use $(Gg\ y^{-1})^a$	In-use stocks $(Gg)^b$	Uses
Lanthanum	21.9	86	Catalysts, batteries, metallurgical additives and alloys
Cerium	27.9	144	Catalyst, metallurgy, glass additives
Praseodymium	4.1	50	Permanent magnets in: cars, electric toys, wind turbines, computers and speakers; glass manufacture
Neodymium	14.8	137	Permanent magnets in: cars, electric toys, wind turbines, computers and speakers
Samarium	2.1	3.3	Defence applications, magnets
Europium	0.3	0.4	Lighting, liquid crystal displays, plasma
Gadolinium	2.2	3.6	Permanent magnets in: cars, electric toys, wind turbines, computers and speakers; ceramics
Terbium	0.3	0.7	Permanent magnets in: cars, electric toys, wind turbines, computers and speakers
Dysprosium	1.7	8.6	Permanent magnets in: cars, electric toys, wind turbines, computers and speakers
Holmium	0.3	2.1	Magnetic applications
Erbium	0.9	3.9	Fibre optics, lasers and optical glasses
Thulium	0.2	0.2	X-ray applications and lasers
Ytterbium	0.7	0.7	Lasers
Lutetium	0.1	0.6	Scintillator in computerised tomography
Yttrium	12.3	6.9	Lighting, liquid crystal displays, plasma

^{a}'Flow into use' is the amount of the element being extracted from ore annually and being used in products.
^{b}In-use stocks is the estimated total amount of that element currently in products such as mobile phones, cars or batteries that are being used or have not yet been thrown away.

for several years whilst the demand for the elements will increase, resulting in continued exploitation of virgin resources.

9.2.1.2 Precious and Speciality Metals

In terms of the precious and speciality metals, very little data is available. However, Japan has shown interest in the potential of its EOL waste as an 'urban mine' for speciality metals.[16] Researchers there have suggested that greater quantities of gold and silver are to be found globally above ground in 'urban mines' than below ground in reserves, with similar quantities of copper above and below ground.[10] They have estimated the amount of rare and precious metals accumulated in Japan in consumer and industrial goods and waste and have calculated that the country has accrued approximately 16% of the world's reserves of gold, (about 6800 tonnes), 22% silver (60 000 tonnes), 61% indium, 11% tin and 10% tantalum.[11] Halada[10] also calculated that the total accumulation for many elements, including gold, silver, copper, cobalt, PGM, vanadium, lithium, tantalum and indium was several times that of the annual global consumption. Overall, per capita in-stock estimates of precious metals and speciality metals in developed countries for which there are data, along with their uses, are shown in Table 9.2. What is clear is that much more data is needed truly to understand the amounts of these elements in-stock and their flows.

9.2.2 Landfill Stockpiles

Landfills are a potentially large source of elements. Unfortunately, information available about the elemental content of landfills is extremely sparse. Rough estimates can be made based on calculations about the total amount of product discarded in the past, its elemental content and the level of recycling.[13] Muller *et al.*[18] calculated that 850 Tg (10^{12}) of iron was present in US landfills, about two-fifths of the economically recoverable iron reserves, whilst Kapur estimated that 225 Tg of copper could be present in landfills globally.[19] These amounts are large but are likely to be highly dispersed, requiring new technologies and techniques if they are to be recovered.

9.3 Source-separation and Recycling of Metals from EOL Waste

Previous information has shown that developing closed-loop lifecycles via source-separation and recycling of end-of-cycle waste is vital to ensure elemental sustainability, particularly for several elements that will be entirely depleted if this is not achieved. Frequently, when demand and costs increase for a particular element, alternative elements with similar properties can be used; however, this is only a short-term solution to the challenge of elemental sustainability. The recovery and recycling of metals from wastes and EOL products is not new. Current aluminium recycling supports 49% of aluminium consumption in the USA.[20] It saves 95% of the energy and generates only 5% CO_2 compared to the mining and electrolysis of alumina from bauxite ore.[21] However, many challenges remain to extend recycling systems to more elements and to improve current recycling rates. Particularly important is improving 'functional recycling' rates, such that the functionality of each metal remains, rather than it being incorporated into a larger bulk material as a 'tramp' element or impurity. This frequently occurs during the recycling of copper alloys in electronic devices containing beryllium, with the latter usually becoming an impurity in the copper recyclate.[17] In this section the focus is maintained on metals, particularly precious and speciality metals due to their

Table 9.2 Per capita in-stock metal quantities of precious and speciality metals and their uses.[13,17]

Metal	More developed country per capita stock	Uses
Ferrous elements		
Chromium	7–50 kg	Cutlery, heat exchangers, car exhaust systems, elevators
Manganese	100 kg	High strength and structural steel, industrial machinery
Nickel	2–4 kg	Superalloys, stainless steel, electronics, industrial machinery
Molybdenum	3 kg	High performance stainless steel, superalloys, catalysts
Non-ferrous		
Magnesium	5 kg	Transport castings and alloys
Titanium	13 kg	Carbides, metal alloys, paints
Cobalt	1 kg	Superalloys, catalysts, batteries, blades and disks
Tin	3 kg	Cans, corrosion prevention, solders
Precious metals		
Gold	35–90 g	Jewellery, dental, electronics
Iridium		Electrochemistry and spark plugs
Palladium	1–4 g	Catalytic converters, electronics, dental
Platinum	1–3 g	Catalytic converters, jewellery, fuel cells
Rhodium	0.2 g	Catalytic converters, glass manufacture, fuel cells
Ruthenium		Hard disk drives, catalysts and electrochemistry
Silver	13 g	Solders, catalysts, batteries, glass/mirrors, jewellery, tableware, photography
Speciality elements		
Antimony	1 kg	Flame retardant, lead acid batteries, ceramics and glass
Arsenic		Semiconductors, alloying, oxide used as wood preservative and glass manufacture
Barium		Drilling fluid ($BaSO_4$) and filler in plastic, paints and rubber
Boron		Glass, ceramics, magnets
Beryllium		Electronics
Cadmium	80 g	Batteries, paints and coatings
Gallium		LEDs, diodes, solar cells
Germanium		Night vision infrared lenses, PET catalysts, fibre optics, solar cell concentrators
Indium		Coating in flat panel displays
Lithium		Batteries
Mercury	10 kg	Gold production, dental amalgams, lighting
Rhenium		Superalloys for gas turbines and catalysts
Scandium		Aluminium alloys
Selenium		Glass manufacture, manganese production, LEDs, photovoltaics and infrared optics

Table 9.2 *(Continued)*

Metal	More developed country per capita stock	Uses
Strontium		Pyrotechnics, ceramic magnets for electronics
Tantalum		Capacitors in electronics
Tellurium		Steel additives, solar cells, thermoelectrics
Thallium		Medical equipment
Tungsten	1 kg	Lighting and cutting tools
Zirconium		Nuclear reactors

technological importance, finite nature, low concentrations and mixed dispersion in many EOL products.

9.3.1 Current Recycling Rates

Surprisingly, considering the importance of metals for technological advancement, relatively little information is available on how efficiently non-renewable resources are recycled and how they flow from extraction to use and ultimately to disposal. Recently the Global Metals Flows Group of UNEP's International Resource Panel evaluated the global recycling rate information for 69 metals, mainly based on order of magnitude expert assessments.[17]

For most ferrous metals the end of life recycling rate (EOL-RR) was estimated to be over 50%, for non-ferrous metals the levels were mainly between 30–60%, for precious metals between 40–70% and for speciality metals the majority of recycling rates were below 1% (Table 9.3).[17] The influence that these levels of recycling have on the recycled content of new products made using these elements and their replacement of virgin metals, also depends on the change in demand for the element in question and the lifetime of the product that the element is in, as this causes a delay in the availability of elements from a product for recycling and reuse (Figure 9.6).

Recycling of elements will never reach 100% owing to thermodynamic considerations,[22] however, the data available on current recycling rates of consumer products shows that there is still a long way to go to 'close the loop' for even the most common and most precious elements. Overall, the effectiveness of recycling depends on economics, technology and social acceptance.

9.3.2 Technology

Metallurgical processes have been developed for the extraction and separation of ores with particular elemental combinations that are found naturally in the earth's crust.[22] However, these combinations differ from those that are found in EOL products, which are usually chosen based on their technical properties, cost and availability. The growing number of element combinations and the miniaturisation of electronic components make recycling EOL products

Table 9.3 Recycling rates for elements and recycling content of new products.[17]

Metal	Old Scrap Rate (%)	Recycled Content of New Products (%)	EOL Recycling Rate (%)	Notes
Ferrous elements				
Chromium	60–72	18–20	85–95	
Manganese	30–70	37	52	
Iron	50–65	30–50	70–90	
Nickel	65–90	30–40	55–65	
Niobium	45–55	22	50–55	
Molybdenum	35–65	33	30	
Vanadium			<1	
Non-ferrous[a]				
Magnesium	42	33	39	
Aluminium	40–50	30–35	40–70	
Titanium	11	51	91	
Cobalt	50	32	68	
Copper	24–78	20–37	42–53	
Zinc	19–71	18–27	20–60	
Tin	50	22	75	
Lead	95	40–60	40–95	c
Precious metals[b]				
Gold	75–80	30	15–96	d
Iridium	>80	15–20	20–30	e
Palladium	>80	50	60–70	e,f,g
Platinum	60 – >80	16–50	60–70	e,f
Rhodium	>80	40	50–60	e,g
Ruthenium	<20	50–60	5–15	e
Silver	75 – >80	20–30	30–95	d
Osmium	<1	<1	<1	
Speciality elements				
Antimony	<10–80	<10–25	<5–85	h
Arsenic	<1	<1	<1	i
Barium				i
Bismuth	<1		<1	i
Boron			<1	i
Beryllium	14–75	10–25	<1–7	j
Cadmium	76	25–75	15	k
Gallium	<1	25–50	<1	l
Germanium	0–40	35–50	<1–76	l
Indium	1	25–50	<1	l
Lithium	<1	<1	<1	i
Mercury	97	25–50	1–60	k
Rhenium	<50	10–25	>50	m
Scandium			<1	i

Table 9.3   (*Continued*)

Metal	Old Scrap Rate (%)	Recycled Content of New Products (%)	EOL Recycling Rate (%)	Notes
Selenium		1–10	<5	*i*
Strontium			<1	*i*
Tantalum	<50	10–25	<1–35	*l*
Tellurium			<1	*i*
Thallium	0	0	0	*i*
Tungsten	80	46	10–65	
Yttrium	0	0	0	*i*
Zirconium		1–10	<1	
Rare-earth elements				
Lanthanum		1–10	<1	*i*
Cerium		1–10	<1	*i*
Praseodymium		1–10	<1	*i*
Neodymium		1–10	<1	*i*
Samarium		<1	<1	*i*
Europium		<1	<1	*i*
Gadolinium		1–10	<1	*i*
Terbium		<1	<1	*i*
Dysprosium		1–10	<1	*i*
Holmium		<1	<1	*i*
Erbium		<1	<1	*i*
Thulium		<1	<1	*i*
Ytterbium		<1	<1	*i*
Lutetium		<1	<1	*i*
Hafnium			<1	*i*

[a]Metals widely used and sufficiently valuable to ensure recycling and reuse are quite high.

[b]EOL-RR for the precious metals does not include jewellery. Generally highly valuable so usually recycled where possible,collection costs may be determining factor.

[c]Mainly used in industrial and large vehicle batteries so recycled in commercial and industrial recycling chains,also collection and recycling cost high as is a hazardous element.

[d]EOL-RR depends on whether jewellery is included or not – lower estimates are where jewellery is excluded. Recycling is highest for industrial applications and jewellery.

[e]Recycling highest for industrial applications.

[f]Recycling highest for jewellery.

[g]Recycling highest for catalytic converters.

[h]Usually reported for more metal,but about 70% used in oxide form which is rarely recycled.

[i]Used in small quantities in complex products, *e.g.* computer chips or mixed alloys. Speciality metals often incorporated into base metal during recycling and functionality is lost.

[j]Functional recycling level low as although materials are collected it usually becomes a tramp metal or is transferred to slag during copper recycling.

[k]Collection and recycling high as hazardous element.

[l]Large amount of new scrap produced during manufacture which is recycled into the final product. Low EOL-RR due to challenge of collecting material, low metal content and challenge of recycling.

[m]Used in many industrial applications and has a high value therefore collection and recycling level is high. Demand for rhenium is growing reducing the percentage of recycled content in use.

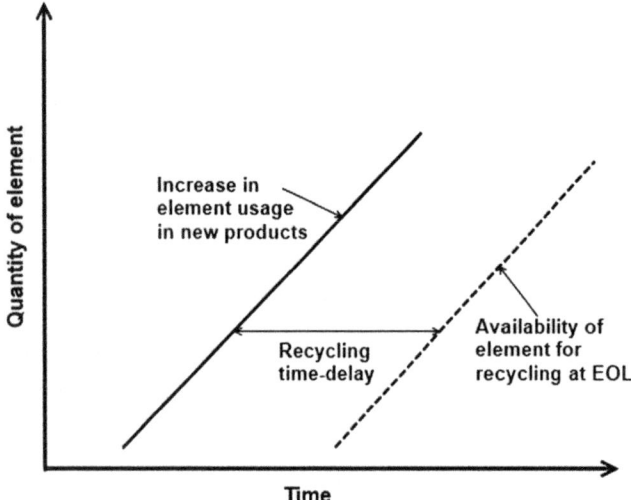

Figure 9.6 Delay in element availability for recycling depending on product lifetime.

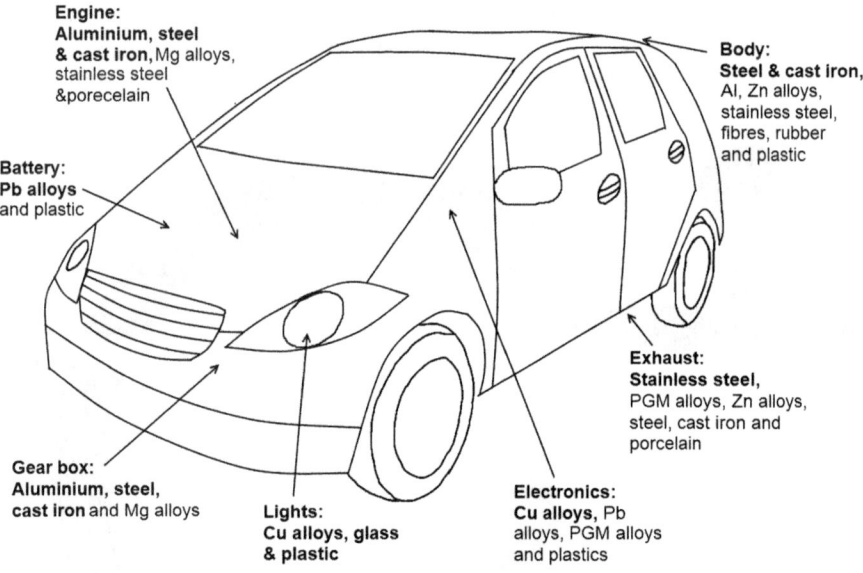

Figure 9.7 Location and combination of different elements in a typical passenger vehicle. Words in bold signify the main element or component.[22]

increasingly complex. This is particularly so when precious elements are diffused throughout a product at extremely low concentrations or when both valuable and hazardous substances are interconnected.[23] Vehicles are a good example for demonstrating the complexity and challenges involved in creating closed-loop recycling systems with many different parts containing different mixtures of elements (Figure 9.7).

9.3.2.1   Contamination

When EOL products are sent for recycling they are usually only partially dismantled to separate valuable distinct fractions of recyclables before shredding to break the materials down and liberate the different elements. These processes, as well as the initial alloys used, can result in contamination of the recyclates. This is particularly the case for EOL vehicles where shredded particles are physically separated into ferrous materials, aluminium, copper, zinc, stainless steel and automotive shredder residue (ASR), mainly composed of non-metallic materials.[24] These separate streams will then enter metallurgical recycling processes; however, they will all be contaminated to some extent owing to imperfect separation processes and shredding, particularly at joints connecting different materials.[25] Traditional metallurgical methods have been developed to separate metals from natural ores and thus the differing metal combinations in an EOL product from those in natural ores can cause significant technical recycling challenges.

The ultimate result of contamination is therefore that these elements will be lost from the functional elemental lifecycle. The quality of the contaminated elements will be reduced after each lifecycle, downgrading its potential applications unless it is diluted with higher purity primary products (Figure 9.8).

In order to overcome this, new systems and methods are needed based on knowledge of the elements in EOL products so that full disassembly of

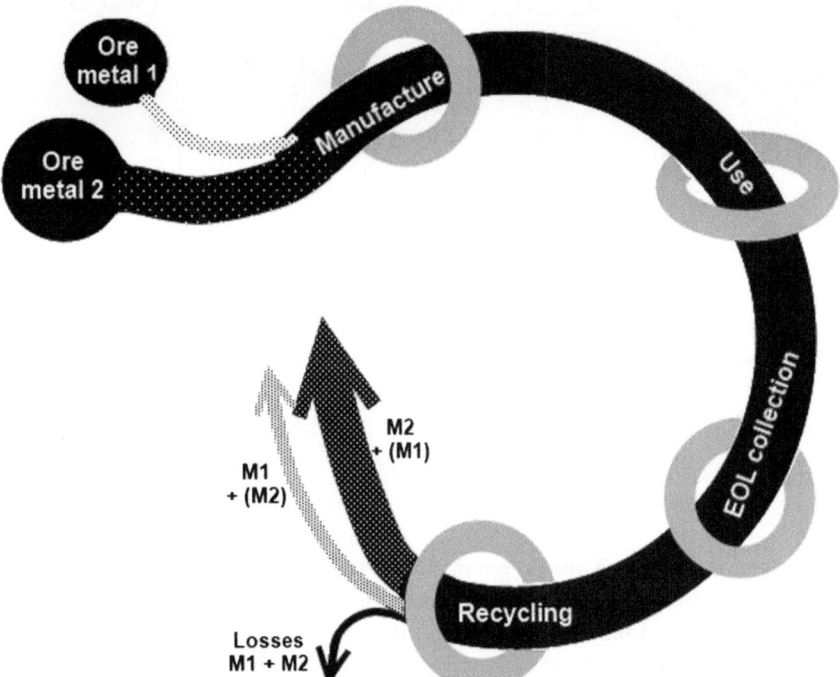

Figure 9.8   Contamination of metals by other elements during the lifecycle.

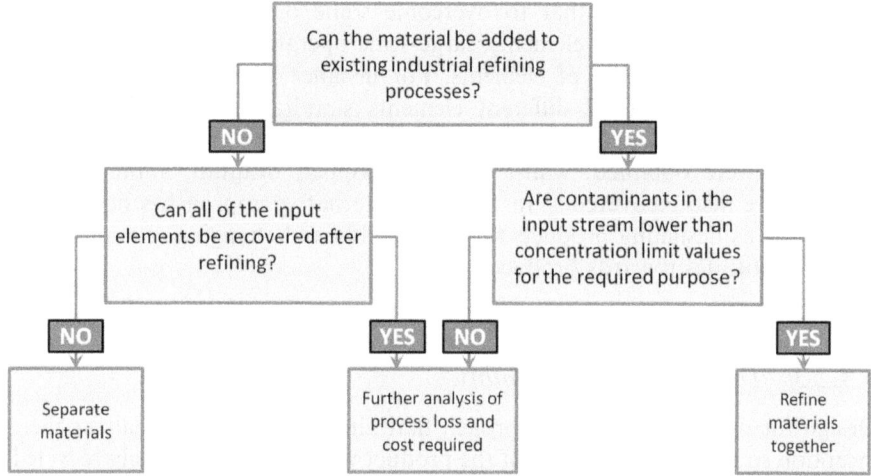

Figure 9.9   Questionnaire to determine if two materials can be recycled together.

products occurs initially, when separation of the elements will not be possible during metallurgical processing. Castro *et al.*[22] have suggested that three questions need to be asked to determine if two materials can be recycled together or not, Figure 9.9 shows the questionnaire process.

For instance in a car with an engine made mainly of aluminium (Figure 9.7), every effort should be made to separate any iron such as bolts which will be difficult to remove from the molten metal. On the other hand, aluminium contamination of the steel body will not result in contamination of the molten steel, as aluminium requires a much higher temperature to be processed. However, the aluminium will be lost in the slag. Meanwhile, platinum group metals (PGMs) may be combined with copper scrap, in which they will dissolve easily and can be recovered by electrolytic refining.[22]

9.3.2.2   Precious and Speciality Metals

As hinted above, particular challenges exist for the recycling and recovery of precious and special metals from post-consumer products. Although levels of recycling from many industrial applications, such as spent catalysts, can be greater than 90% levels, recovery from consumer products such as cars and electronics only reaches around 40–70% and is down to less than 1% for some important elements such as indium used in LCD screens.[17]

The recovery of some of these elements is problematic owing to their concentrations at ppm levels, such as PGMs in circuit boards or indium in LCD screens, or their distribution throughout an entire product, such as car electronics which are usually lost during shredding.[23] In contrast, autocatalysts can be cut from the exhaust and fed directly into the appropriate recycling stream, although even with this easily separable component only about 50% of the PGMs are recovered, mainly owing to lack of collection of the catalytic converters.[26]

It has been suggested that to overcome some of the challenges of low concentration distributive elements, large-scale operations are required which can recover a wide range of elements. For instance Umicore has developed technology to recover 17 different elements simultaneously from complex feedstocks including circuit boards and catalysts. For precious metals, yields of over 95% were obtained, whilst tin, lead, copper, bismuth, antimony and indium were also recovered.[23] In addition, alternative approaches need to be taken such as designing products for disassembly and changing consumer and industry attitudes towards products.

9.3.2.3   Design for Disassembly

Design for disassembly is an approach that aims to design materials at their inception so that different parts of the product which cannot be easily recycled or reused together can be easily separated to aid the development of a closed-loop materials system. As discussed above, particular problems are present when two materials that need to be recycled in different ways are bonded together or are difficult to separate, such as different types of plastics in a single toy or mixed cotton and synthetic polymer clothing.

Several simple steps have been suggested to improve the potential for the disassembly of products at their end of life or for easier repair. The main considerations are:

- use parts from pure materials
- use fewer components and component types and integrate them to reduce he time and cost for disassembly and to reduce the amount of materials used
- use batteries and electronics that are easy to separate
- use standardised fasteners such as screws and bolts throughout a product and make them accessible. Making these magnetic can help to improve disassembly and their separation for reuse or recycling. Alternatively, not using fasteners at all but using plastic clips or tabs enables materials to be separated quickly. Glues and bonds should be avoided except where materials can be recycled together.
- label parts so that whoever is doing the repair or recycling knows which stream it should go in so as not to contaminate other materials
- use modular components which will particularly help with upgrades and repairs.

An example of design for disassembly is the suggested redesign of the location of the wire harness in cars and the collection of wires that link the electronic devices and are rich in copper. Currently, these are considered too labour intensive to remove and generally end up in the automotive shredder waste. However, some research has suggested that a different routing of the wire harness through the car or altering the attachment of the harness to the vehicle would make it easier to remove. This could enable a greater proportion of the copper from this component, around 10 kg per car, to be recovered.[27]

9.3.3 Improving Collection Rates

Closed recycling cycles are much more typical for manufacturing and industrial processes than for households owing to fewer 'owners' handling larger quantities of materials, reducing the dissipation of products and increasing the direct economic incentives for recycling.[17] For the recycling of domestic EOL products, the initial collection activity is usually the least efficient step in the process, with efficient collection systems being a prerequisite for elemental recovery.[26] Studies generally show that the main factors that affect household recycling levels are the availability of convenient, reliable and simple (*i.e.* less sorting) recycling services, knowledge, time and storage space.[28,29] In addition, concern for the environment and the local community and awareness about the impact of waste are important to overcome the energy and time required to sort recyclables and set-up recycling systems in the home, with changes in societal norms also being important, in other words recycling becomes the normal behaviour.[29] Economic incentives have been found to increase recycling in communities on low incomes, although these do not impact on the activities of people with higher incomes.[28]

Some items such as mobile phones or computers are often stored at home (estimated to be 44% of all devices) and do not enter into reuse or recovery systems (only 3%).[26] One potential approach to improve the level of recycling and to change the consumption relationship is to switch from 'selling' products to 'leasing' of them, with the manufacture retaining ownership of the item and taking it back after use for refurbishment or recycling. Many consumers lack awareness about the value of the elements in the products they use and although these are often at low levels per item, the large numbers of items overall represent a significant elemental resource.

Furthermore, weight-based recycling quotas, promoting efforts for recycling heavy materials such as glass, can have a negative impact on the recovery of trace speciality and precious elements which may be the most valuable and the most important from a sustainability perspective to recover.[23]

More knowledge is also needed about the flow of EOL products between different regions of the world, as the trade in recyclates and the export of second hand goods to developing countries increases. It is estimated that 50% of used IT electronics leave Europe by different means, whilst only 504 000 of Germany's 3.2 million EOL passenger cars were recycled in Germany.[26] Products that have been exported may be reused, extending their lifetime, however, knowledge and control of their eventual fate is currently lost with the result that worse waste disposal or recycling practices may be used resulting in a greater loss of elements from a circular material economy.

9.4 Excavation of Existing Landfill Sites 'Landfill Mining'

As mentioned in Section 9.2.2, municipal landfills could offer a potential source of a range of elements and as such the concept of landfill mining, excavating waste from landfills and subsequent extraction of resources, has been suggested as a means to exploit this supply. The concept of landfill mining can be traced back as far as 1953 where it was used as a way to obtain fertilisers for orchards.[30] Since then interest has been sporadic, reflected by the fluctuation in numbers of publications in the area over the last couple of decades (Figure 9.10). Interest appeared to peak in the mid-1990s but has since dropped off to only 1–2 studies each year. There are a number of potential reasons for this lack of interest, such as economics or less demand for landfill space owing to the use of alternative waste management technologies. However, the literature indicates that the most likely cause is the difficulty of obtaining useable, recyclable products from landfill sites.[31]

Theoretically, considering the vast array of wastes disposed of almost indiscriminately in landfills, they should offer a whole wealth of valuable materials, particularly with the aim of creating a circular economy (Figure 9.11). Unfortunately, in practice there is a significant lack of certainty. The specific contents of individual landfill sites are hard to ascertain, especially for those that have been used for a long period or were closed up some time ago, whilst there is insufficient information about the environmental or atmospheric pollution that may occur by opening up and disturbing sites. Finally, the development of suitable technologies to enable landfill mining and separation, recovery and reuse of the excavated contents is still in its infancy. Much more work is required to demonstrate the efficiency, capacity and suitability of different technologies.[31]

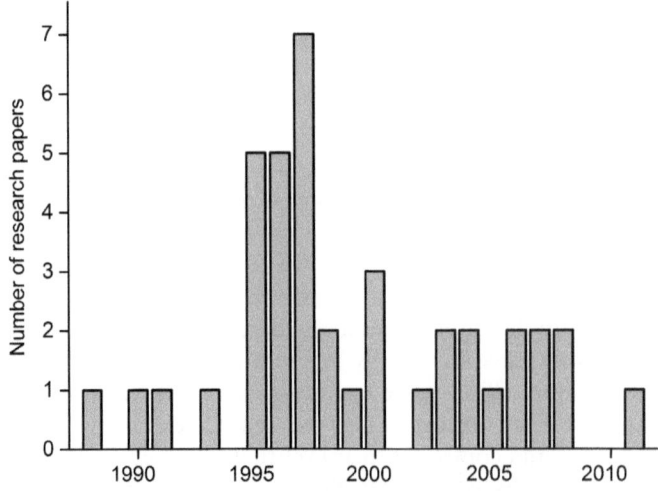

Figure 9.10   Number of research papers focusing on landfill mining from 1988–2011.[31]

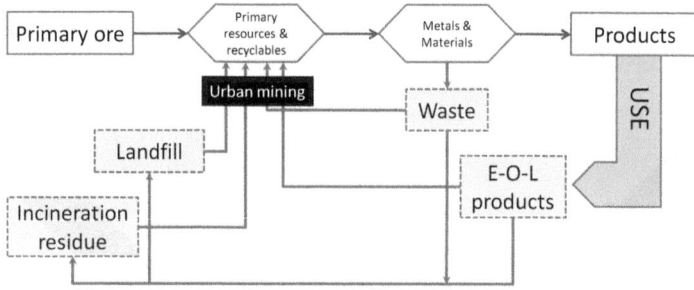

Figure 9.11   How landfill mining fits into a circular economy.[32]

Table 9.4   Landfill mining or management concepts currently under development.[33]

Concept	Definition	Ref.
Enhanced landfill mining (ELFM)	Accessing the resources and value in landfills as both resources for materials and energy.	
Enhanced biodegradation (& bioreactor)	Technology for cost-effective remediation for contaminated soil and water.	34–37
Sustainable landfill	Methods to minimise the pollution potential of landfills. Using the bioreactor concept pollutants are broken down, removed or immobilised within landfills.	33,38
Natural cap/catch	An alternative to the standard landfill cover (soil and synthetic foil) by a natural layer of living organic material (natural cap). The principle of natural catch is to improve the natural attenuation of seepage from waste dumps by cultivating wetlands.	39,40
Temporary storage place	Fills the gap between ELFM and resource recovery concepts by providing an environmentally and structurally safe area to store MSW for future materials recovery.	32

Nevertheless, new landfill mining strategies are being developed. Table 9.4 gives details of five landfill mining or management schemes currently undergoing research and development. The relevance and future use of these techniques depends closely on the specific nature of the landfill being mined with factors such as its size, location, age, composition and past records all affecting the appropriateness of different concepts.[33]

Despite the difficulties in implementing landfill mining, there is still the potential to recover metals without the necessity for excavation. A range of emissions are given off by current and sealed landfill sites, the most significant being a liquid leachate.

9.4.1   Potential Metal Recovery from Landfill Leachate

Landfill leachate is defined as the aqueous effluent generated as a consequence of rainwater percolation through waste.[41] Leachate production increases for

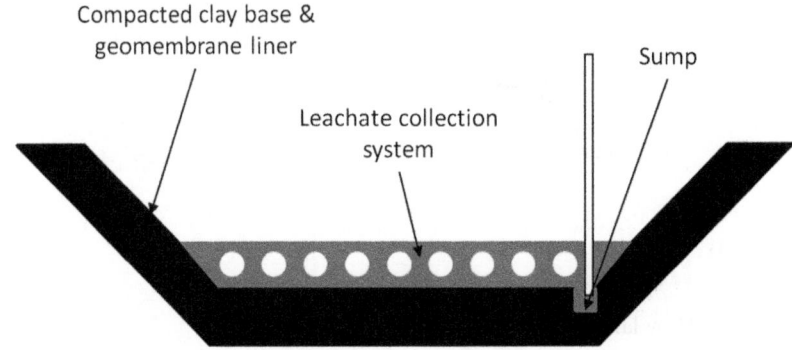

Figure 9.12   Structure of the base of a sanitary landfill showing leachate collection system.[43]

landfills that are uncapped and those with less compacted waste, as water can penetrate more easily. Landfill leachates contain a large number of compounds as a result of the biological and chemical breakdown of the refuse in the landfill.[42] The composition is greatly dependent on the type of waste, the location and the age of the landfill. Figure 9.12 illustrates the structure of a standard sanitary landfill, with leachate being collected at the bottom of the landfill where it flows into the sump from which it can be piped to the surface.

At present research into the characterisation of landfill leachate is being carried out with a view to preventing environmental contamination by treating the leachate to lower its toxicity by removal of a range of organic, inorganic and biological substances, rather than to extract and concentrate the metals. However, the data collected by this work can be used to give an impression of the potential metal content of the leachate with a view to their extraction in the future. Oman and Junestedt[44] carried out a very thorough characterisation of landfill leachate samples from 12 Swedish landfill sites and have successfully identified 400 compounds, including 40 metals at a range of concentrations, contained within it (Table 9.5).

As this data demonstrates, the concentrations of metals in leachate are not trivial with significant concentrations across all classes of metals. In order to recover the metals from this highly hazardous source, considerable work is required to develop suitable techniques and technologies. However, the wide range of metals present in the leachate could point to a future economic and accessible route to the resources contained in landfills.

9.5   Use of Ash from Incineration Plants

The definition of incineration is the controlled burning of solid, liquid or gaseous waste. Owing to the increases in MSW being produced, incineration is becoming a part of modern municipal waste management, especially in industrialised countries and is dependent on their individual public policies.[43] The primary purpose and advantage of municipal solid waste incineration

Table 9.5   Metals found in landfill leachate and sediment samples.[44]

| Metal | Leachate sediment (mg kg^{-1}) | | | |
	Min	Max	Min	Max
Ferrous elements				
Chromium	1.4	1500	4.0	124
Manganese	30	1400 000	954	3000
Iron	160	5500 000	13 500	285 000
Nickel	10	13 000	2.9	68
Niobium	0.04	0.9	14	0
Molybdenum	0.04	223	83	37
Vanadium	2.0	23	102	69
Non-ferrous				
Magnesium	13 800	15 000 000	1680	6900
Aluminium	24	579	27 400	7430
Cobalt	1.7	1500	0	26
Copper	5.0	10 000	1.9	1890
Zinc	13	1000 000	59	1890
Tin	0	3.5	95	32
Lead	1.0	5000	8.7	500
Precious metals				
Palladium	0.04	0.2	1.3	0.3
Platinum	0.04	0.02	0.05	0.002
Silver	0.04	0.3	1.6	0.7
Speciality elements				
Antimony	0.04	6.0	15	0
Arsenic	10	1000	0.9	258
Barium	2.0	1370	676	2950
Bismuth	0.02	0.1	0.7	0.2
Cadmium	0	400	0.1	11
Germanium	0.04	0.3	8.2	0.1
Indium	0.04	0.06	0.18	0.06
Lithium	0.4	622	111	11
Mercury	0.05	160	0	2.5
Scandium	0.3	1.0	24	14
Selenium	0.02	113	7.6	0.6
Strontium	0.2	1430	523	356
Tantalum	0.01	0.01	0.40	0.22
Tellurium	0.04	0.05	0.06	0.04
Thallium	0	0.28	0	0
Tungsten	0	2.9	0	0
Yttrium	1.0	6.6	146	50
Zirconium	1.0	10	291	160
Rare-earth elements				
Lanthanum	0.04	8.3	134	14
Cerium	1.4	1500	4.0	124
Neodymium	0.04	0.9	0	14
Samarium	0.2	1430	523	356
Terbium	0.04	0.05	0.06	0.04

(MSWI) is the resulting reduction in the volume (up to 90%) and mass (up to 70%) of the waste.[45,46] The recovery of heat energy produced by incineration to generate electricity offers a secondary advertised benefit of this process (however, See section 9.5.3 on the environmental impact).[43,45]

Various incinerator designs are used but most are based on the mass-burn incinerator as the most straightforward incineration technology available. This technology requires very little processing of the MSW received, generally only the removal of large bulky items (*e.g.* white goods), bulky combustible items (*e.g.* mattresses) and hazardous waste. The process consists of three main parts: incineration, energy recovery and air pollution control, a schematic diagram of a typical mass-burn incinerator is shown in Figure 9.13.

During MSWI a number of residues (or secondary wastes) are produced that can be categorised as:

- flue gases (*e.g.* SO_2, NO_x, HCl, H_2O)
- particulate matter carried by the gas stream, termed 'fly ash'
- incineration residue or 'bottom ash'.

Studies of waste input and subsequent output into these residues as a result of incineration show that any inorganic, metal-containing waste fed into the incinerator is contained in the bottom and fly ash fractions post incineration.[47] These residual ashes can be generated on a large scale, for example, in Sweden waste incinerators annually produce about 700 000 tonnes of bottom ash and 200 000 tonnes of fly ash.[48] These ash fractions can contain a vast and varying mixture of potentially hazardous substances making their disposal complex, as they present a threat if accidentally released into the environment. The most common method of disposal is land-filling, with or without further treatment, depending on the requirements of the country where they are produced, although this is not an ideal solution. Importantly, apart from containing hazardous substances, MSWI ash can also contain significant amounts of

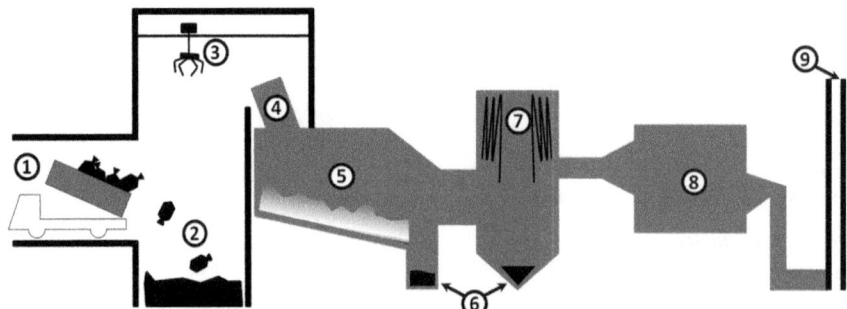

Figure 9.13 Schematic diagram of a typical mass-burn incinerator: (1) tipping area, (2) storage area, (3) crane, (4) feeding chute, (5) combustion chamber, (6) bottom ash recovery, (7) boiler, (8) air pollution control, (9) flue (stack).[47]

valuable metals, such as Cu, Zn, Ag and Au leading to interest in its exploitation and use as an 'artificial ore'.[48,49]

9.5.1 Characterisation of and Metal Concentrations in MSWI Ash

Results from various studies into the metal content of bottom and fly ash produced from MSWI show the presence of a wide range of ferrous, non-ferrous, precious, speciality and rare earth metals (Table 9.6). What is apparent is that the concentrations of different metals can fluctuate extensively depending on the source of the ash, which indicates that the metal content of the ash is strongly influenced by the MSW feedstock, which in turn is largely influenced by locality and municipality of the incineration site.[50] The concentration of metals in the bottom ash compared to fly ash also differs between incinerator plants, owing to the use of different incineration conditions which influence the partitioning of metals.[47] The fate of metals during incineration is determined by a number of variables:

1 MSW feedstock composition
2 physical–chemical behaviour of the individual metal
3 operating conditions of the incinerator.

What this data show is that the metal content of ash cannot be generalised, rather it must be determined on a case by case basis. This is problematic as MSWI residues cannot be relied upon to be a constant, continuous feedstock of metals and different recovery methods are potentially needed for each source.

9.5.1.1 Precious and Speciality Metals

The concentrations of precious and special metals in MSWI ash on first glance appear to be very small and potentially insignificant (palladium levels of only 0.03 mg kg^{-1}). However, a comparison shows that palladium is present in fly ash at concentrations approximately 20 000 times higher than in the continental crust (Figure 9.14 and Figure 9.15). This is because incineration concentrates these metals which are present only in small amounts in EOL products.[50] Yet, as discussed regarding the recycling of metals, owing to the extremely heterogeneous nature of MSWI wastes, the separation, recovery and reuse of these elements can be problematic and varying amounts of speciation and contamination are highly likely owing to the presence of so many other elements.

9.5.2 Techniques for Treatment and Metal Extraction from MSWI Ash

Unlike other materials such as plastics, metal properties can potentially be restored via recycling regardless of the chemical or physical form that they are in. However, this is not always a simple or economical task. The success of

Table 9.6 Concentrations of elements indentified in MSWI bottom and fly ash.

Metal	Concentration range (mg kg^{-1}) Bottom ash	Ref.	Concentration range (mg kg^{-1}) Fly ash	Ref.
Ferrous elements				
Chromium	13–1400	45,51–53	13–1900	45,51,53
Manganese	0.8–8500	45,51,52	<0.7–3100	45,51
Iron	32 000–84 000	51–53	420–10 000	51,53
Nickel	9.0–510	45,51,53	6.5–2000	45,51,53
Niobium	2.3–ND	53	0.2–ND	53
Molybdenum	7.9–33	51,53	0.7–47	51,53
Vanadium	10–90	45,51–53	0.7–150	45,51,53
Non-ferrous				
Magnesium	9700–12 000	51,52	8200–14 000	51
Aluminium	16 000–85 000	51–53	663–47 000	52,53
Cobalt	9.9–700	45,51–53	1.1–1700	45,51,53
Copper	80–25 000	45,51–53	45–4000	45,50,51,53
Zinc	200–20 000	45,51–53	605–150 000	45,50,51,53
Tin	31–1300	45,51,53	30–8200	45,50,51,53
Lead	98–6500	45,51,53	200–19 000	45,50,51,53
Precious metals				
Gold	0.2–19	52–54	ND–ND	–
Palladium	0.03–ND	51	3.6–18	50,51
Platinum	0.1–ND	51,53	<1–ND	51
Silver	2–62	45,51–54	1.2–700	45,50,51,53
Speciality elements				
Antimony	7.6–120	51,52	170–2600	50,51
Arsenic	1.3–230	45,52	15–751	45
Barium	47–2700	45,52,53	37–9000	45,53
Bismuth	1.2–57	51,53	1.6–142	50,51,53
Cadmium	0.3–61	45,51,53	5–2200	45,51,53
Gallium	1.9–24	51,53	0.3–164	50,51,53
Germanium	0.2–1.5	51,53	<1–27	50,51,53
Hafnium	1.8–2.7	52,53	ND–ND	–
Indium	0.2–2	51,53	0.1–23	50,51,53
Lithium	8.1–ND	53	0.9–ND	53
Mercury	<0.01–3	45	0.8–73	45
Scandium	0.9–2.1	52,53	0.1–ND	53
Selenium	0.2–ND	53	0.2–ND	53
Strontium	122–ND	53	7.8–ND	53
Tantalum	1.1–5.3	51–53	0.1–43	52,53
Tellurium	<1–1.2	51	1.8–12	50,51
Thallium	<1–ND	51	<1–2.0	51
Tungsten	11–22	51,52	6.0–9.0	51
Yttrium	4.8–ND	53	3.0–ND	53
Zirconium	65–81	51,53	27–57	51,53
Rare-earth elements				
Lanthanum	24–ND	52	ND–ND	—
Cerium	35–ND	52	ND–ND	—

Table 9.6 (*Continued*)

Metal	Concentration range (mg kg^{-1}) Bottom ash	Ref.	Concentration range (mg kg^{-1}) Fly ash	Ref.
Praseodymium	1.9–ND	53	ND–ND	—
Neodymium	7.1–ND	53	0.2–ND	53
Samarium	1.7–ND	52	ND–ND	—
Terbium	0.64–ND	52	ND–ND	—

ND, not determined.

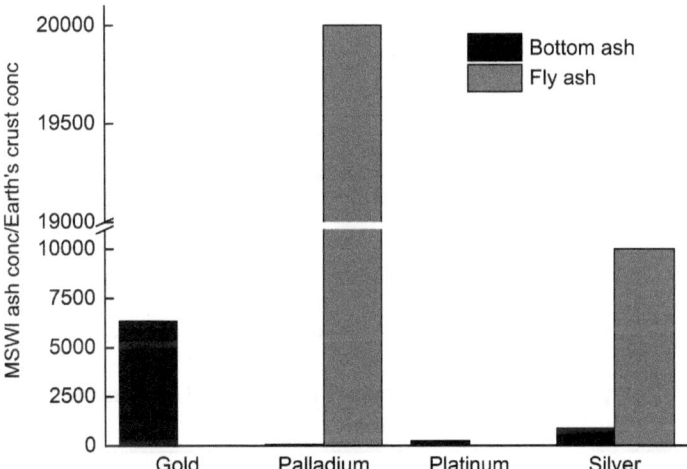

Figure 9.14 Ratio of the precious metal concentration measured in MSW and Taylor's concentration values in bottom and fly ash (ther ratio based on maximum concentration values achieved for each metal[55]).

secondary metal markets is entirely dependent on the cost incurred in retrieving and processing the elements from waste streams.[56] There are several technologies that are currently available and many more that are being developed in an attempt to make metal retrieval from ash both scientifically and financially possible. The technique chosen for the recovery of metals from MSWI ash depends on the characteristics of the residue and what follows is a description of different separation techniques and where they can be best employed.

9.5.2.1 Mechanical Separation Technique: Melting MSWI Ash

Remelting of MSWI residues has three main objectives (1) extraction of metals, (2) destruction of dioxins and (3) volume reduction of ash.[57] This is not a technique widely used in Europe. However, in 2009 approximately 10% of MSWI ash produced in Japan was treated using melting.[58] Electric, burner and blast melting type systems are available for MSWI melting. For the purpose of metal recovery, the electric melting system is the most appropriate as it has the highest capacity, lowest energy requirements and lowest operating costs

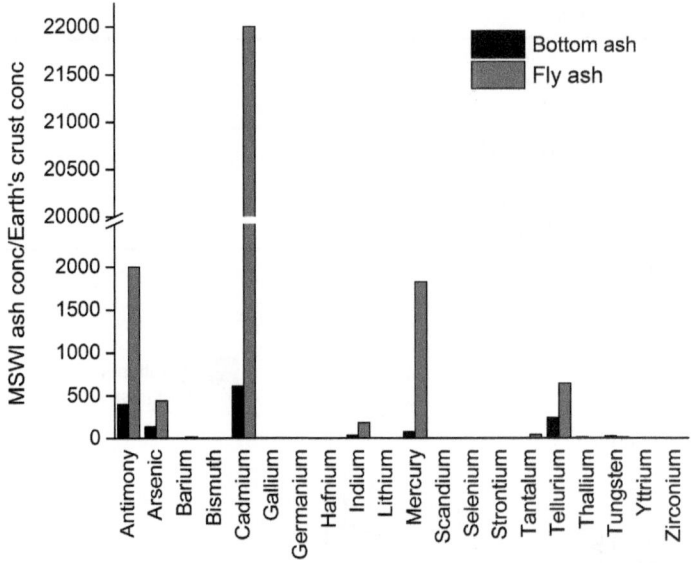

Figure 9.15 Ratio of the speciality element concentration measured in MSW and Taylor's concentration values in bottom and fly ash (ratio based on maximum concentration values achieved for each metal).

Table 9.7 Energy and cost requirements for different melting techniques for treatment of MSWI in Japan.[57]

Melting System	Furnace Type	Average Capacity (t d⁻¹)	Average Energy Needs (kWh t⁻¹)	Average Operation Costs (10³ ¥ t⁻¹)
Electric	Electric arc	655	675 ± 75	ND
	Plasma arc	37	1050 ± 250	30
	Electric resistance	74	$650 \pm ND$	<14
Burner	Reflecting surface	37	8550 ± 350	24
	Rotating surface	84	3900 ± 900	15
	Swirling flow	ND	2900 ± 400	ND
Blast	Coke bed	160	2700 ± 200	ND

compared with the other techniques (Table 9.7). The three most common types of furnaces used for electric melting are illustrated in Figure 9.16 and are described below:

1 Electric arc furnace: this furnace transforms electrical energy to heat in the form of an electric arc to melt the slag (ash) fed into the furnace. The arc is formed between the graphite electrode and the melted metals collected at the bottom of the furnace.[59]
2 Plasma arc furnace: in this case copper alloy rear electrodes are used to generate plasma to heat the slag (ash).[57] Nitrogen gas is commonly used to

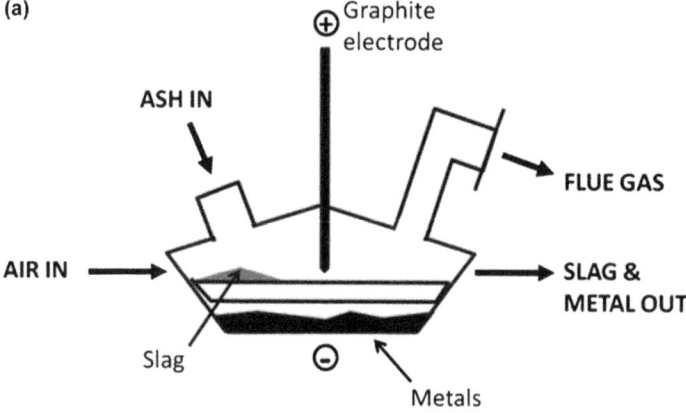

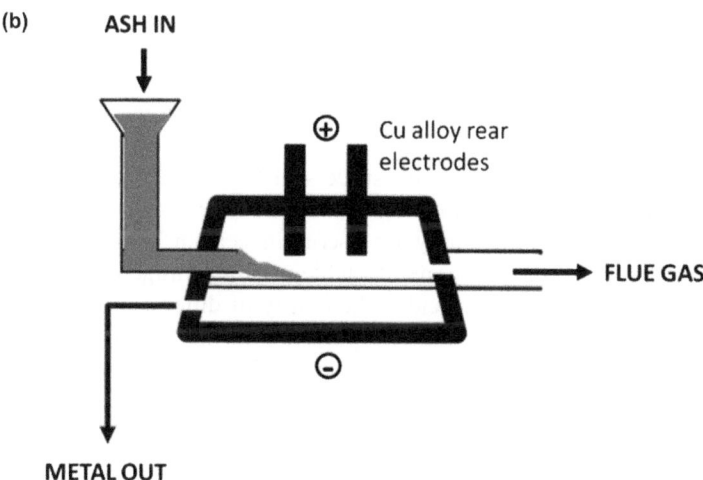

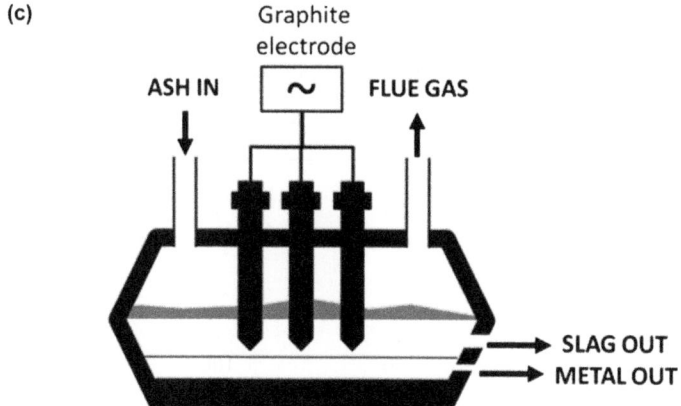

Figure 9.16 Furnace types for electric melting systems (a) electric arc furnace, (b) plasma arc furnace, (c) electric resistance furnace.[57]

produce the plasma and is maintained at a constant temperature of $1400 \pm 50\,^{\circ}C$.[60]

3 Electric resistance furnace: here graphite electrodes are inserted directly into the melting bath to conduct current and melt the metals. MSWI is used to insulate the bath and to reduce heat dissipation.[57]

9.5.2.2   Chemical Separation Technique: Acid Leaching of Metals from MSWI Ash

This is the most widespread method of metal removal from MSWI ash.[56] Metals are leached from the ash using an appropriate acid followed by recovery of the leached metal, usually in the form of precipitates. This technique offers a number of advantages over melting treatments. First, if the concentration of chlorine in the ash is too high melting cannot be carried out as chlorine can inhibit the process. Second, melting is more energy intensive and expensive than leaching and emits considerable amounts of CO_2. Finally, leaching is a relatively rapid way of extracting metals, enabling faster processing of the ash.[61,62]

In general, acid leaching is more successful than alkali leaching, although the latter can be used to remove lead and zinc selectively from ash leaving other metals behind, which may be useful in certain situations. Table 9.8 shows leaching of some heavy metals achieved by a range of acids (sulfuric, hydrochloric, acetic and citric acid). The ability of different acids to solubilise metals in varying amounts makes separation of the metals possible. This process is simple and can be economically carried out on a large scale, although it does result in considerable amounts of hazardous, acidic waste being produced.

Table 9.8   Leaching metals from MSWI ash using a range of acids.

	Acid Used for Metal Extraction			
	H_2SO_4[61,63]	HCl[61,63]	Acetic Acid[61,63]	Citric acid[62]
Metal	Leaching (%)			
Ferrous elements				
Chromium	25	20	18	—
Iron	57	36	23	67
Non-ferrous				
Magnesium	65	53	47	—
Aluminium	73	43	88	100
Copper	50	50	100	100
Zinc	68	68	100	100
Lead	15	51	93	97
Speciality elements				
Cadmium	71	72	73	—

In the last decade interest in bio-derived chemicals has increased. In the case of MSWI ash leaching, this has led to attention being given to the use of acids, such as citric acid, which can be extracted from many biomass sources, *e.g.* citrus fruits. Citric acid has been shown to be a successful metal leaching agent compared to other acids (Table 9.8) which has been attributed to its ability to chelate metals owing to the presence of three carboxyl groups and one hydroxyl group in its structure. Unlike alternative synthetic chelating agents like EDTA, citric acid is biodegradable and environmentally benign. Citric acid has also been shown to be more efficient than other acids for metal leaching, thus reducing the amount that is required for efficient metal extraction and reducing the amount of acidic waste produced by the process.[62] In the future there may be an increase in chemicals such as this being employed in waste treatment to improve the sustainability of metal-leaching processes.

9.5.2.2.1 Precious and Speciality Metal Leaching. Information and data relating to the recovery of precious and special metals from MSWI ash is sparse. Jung and Osako[58] carried out a study looking at the effect of pH on the leaching of Ag, Bi, Ga, Ge, In, Pd, Sb, Sn, Te and Tl from incineration ash in a pH range of 2–13. In this study, Ag, Ge, Sb, In and Te were leached more effectively under acidic conditions with only minimal recovery in neutral and alkaline solutions. Bi, Sn and Ga were leachable under extremely alkali or acidic conditions but not in between. Pd and Tl leaching was relatively unaffected by pH, however, although large amounts of Tl were recovered, Pd extraction was low.[58] This work indicates some success in precious metal leaching from incineration residues, however, significant further work is necessary to prove its feasibility.

9.5.2.3 Biological Separation Technique: Bioleaching

Over the last four decades biotechnology has also been investigated as an alternative method for resource recovery.[56] Bioleaching is of particular interest, based on the interaction between metals and microorganisms, enabling metal recycling using similar processes to those occurring in natural biogeochemical cycles.[56] Sulfur-oxidising and iron-oxidising bacteria have been studied, especially strains of *Thiobacilli,* commonly *T. thiooxidans* and *T. ferrooxidans*. These bacteria produce highly acidic solutions (approx pH 1) which dissolve the majority of metals whilst the bacteria are tolerant of high concentrations of toxic and heavy metals.[64] A range of metals including Al, Cd, Cu and Zn have been successfully recovered from MSWI ash using bioleaching with different microorganisms showing specificity for different elements (Figure 9.17). Unfortunately, no studies on the recovery of precious and special metals using bioleaching are known. Despite the first efforts in bioleaching being made more than 30 years ago, this technique has still not been widely deployed commercially; further research is required to demonstrate its feasibility on a large scale.[65]

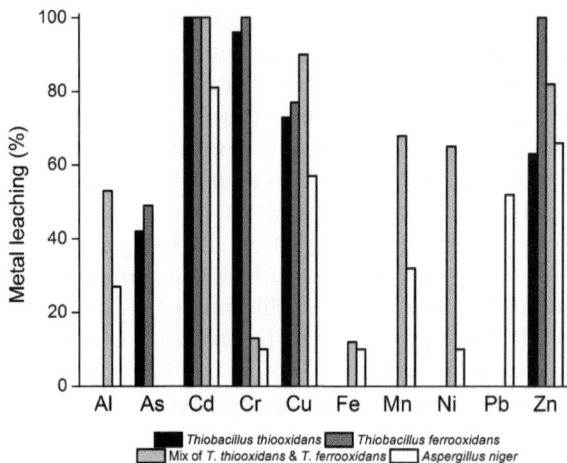

Figure 9.17 Metal bioleaching from MSWI ash using *Thiobacillus thiooxidans* (black),[49] *Thiobacillus ferrooxidans* (dark grey),[49] a mix of *T. Thiooxidans* & *T. Ferrooxidans* (light grey)[65] and *Aspergillus niger* (white).[66]

9.5.3   Environmental Impact of MSWI

Despite some of the advantages that MSWI can offer and its deployment in many countries, there are several negative aspects and environmental problems associated with its use. Waste incineration is responsible for the release of a wide variety of toxic chemicals, mainly as gases, such as dioxins, furans, sulfur dioxide, hydrochloride, nitrogen oxide and carbon monoxide.[67] These gases are closely associated with environmental impacts Like climate change and acid rain and can have extremely adverse effects on human and animal health.[68] The burning of materials such as plastics, contained in the waste, is largely responsible for these emissions, as these materials were never designed to be burnt safely.[69]

MSWI can also be responsible for the emission of large concentrations of heavy metals, *e.g.* cadmium, lead and mercury, in flue gases, dependent on the MSW source. The metals volatilise during incineration and attach to soot particles that are emitted along with the gas. In Germany, the concentrations of heavy metals in tree leaves surrounding incineration plants can reach such high levels that they have to be stripped and dealt with as hazardous waste.[69]

Another problem associated with MSWI is the loss of resources, namely the loss of recyclable materials, such as, paper, wood and plastic. During their combustion, the carbon contained within them is released and emitted into the atmosphere increasing the quantities of climate-change inducing CO_2. From this perspective, land-filling products like plastics made from fossil fuels rather than incinerating them may offer an advantage as it acts as a form of carbon storage. If this was carried out in a sensible way, that is, collecting similar materials together in one landfill, there is also the potential that the products could be retrieved at a later date and recycled. In addition, as discussed earlier, separating out specific streams of metal-containing waste enables more selective recovery processes to be applied which should result in a greater elemental recovery than treating the mixed and potentially hazardous incineration ashes. Overall, whilst MSWI may solve the problem of volumes of waste, the negative impact on the environment and resource recovery may negate its benefits.

9.6 Urban Mining in Practice: Platinum and Palladium Recovery from Roadside Dust

Automobile catalysts or catalytic converters are the largest annual users of platinum and palladium. Catalytic converters were first introduced around 1975 and are used to reduce harmful emissions of gases, such as uncombusted hydrocarbons, NO_x and CO, from exhaust fumes (Figure 9.18).[70] The precious metals Pt, Pd and Rh are supported on a ceramic honeycomb within the converter and, as the name suggests, catalyse the oxidation or reduction of the gases. Ever since their introduction there have been concerns related to the magnitude and relevance of possible metal emissions from the converters owing to the mechanical and thermal impact on the active layer during use.[71]

Three-way catalytic converters contain approximately 0.08% platinum, 0.04% palladium and 0.007% rhodium, although presently the metal composition is changing owing to increases in the price of platinum and it is being replaced by the cheaper metal palladium.[72,73] Studies have shown that up to 80% of these PGMs are released from the catalytic converters during their lifetime (which with modern converters is usually the lifetime of a car), the majority of the metal is emitted in metallic form and is primarily deposited in the dust and soil along the roadside.[74] From here the metals can be further dispersed into the wider environment, for example, rain can wash the metals into drains which eventually lead out to the sea. It is not well known what impact the presence of these metals will have on the environment, wildlife or even humans. However, what is clear is that these incredibly valuable and useful metals are being heavily diluted and dispersed into the environment.

9.6.1 Levels of Pt and Pd in Environmental Matrices

All over the world road sweeping is carried out by local government or private companies to collect rubbish and debris from roadsides and prevent blockage of street drains. Another minor result of this practice is that small particulate matter, such as dust, is collected during sweeping. Characterisation of this dust has shown concentrations of PGMs can be as much as 1 ppm.[75]

A wider study by Helmers *et al.*[71] investigated the Pt and Pd concentrations of various environmental matrices surrounding roads and combined this with literature data to give a representative idea of contamination levels (Table 9.9). The results show that Pt and Pd are indeed distributed into the surrounding environment and are accumulating at levels much higher than would naturally occur. However what is promising from their study is that the highest concentrations of the metals are in the dust and have the potential to be recovered.

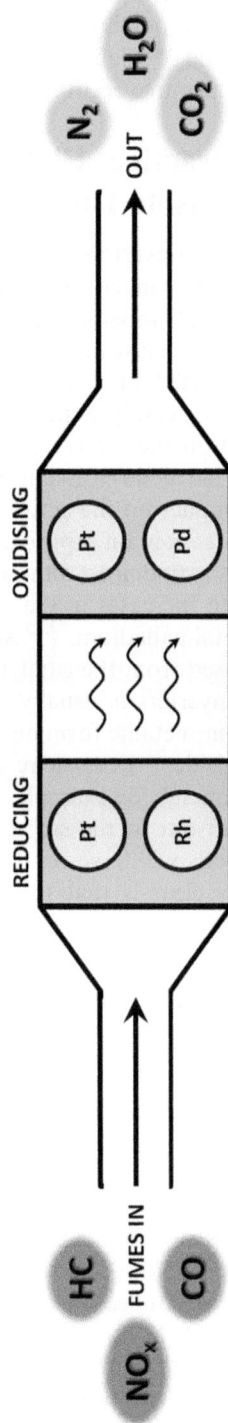

Figure 9.18 Diagram of a catalytic converter.

Table 9.9 Concentration ranges of Pt and Pt in various polluted matrices surrounding roads.[71,74]

Matrix	Pt	Pd	Unit
Grass	3.4–7.7	<0.5–0.6	mg kg^{-1}
Rain	<5–ND	<5–ND	ng l^{-1}
Soil	10–253	2–77	µg kg^{-1}
Dust	51–360	5–2465	µg kg^{-1}
Sludge	<2–220	<1–260	µg kg^{-1}
Car Exhaust	120–700	0.3–480	ng m^{-3}
Background (continental crust)	0.4	0.4	µg kg^{-1}

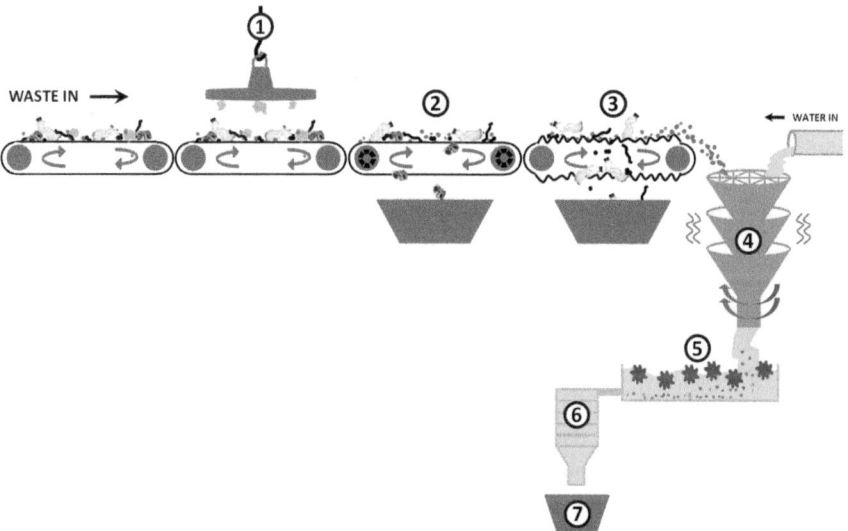

Figure 9.19 Representation of the palladium recovery process being developed by Veolia. (1) Magnets to remove steel and iron, (2) eddy current separator to remove aluminium cans, (3) vibration conveyor belts to remove plastics, twigs and grit, (4) water added followed by filter, shake spin and sieve to remove grit and dust, (5) 'smart sponges' used to adsorb oil, (6) membrane separation to catch dust particles, (7) palladium collection.[75,77]

9.6.2 Novel Processing Techniques for the Recovery of Pt and Pd

At present, once roadside dust has been collected, it is disposed of in landfill with no attempt made to retrieve the PGMs. Yet, in 2012 a patent for the design of a mechanical processing method was filed for the separation of roadside dust.[75] Also Veolia, the world's largest environmental services company, has released plans for a facility dedicated to the treatment of roadside dust and the retrieval of palladium, aimed to be open in 2013.[76] A schematic representation of their planned process is shown in Figure 9.19, which involves a variety of steps not only to separate the palladium but also other valuable and recyclable materials from the waste. Veolia is hoping to retrieve around £80 000 worth of palladium per year from their treatment plant plus other profitable materials such as steel, iron and aluminium.[77]

This is a fantastic example of how a potentially toxic waste that would cost money to be disposed of in landfill can be effectively 'urban mined' to recover valuable resources. Hopefully, similar developments will occur in the near future.

9.7   Conclusions

'Waste' is an expanding challenge owing to the current linear, throw-away material economy and increasing consumption across the world. Of critical importance is changing societal attitudes towards waste, to value the resources contained within it and to reduce, reuse and recycle the elements as a vital step towards 'closing the loop' and preventing future resource crises and environmental degradation.

To achieve this, several challenges must be addressed that have been highlighted throughout this chapter. We need better knowledge and analysis of the flow of resources into products and their flow into waste. The biggest losses occur during the collection of waste and this must be improved and prioritised, with waste treated not as one indistinct mass but as a combination of components that must be separated and treated individually to extract the most elemental and economic value. Low volume precious and speciality elements must not be lost at the expense of weight-based recycling quotas.

Changing the dynamics surrrounding ownership of products by valuing the resources at their end-of-life, as well as changing societal norms and perceptions of waste may also help to improve collection rates. Better labelling of products and design-for-disassembly will enable greater ease of recycling of collected waste, whilst new improved techniques for the recovery of combinations of elements not present in virgin ores are necessary.

Many avenues are being researched to valorise past and present waste, such as landfill mining, urban mining from roadside dust and recycling computer and mobile circuit boards on a large-scale. However, we are still only scratching the surface of what is possible in terms of, for instance, the 22% of silver reserves estimated to be above ground in products in Japan. If we want to ensure 'elemental sustainability' so that supplies of these elements remain for future generations, whilst reducing the environmental burden of resource extraction in ever more sensitive environments, it is vital that we focus on 'closing the loop' for wastes before the problem of elemental scarcity and increasing cost is on our doorstep, knocking at the door.

References

1. M. Fischer-Kowalski and M. Siwilling, *UNEP International Resource Panel*, Paris, 2011.
2. E. Matthews, C. Amann, S. Bringezu, M. Fischer-Kowalski, W. Huttler, R. Kleijn, Y. Moriguchi, C. Ottke, E. Rodenburg, D. Rogich, H. Schandl, H. Schutz, E. Van der Voet and H. Weisz, *Weight of Nations: material*

outflows from industrial economies, World Resources Institute, Washington DC, 2000.

3. *Resource Efficiency: Economics and Outlook (REEO) for Asia and the Pacific*, United Nations Environment Programme, Bangkok, 2011.
4. C. Hagelueken and C. Meskers, *Technology Challenges to Recover Precious and Special Metals from Complex Products*, R'09 World Congress, Davos, Switzerland, 2009.
5. F. Krausmann, S. Gingrich, N. Eisenmenger, K.-H. Erb, H. Haberl and M. Fischer-Kowalski, *Ecol. Econ.*, 2009, **68**, 2696.
6. A. Maddison, *Statistics on World Population*, GDP and Per Capita GDP, 1-2008 AD, http://www.ggdc.net/MADDISON/oriindex.htm, Accessed 1st June 2013.
7. K. Halada, M. Shimada and K. Ijima, *Mater. Trans.*, 2008, **49**, 402.
8. E. Baker, E. Bournay, A. Harayama and P. Rekacewicz, *Vital Waste Graphics*, UNEP, 2004.
9. L. F. Mortensen and H. Mountford, *OECD Environmental Outlook*, OECD, Paris, 2001.
10. K. Halada, http://www.nims.go.jp/genso/lecture/0ej00700000030pw-att/0ej00700000034on.pdf, Accessed 23rd May 2013.
11. K. Halada, K. Ijima, M. Shimada and N. Katagiri, *J. Jpn. Inst. Met.*, 2009, **73**, 151.
12. S. Hashimoto, H. Tanikawa and Y. Moriguchi, *Waste Manage.*, 2007, **27**, 1725.
13. T. E. Graedel, A. Dubreuil, M. Gerst, S. Hashimoto, Y. Moriguchi, D. Müller, C. Pena, J. Rauch, T. Sinkala and G. Sonnemann, *UNEP Metal Stocks in Society – A Status Report, A Report of the Working Group on the Global Metal Flows of the International Resource Panel*, 2010.
14. M. D. Gerst and T. E. Graedel, *Environ. Sci. Technol.*, 2008, **42**, 7038.
15. X. Du and T. E. Graedel, *Environ. Sci. Technol.*, 2011, **45**, 4096.
16. C.-H. Jung and M. Osako, *Resour., Conserv. Recycl.*, 2009, **53**, 301.
17. T. E. Graedel, J. Allwood, J.-P. Birat, B. K. Reck, S. F. Sibley, G. Sonnemann, M. Buchert and C. Hagelucken, *Recycling Rates of Metals – A Status Report, A Report of the Working Group on the Global Metal Flows of the International Resource Panel, UNEP*, 2011.
18. D. B. Muller, T. Wang, B. Duval and T. E. Graedel, *Proc. Natl. Acad. Sci. USA*, 2006, **103**, 16111.
19. A. Kapur, PhD, The Future of the Red Metal: Scenario Analysis and Implications for Policy, *Yale University*, 2004.
20. C. Rhodes, *Chem. Ind.*, 2008, **16**, 21.
21. J. Cui and L. Zhang, *J. Hazard. Mater.*, 2008, **158**, 228.
22. M. B. G. Castro, J. A. M. Remmerswaal, M. A. Reuter and U. J. M. Boin, *Resour., Conserv. Recycl.*, 2004, **43**, 1.
23. C. Hagelucken, *The Challenge of Open Cycles*, R'07 8th World Congress, Davos, Switzerland, 2007.
24. M. B. Castro, J. A. M. Remmerswaal, J. C. Brezet, A. van Schaik and M. A. Reuter, *Int. J. Min. Process.*, 2005, **75**, 255.
25. M. B. G. Castro, J. A. M. Remmerswaal, J. C. Brezet and M. A. Reuter, *Resour., Conserv. Recycl.*, 2007, **52**, 219.

26. C. Hagelucken and C. Meskers, in *Strüngmann Forum Report: Linkages of Sustainability*, eds. T. E. Graedel and E. van der Voet, MIT Press, Cambridge, 2010, pp. 163–197.
27. E. Brahmst, *Copper in End-of-Life Vehicle Recycling*, Centre for Automotive Research, 2006.
28. M. Martin, I. D. Williams and M. Clark, *Resour., Conserv. Recycl.*, 2006, **48**, 357.
29. M. Tonglet, P. S. Phillips and M. P. Bates, *Resour., Conserv. Recycl.*, 2004, **42**, 27.
30. G. M. Savage, C. G. Colueke and E. L. Von Stein, *Biocycle*, 1993, **34**, 58.
31. J. Krook, N. Svensson and M. Eklund, *Waste Manage.*, 2012, **32**, 513.
32. P. J. Jones, T. Van Gerven, K. Van Acker, D. Geysen, K. Binnemans, J. Fransaer, B. Blanpain, B. Mishra and D. Apelian, *JOM*, 2011, **63**, 14.
33. P. J. Jones, D. Geysen, Y. Tielemans, S. Van Passel, Y. Pontikes, B. Blanpain, M. Quaghebeur and N. Hoekstra, *J. Cleaner Prod.*, 2012, **30**, 1–14.
34. N. Hoekstra, A. Langenhoff, T. Verheij, J. Dijkhuis and H. Slenders, eds., *Enhanced Microbial Degradation of Chloroethenes in a Bioscreen*, IAHS, Oxfordshire, 2005.
35. N. Hoekstra and A. Langenhoff, *Proceedings of the 3rd International Symposium on Permeable Reactive Barriers and Reactive Zones*, 2007, 59–62.
36. A. D. Read, M. Hudgins, P. Phillips and J. Morris, *Resour. Conserv. Recycl.*, 2011, **32**, 115.
37. C. Rich, J. Gronow and N. Voulvoulis, *Waste Manage.*, 2008, **28**, 1039.
38. H. Woelder, H. Hermkes, H. Oonk and L. Luning, *Sardinia Proceedings 11th International Waste Management and Landfill Symposium*, 2007, 1.
39. C. H. Clemens, M. Van Der Wijk, P. Stook and J. C. N. Van Der Pal, in *Proceedings Consoil, 11th International UFZ-Deltared/TNO Conference on Soil–Water Systems*, Salzburg, Austria, 2010.
40. R. Dijcker, W. Den Engst and T. F. Praamstra, in *Proceedings Natural Cap Conference* Amsterdam, The Netherlands, 2011.
41. S. Renou, J. G. Givuadan, S. Poulain, F. Dirassouyan and P. Moulin, *J. Hazard. Mater.*, 2008, **150**, 468.
42. P. Kjeldsen, M. A. Barlaz, A. P. Rooker, A. Baun, A. Ledin and T. H. Christensen, *Crit. Rev. Environ. Sci. Technol.*, 2002, **32**, 297.
43. J. Pichtel, *Waste management practices: municipal, hazardous and industrial*, CRC Press, 2005.
44. C. B. Oman and C. Junestedt, *Waste Manage.*, 2008, **28**, 1876.
45. M. A. Salam and R. C. Burk, *Water, Air, Soil Pollut.*, 2010, **210**, 101.
46. G. Li and Y. Hu, *Mechanic Automation and Control Engineering (MACE)*, 2010 International Conference, IEEE, Wuhan, China, 2010.
47. P. H. Brunner and H. Monch, *Waste Manage. Res.*, 1986, **4**, 105.
48. S. Karlsson, P. Carlsson, D. Aberg, K. K. Fedje, J. Krook and B.-M. Steenari, *Linnaeus ECO-TECH*, 2010, **10**, 22–24.
49. T. Ishigaki, A. Nakanishi, M. Tateda, M. Ike and M. Fujita, *Chemosphere*, 2005, **60**, 1087.
50. C.-H. Jung and M. Osako, *Resour., Conserv. Recycl.*, 2009, **53**, 301.
51. C.-H. Jung and M. Osako, *Chemosphere*, 2007, **69**, 279.

52. Y.-S. Chung, J.-H. Moon, S.-H. Kim, S.-H. Kang and Y.-J. Kim, *J. Radioanal. Nucl. Chem. Art.*, 2007, **271**, 339.
53. L. S. Morf, R. Gloor, O. Hagg, M. Haupt, S. Skutan, F. D. Lorenzo and D. Boni, *Waste Manage.*, 2013, **3**, 634.
54. Y. Hu, M. Bakker, G. Brem and G. Chen, *Waste Manage.*, 2011, **31**, 259.
55. K. H. Wedepohl, *Geochim. Cosmochim. Acta*, 1995, **59**, 1217.
56. U. U. Jadhav and H. Hocheng, *J.Achiev. Mater. Manuf. Eng.*, 2012, **54**, 159.
57. H. Ecke, H. Sakanakura, T. Matsuta, N. Tanaka and A. Lagerkvist, *Waste Manage. Res.*, 2000, **18**, 41.
58. C.-H. Jung and M. Osako, *Waste Manage.*, 2009, **29**, 1532.
59. I. Vervenne, *3rd IEEE Benelux Young Researchers Symposium in Electrical Power Engineering*, Ghent, Belgium, 2006.
60. H.-S. Park, *Thermophys. Aeromech.*, 2011, **18**, 313.
61. S. Nagib and K. Inoue, *Hydrometallurgy*, 2000, **56**, 269.
62. K. Huang, K. Inoue, H. Harada, H. Kawakita and K. Ohto, *J. Mater. Cycle Waste Manage.*, 2011, **13**, 118.
63. J. Xue, W. Wang, Q. Wang, S. Lui, J. Yang and T. Wui, *J. Chem. Technol. Biotechnol.*, 2010, **85**, 1268.
64. W. Krebs, R. Bachofen and H. Brandl, *Hydrometallurgy*, 2001, **59**, 283–290.
65. C. Brombacher, R. Bachofen and H. Brandl, *Appl. Environ.Microbiol.*, 1998, 1237.
66. P. P. Bosshard, R. Bachofen and H. Brandl, *Environ. Sci. Technol.*, 1996, **30**, 3066.
67. Greenpeace, *Pollution and Health Impacts of Waste Incinerators*, http://www.greenpeace.org.uk/MultimediaFiles/Live/FullReport/3809.PDF, Accessed 25 January 2013.
68. M. Allsopp, P. Costner and P. Johnston, *Environ. Sci. Pollut. Res.*, 2001, **8.2**, 141.
69. M. Braungart and W. McDonough, *Cradle to Cradle: Re-making the way we make things*, Jonathan Cape, London, 2008.
70. V. F. Hodge and M. O. Stallard, *Environ. Sci. Technol.*, 1986, **20**, 1058.
71. E. Helmers, M. Schwarzer and M. Schuster, *Environ. Sci. Pollut. Res.*, 1998, **5**, 44.
72. F. Zereini and F. Alt (eds), *Palladium Emissions in the Environment: Analytical Methods, Environmental Assessment and Health Effects*, Springer, Verlag Berlin Heidelberg, 2006.
73. M. E. Farago, P. Kavanagh, R. Blanks, J. Kelly, G. Kazantzis, I. Thornton, P. R. Simpson, J. M. Cook, S. Parry and G. M. Hall, *Fresenius' Journal of Analytical Chemisty*, 1996, **354**, 660.
74. J. D. Whiteley and F. Murray, *Sci. Total Environ.*, 2003, **317**, 121.
75. A. J. Murray, *US Pat.* 2012/0000832 A1, 2012.
76. Veolia, http://www.veolia.com/en/, Accessed 19th November 2012.
77. D. Derbyshire, http://www.davidderbyshire.co.uk/stories240911.html, Accessed 19th November 2012.

Subject Index